Neuromorphic Computing

Edited by Yang (Cindy) Yi and Hongyu An

Published in London, United Kingdom

Neuromorphic Computing
http://dx.doi.org/10.5772/intechopen.106113
Edited by Yang (Cindy) Yi and Hongyu An

Contributors
Doaa Mahmoud Habashy, Mahmoud Yaseen El-Bakry, El-Sayed Ahmed El-Dahshan, Hanem Ibrahim Lebda, Khaled S. Sayed, Fayroz F. Farouk, Zeynab Moradi, Ali Reza Mikaeili Tabrizi, Gokul Krishnan, Yu Cao, Sumit K. Mandal, Alper A. Goksoy, Chaitali Chakrabarti, Jae-sun Seo, Umit Y. Ogras, Zhenyu Wang, Sally Ghanem, Hamid Krim, Honghao Zheng, Yang (Cindy) Yi, Hongyu An, Noah Zins, Yan Zhang, Ritika Mehra, Vrinda Tandon, John Paul Shen, Harideep Nair, Tommaso Zanotti, Paolo Pavan, Francesco Maria Puglisi, Chenchen Liu, Xiang Chen, Bojing Li, Duo Zhong

First published in London, United Kingdom, 2023 by IntechOpen
IntechOpen is the global imprint of INTECHOPEN LIMITED, registered in England and Wales, registration number: 11086078, 5 Princes Gate Court, London, SW7 2QJ, United Kingdom
Printed in Croatia

British Library Cataloguing-in-Publication Data
A catalogue record for this book is available from the British Library

Additional hard and PDF copies can be obtained from orders@intechopen.com

Neuromorphic Computing
Edited by Yang (Cindy) Yi and Hongyu An
p. cm.

This title is part of the Artificial Intelligence Book Series, Volume 17
Topic: Computational Neuroscience
Series Editor: Andries Engelbrecht
Topic Editor: Magnus Johnsson

Print ISBN 978-1-80356-143-1
Online ISBN 978-1-80356-144-8
eBook (PDF) ISBN 978-1-80356-145-5
ISSN 2633-1403

For EU product safety concerns:
IN TECH d.o.o., Prolaz Marije Krucifikse Kozulić 3, 51000 Rijeka, Croatia,
info@intechopen.com or visit our website at intechopen.com.

IntechOpen Book Series

Artificial Intelligence

Volume 17

Aims and Scope of the Series

Artificial Intelligence (AI) is a rapidly developing multidisciplinary research area that aims to solve increasingly complex problems. In today's highly integrated world, AI promises to become a robust and powerful means for obtaining solutions to previously unsolvable problems. This Series is intended for researchers and students alike interested in this fascinating field and its many applications.

Meet the Series Editor

Andries Engelbrecht received the Masters and Ph.D. degrees in Computer Science from the University of Stellenbosch, South Africa, in 1994 and 1999 respectively. He is currently appointed as the Voigt Chair in Data Science in the Department of Industrial Engineering, with a joint appointment as Professor in the Computer Science Division, Stellenbosch University. Prior to his appointment at Stellenbosch University, he has been at the University of Pretoria, Department of Computer Science (1998-2018), where he was appointed as South Africa Research Chair in Artifical Intelligence (2007-2018), the head of the Department of Computer Science (2008-2017), and Director of the Institute for Big Data and Data Science (2017-2018). In addition to a number of research articles, he has written two books, Computational Intelligence: An Introduction and Fundamentals of Computational Swarm Intelligence.

Meet the Volume Editors

Yang (Cindy) Yi is a tenured associate professor in the Bradley Department of Electrical and Computer Engineering (ECE), Virginia Polytechnic Institute and State University (Virginia Tech), USA, where she is also the director of the Multifunctional Integrated Circuits & Systems (MICS) Group. Over the past 18 years, she has been working on various research topics in integrated circuits, neuromorphic computing, high performance computing and communication at Texas A&M University, IBM, Freescale, Intel, Texas Instruments (TI), University of Missouri Kansas City, University of Kansas (KU), and Virginia Tech. She has 180 publications in international journals, conference proceedings, and book chapters to her credit. Eleven of her papers have been selected for various best paper awards. Dr. Yi has received numerous awards for her research and service, including 2023 Institute of Electrical and Electronics Engineers (IEEE) Featured Author, 2020 Jeffress Trust Research Award, 2019 Virginia Tech College of Engineering Faculty Fellow and Virginia Tech Institute for Critical Technology and Applied Science (ICTAS) Junior Faculty Award, 2018 National Science Foundation (NSF) CAREER Award, a United States Air Force (USAF) Summer Faculty Fellowship, 2016 University of Kansas (KU) Miller Professional Development Award for Distinguished Research, NSF CISE Research Initiation Initiative (CRII) Award, NSF EPSCoR First Award, USAF Summer Faculty Fellowship, 2015 KU Miller Scholar Award, KU New Faculty General Research Award, Air Force Research Lab (AFRL) Visiting Faculty Research Fellowship, University of Missouri (MU) Academic Program Faculty Scholar, and 2014 MU Outstanding Service Award.

Hongyu An is an Assistant Professor of Electrical and Computer Engineering, Michigan Technological University, USA. He obtained his doctoral degree in Electrical Engineering at Virginia Polytechnic Institute and State University (Virginia Tech), USA. Prior to Virginia Tech, he received his MS and BS in Electrical Engineering at Missouri University of Science and Technology, USA, and Shenyang University of Technology, China, respectively. He was awarded the Bill and LaRue Blackwell thesis/dissertation award in 2020. His areas of interest include neuromorphic computing and engineering, neuromorphic electronic circuit design for artificial intelligence systems, and artificial intelligence for robotics and medical devices.

Contents

Preface

In this book, *Neuromorphic Computing*, we embark on a captivating journey through the cutting-edge realm of brain-inspired computing systems that are revolutionizing the landscape of artificial intelligence and computational paradigms.

Neuromorphic computing, inspired by the intricate workings of the human brain, represents a groundbreaking approach to computational architectures. With an emphasis on mimicking the brain's neural networks, this emerging field has shown remarkable promise in addressing the inherent limitations of traditional computing systems. By harnessing the power of neural networks and synaptic connections, neuromorphic computing opens up new vistas of possibilities in data processing, pattern recognition, decision-making, and beyond.

This book is a comprehensive guide to understanding the principles, methodologies, and applications of neuromorphic computing. This book is for a diverse audience eager to explore the boundless potentials of brain-inspired computing.

Key features of this book include 1) foundational concepts, 2) neuromorphic circuits and systems; 3) emerging devices for neuromorphic computing; 4) deep learning, unsupervised learning, and sparse learning, 5) spiking neural network, recurrent neural network, and temporal neural network, and 6) memory computing.

In crafting this book, our goal is to provide readers with a comprehensive and insightful journey into the realm of neuromorphic computing. We have endeavored to make the content accessible to both beginners and experts, ensuring an enriching experience for everyone.

We extend our heartfelt gratitude to the researchers, scientists, and engineers who have contributed to the field of neuromorphic computing, pushing the boundaries of innovation. Their work continues to inspire and shape the landscape of artificial intelligence and computing systems.

Yang (Cindy) Yi
Bradley Department of Electrical and Computer Engineering,
College of Engineering,
Virginia Tech,
Blacksburg, VA, USA

Hongyu An, Ph.D.
Assistant Professor,
Department of Electrical and Computer Engineering,
Institute of Computing and Cybersystems (ICC),
College of Engineering,
Michigan Technological University,
Houghton, Michigan, USA

Section 1

In Memory Computing for Neuromorphic Computing

Chapter 1

End-to-End Benchmarking of Chiplet-Based In-Memory Computing

Gokul Krishnan, Sumit K. Mandal, Alper A. Goksoy, Zhenyu Wang, Chaitali Chakrabarti, Jae-sun Seo, Umit Y. Ogras and Yu Cao

Abstract

In-memory computing (IMC)-based hardware reduces latency and energy consumption for compute-intensive machine learning (ML) applications. Several SRAM/RRAM-based IMC hardware architectures to accelerate ML applications have been proposed in the literature. However, crossbar-based IMC hardware poses several design challenges. We first discuss the different ML algorithms recently adopted in the literature. We then discuss the hardware implications of ML algorithms. Next, we elucidate the need for IMC architecture and the different components within a conventional IMC architecture. After that, we introduce the need for 2.5D or chiplet-based architectures. We then discuss the different benchmarking simulators proposed for monolithic IMC architectures. Finally, we describe an end-to-end chiplet-based IMC benchmarking simulator, SIAM.

Keywords: in-memory compute, SRAM, RRAM, network-on-chip, network-on-package, convolutional neural networks, artificial intelligence

1. Introduction

1.1 Modern-day AI algorithms

Machine learning (ML) and artificial intelligence (AI) significantly impacted society. AI algorithms, such as deep neural networks (DNNs), achieved accuracy that exceeded human-level perception for various applications, including medical imaging, natural language processing, and computer vision [1–3]. The popularity of AI algorithms was mainly driven by the availability of big datasets (for various applications like image classification and object detection) [1, 4, 5], as well as the increased computing power provided by the next-generation ML hardware accelerators and general-purpose computing platforms. **Figure 1** illustrates the taxonomy of ML algorithms, which could be broadly categorized into supervised and unsupervised

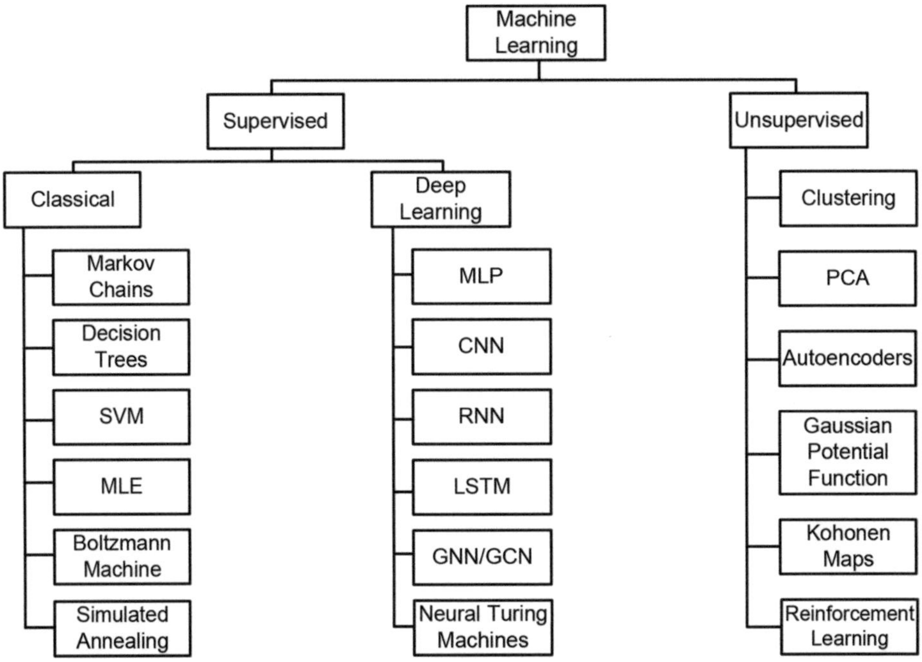

Figure 1.
Taxonomy showing different ML with different learning techniques.

learning. Unsupervised learning involved extracting features from a distribution without data annotation, including selecting samples, denoising data, and clustering data into groups. The unsupervised learning algorithm aimed to find the optimal representation of the data that preserved maximum information about the input data x while ensuring the representation was simpler than the data itself, utilizing constraints such as lower-dimensional, sparse, or independent representation [6, 7]. Some popular unsupervised learning techniques were clustering, principal component analysis (PCA), autoencoders, and Gaussian potential functions.

Supervised learning involves training the ML model using a set of labeled training data and then testing it using a labeled testing set. There are two types of supervised learning: classical approaches and deep learning. Classical approaches use conventional techniques that employ a probabilistic model to determine the next state based on a set of parameters. Some popular classical techniques are decision trees, support vector machines (SVM), Markov chains, and maximum likelihood estimation (MLE) [8–11]. However, classical techniques had limitations such as difficulty in scaling, lack of generalization, and the need for significant data engineering for each algorithm. Classical techniques serve as the foundation for deep learning algorithms, which overcome their limitations. This chapter focuses on deep learning techniques used in supervised learning. Convolutional neural networks (CNNs) stand out for their superior performance in various machine-learning tasks, including computer vision, object detection, and object segmentation. Additionally, recurrent neural networks (RNNs) excel in processing temporal data, while graph convolutional networks (GCNs) combine graphs and neural networks for various applications. The chapter covers recent advancements in CNNs, RNNs, and GCNs, discussing their structures, training methods, and execution efficiency for training and inference operations.

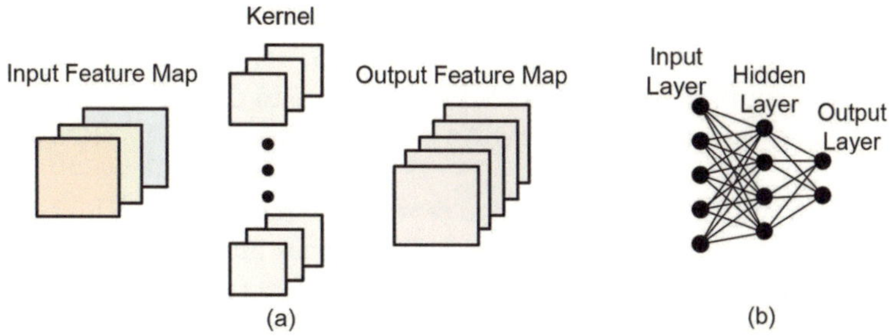

Figure 2.
(a) Convolution layer consisting of the input feature map (IFM), kernel, and the output feature map (OFM), (b) FC layer in a CNN.

Conventional CNNs consist of layers connected either sequentially or with skip connections. Besides convolutional layers, ReLU, pooling, and batch-normalization methods are commonly used to enhance performance. **Figure 2** illustrates the typical structure of a convolutional and fully connected (FC) layer. Sequential layers usually involve a stack of convolution (conv) layers that extract features from the input. Convolution layer kernels may include 7×7, 5×5, 3×3, and 1×1. Additionally, as proposed in MobileNet [12], depthwise convolutions divide a given $N \times N$ convolution into two parts: first, a $N \times 1$ convolution is performed, followed by a $1 \times N$ convolution. Depth-wise convolution yields better accuracy and lower hardware complexity. Pooling layers are periodically utilized to reduce the feature map size and eliminate noisy input. In order to perform classification on extracted features, a set of FC layers or classifier layers are utilized. These layers, along with the Conv layers, have a set of weights that are trained to achieve the best accuracy. Popular CNN structures, such as AlexNet [1], GoogleNet [13], ResNet [14], DenseNet [15], MobileNet [12], and SqueezeNet [16], utilize a combination of convolutional, pooling, and FC layers to achieve high accuracy in a variety of ML tasks. Additionally, CNNs like DenseNet and ResNet utilize skip connections from previous layers to create a highly branched structure, which aims to improve the feature extraction process. These skip connections are present within the conv layers only. In contrast, conventional CNNs face several limitations, such as the vanishing gradient problem, higher hardware costs during training and inference, and over-parameterization [17–19]. To address these issues, network architecture search (NAS) was introduced to automatically find the optimal neural network architecture based on the specific design point for the target application. Different design points, such as higher accuracy, better generalization, higher hardware efficiency, and lower memory access, have been proposed for NAS. Popular techniques like NasNet [20], FBNet [21], AmoebaNet [22], PNAS [23], ECONas [24], and MNasNet [25] have been developed to address these issues.

RNN is a commonly used deep learning technique that offers an effective solution for modeling data with sequential or temporal structures and variable length inputs and outputs in various applications [26–28]. It processes sequential data one element at a time, using a connectionist model that selectively passes information. This enables RNNs to model input and/or output data consisting of a sequence of dependent elements. Additionally, RNNs can simultaneously model both sequential and time dependencies at different scales. They employ a feedforward network that includes edges connecting adjacent time steps, which introduces time to the model. While conventional edges do not have cycles, recurrent edges that connect adjacent time steps can form cycles. Modern RNN architectures can be categorized into two main types. The first is long-short-term memory (LSTM), which includes a memory cell, a computation unit that replaces traditional nodes in the hidden layer of a network [29]. The second variant of RNNs is bi-directional RNNs (BRNNs), as proposed in [30].

As more applications rely on graphs to represent data, using CNNs and RNNs to capture hidden patterns within Euclidean data becomes limited. For instance, in e-commerce, a graph-based learning system can use interactions between users and products to recommend highly accurate products. However, graphs' complexity and irregularities pose significant challenges to existing DNNs. To address this issue, Graph Neural Networks (GNNs) were introduced and categorized into three types: Recurrent GNNs (RecGNNs) [31–33], Convolutional GNNs (CGNNs) [34–36], and Graph Autoencoders (GAE) [37–39]. RecGNNs use recurrent neural architectures to learn node representations, while CGNNs generalize convolution operations to graph data by aggregating features from neighboring nodes. GAEs map nodes into a latent feature space, preserving the node's topological information with a low-dimensional vector. Finally, GAEs learn network embeddings using an encoder and a decoder to enforce the preservation of graph topological information.

1.2 Hardware implications of DNNs

State-of-the-art DNNs, such as CNNs, RNNs, and GCNs, have diverse structures that lead to significant demands on compute and memory resources. Achieving higher accuracy with these machine-learning models requires increased computational complexity and model size, which, in turn, require more memory to store both the weights and activations. The increased model size and complexity also lead to a larger volume of on-chip data movement. For instance, the ImageNet dataset's [1] ResNet-50 [14] requires 50 MB of memory and 4 GFLOPs per inference, while DenseNet-121 [15] requires 110 MB of memory and 8 GFLOPs per inference. Conventional architectures that separate memory and computation lead to a considerable number of external memory accesses, which reduce energy efficiency and performance. The average cost of an external memory access is 1000 times higher than the energy required for computations [40]. When considering the total energy spent on performing inference for VGG-16 and ResNet-50 using conventional von Neumann architectures, a floating-point 32-bit (FP-32) multiplication results in 3.2pJ, while an FP-32 add requires 0.9pJ in the 45 nm technology node [41]. Therefore, performing inference for one image consumes 65 mJ of energy using the VGG-16 CNN, while ResNet-50 consumes 16 mJ. Scaling the computation energy up, for 1000 inference performs, VGG-16 takes 65 J while ResNet-50 consumes 16 J of energy. In conclusion, the higher accuracy achieved by DNNs results in higher computational complexity, increased memory requirements, more off-chip memory access, and lower energy efficiency.

This chapter delves into in-memory computing (IMC) as an alternative to traditional von-Neumann architecture, which offers improved energy efficiency, better performance, and reduced off-chip memory access. IMC has emerged as a promising solution to tackle the memory access, energy efficiency, and performance bottlenecks encountered by DNN applications. Hardware architectures for IMC, such as those based on SRAM and nanoscale nonvolatile memory (e.g., resistive RAM or RRAM), provide a dense and parallel structure to achieve high performance and energy efficiency [42–57]. Additionally, the chapter introduces chiplet-based IMC architectures, as well as a benchmarking simulator for this architecture. The SIAM simulator and the associated architecture enabled through SIAM are described in detail. Overall, this chapter explores various benchmarking simulators for IMC architectures.

2. IMC architectures

In the preceding section, we examined the hardware implications of modern DNNs, specifically the memory and computation complexity associated with von Neumann architectures. For instance, dense structures such as DenseNet require roughly 2.7×10^7 off-chip memory accesses to process an image frame [15]. This elevated number of off-chip memory access negatively impacts the energy efficiency of the overall system. IMC architectures provide a promising alternative to traditional von Neumann architectures. **Figure 3** depicts a generic block diagram of an IMC architecture with RRAM/SRAM memory cells. IMC uses analog- or digital-domain computation to carry out multiply-and-accumulate (MAC) operations. The crossbar-based IMC structure efficiently combines memory access and analog-domain computation into a single unit, resulting in faster execution of DNN workloads. The superior energy efficiency is primarily due to a full-custom design, higher density, and higher memory bandwidth [44, 45, 58]. Consequently, IMC-based systems are becoming increasingly popular for implementing compute- and memory-intensive AI applications. This section will examine various IMC architectures in depth using both SRAM and RRAM memory cells.

2.1 RRAM/SRAM-based IMC architectures

2.1.1 RRAM Device

RRAM-based IMC architectures consist of an RRAM memory cell at each crosspoint within the IMC crossbar array. RRAM is a two-terminal device with programmable resistance representing the neural network's weights. It has high integration density, fast read speed, high memory accessing bandwidth, and good compatibility with CMOS fabrication technology. For example, the RRAM device stack can include a TiN bottom electrode, HfO_2 mem-resistive switching layer, a PVD Ti oxygen exchange layer (OEL), and $\sim$ 40 nm TiN top electrode [59, 60]. This specific stack is implemented between the M1 and M2 metallization layers using a FEOL-compatible process flow.

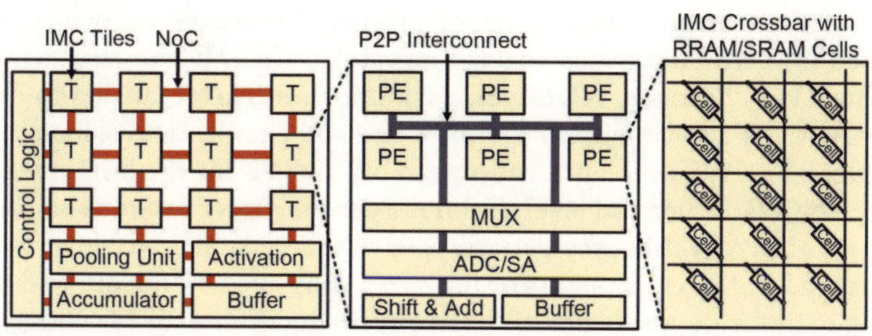

Figure 3.
Generic block diagram of an IMC architecture for DNN acceleration. It consists of an array of IMC tiles connected by an NoC with each tile consisting of a number of IMC arrays [44, 45].

Each RRAM cell can be characterized by the number of resistance levels accessed within them. Broadly, RRAM can be classified into single-level cells (SLC) and multi-level cells (MLC). SLC only has two resistance levels, that is, they can store only binary data. On the other hand, MLC cells have multiple resistance levels that represent higher precision data. The number of available resistance levels is governed by the ratio of the off resistance (R_{off}) to the on resistance (R_{on}) ratio [61]. The ratio provides the range of resistances accessible for the given RRAM device. The overall resistance range can be divided into two main states, a low resistance state (LRS) and a high resistance state (HRS). LRS deals with the lower spectrum within the resistance band, while the HRS deals with the upper band of resistance of the RRAM device.

To program the RRAM device, a series of steps need to be followed [62]. First, the RRAM device is formed by applying a large voltage across the two terminals. This process breaks the barrier and allows for electron flow across the terminals. Next, the RRAM is programmed to the required resistance by passing a specific current (compliance current) through the two electrodes. The RRAM can be programmed at different resistances depending on the compliance current. Furthermore, different resistance levels can be achieved depending on the RRAM device (SLC or MLC). Finally, once the RRAM device is programmed, we can perform a read by applying a voltage across the device electrodes. For the RRAM device proposed in [59, 60], a read voltage of up to 0.4 V can be sustained by the RRAM device. Applying a higher voltage damages the device or goes into the write state, changing the programmed resistance level.

2.1.2 IMC Architecture

Studies involving crossbar architectures have demonstrated that a 100× to 1000× improvement in energy efficiency is achieved as compared to traditional CPU and GPU architectures [44, 45, 47, 49, 50, 52, 63–67]. **Figure 3** shows the block diagram of a IMC architecture with an RRAM/SRAM memory cell. The architecture consists of an array of IMC tiles connected by a network-on-chip (NoC). The architecture also consists of a global pooling unit, nonlinear activation unit, accumulator, and input/output buffers. A global control logic performs the architecture's overall handling of the blocks.

Each tile consists of an array of processing elements (PEs), where each PE is an IMC crossbar array with either an SRAM or an RRAM cell. Each IMC crossbar array consists of a set of peripheral circuits that enable the MAC computations.

Figure 4 shows the generic block diagram for a single RRAM-based IMC crossbar array. In the case of RRAM IMC, a transistor connects the gate to the wordline (WL) of the IMC crossbar array [60]. The WL connects to the access transistors for the SRAM-based IMC with a conventional 6 T structure. The IMC crossbar arrays consist of a wordline (WL) decoder, WL driver, a column multiplexer, analog-to-digital converter (ADC) or a sense amplifier, shifter and add circuit, control logic, and input/output buffers. The WL decoder turns on and off the WL for the IMC crossbar array. Meanwhile, the WL driver and level shifter ensure the driver can turn on the memory cell. Next, for an N × N IMC crossbar array, M columns are shared across the read-out circuit. The read-out circuit consists of the ADC, shift and add circuit, and the precharge circuit for the read operation. To enable the sharing of M columns, a column multiplexer is used. Finally, a custom control logic is utilized to drive the control signals during the operation of the IMC crossbar array. We will now go over the operation for both the SRAM and RRAM-based IMC architectures. First, we will

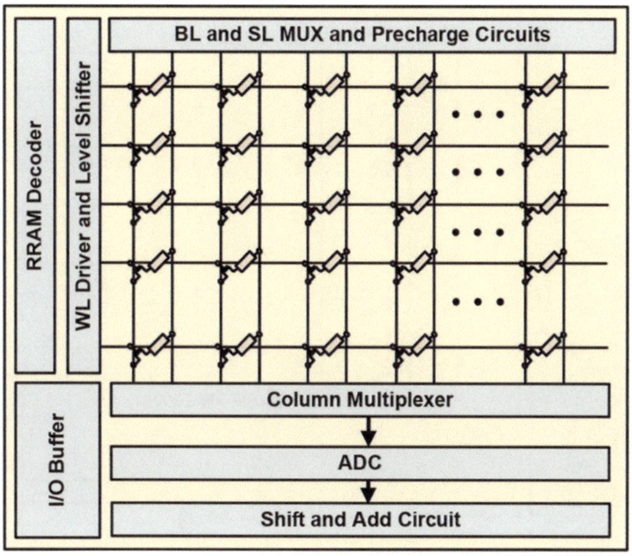

Figure 4.
Block diagram of a RRAM-based IMC crossbar array. An array of RRAM cells form the IMC crossbar array. Peripheral circuits such as bitline (BL)/select line (SL)/column multiplexer (MUX), precharge circuit, wordline (WL) decoder and driver, buffers, level shifters, ADC, and shift and add circuit complete the RRAM-based IMC.

detail the working of the RRAM-based IMC architecture. **Figure 4** shows the generic block diagram for a single RRAM-based IMC crossbar array. The RRAM devices are programmed by connecting the two terminals to a given voltage. To facilitate this, the terminals are connected to the bitline (BL) and the selectline (SL). By applying a voltage across the BL and the SL, forming, programming, and reading operations are performed cell-by-cell. Each cell is chosen during the write state of the IMC, and then the write is performed. During the compute state, the RRAM undergoes the read operation. Two kinds of read-out are performed, parallel and serial. During the parallel read-out, all/multiple WLs are turned on simultaneously, and the output is accumulated across the BL. Two kinds of input schemes are employed for single and multi-bit inputs. The first method uses a digital-to-analog converter (DAC) to convert the input vector to an analog voltage and performs the computation in the charge domain [44]. The second method is to perform bit-serial computing, where each bit in the input vector is computed one at a time. Each input vector's bit significance is handled using a shift and add circuit [45, 49, 63].

Depending on the resistance stored in the RRAM, an output current/charge is generated by the product of the voltage and resistance (conductance). This operation is analogous to the multiply with the MAC. This current/charge is then accumulated across all rows for a given column to perform the addition in the MAC. In the case of the serial read-out, row-wise access of the IMC array is performed for MAC computations. Overall, the final MAC output is generated by accumulating across all rows of the IMC crossbar array.

Figure 5 shows the generic block diagram for a single SRAM-based IMC crossbar array. Next, we will discuss the operation for an SRAM-based IMC architecture [48, 49, 51, 52, 68–70]. Depending on the SRAM bit-cell type and the degree of parallelism, the IMC design can be largely divided into three categories [71]: 6 T bit-cell with parallel computing, 6 T bit-cell with local computing, and (6 T + extra-T) bit-cell with parallel computing. Initially, SRAM-based IMC architecture employed the

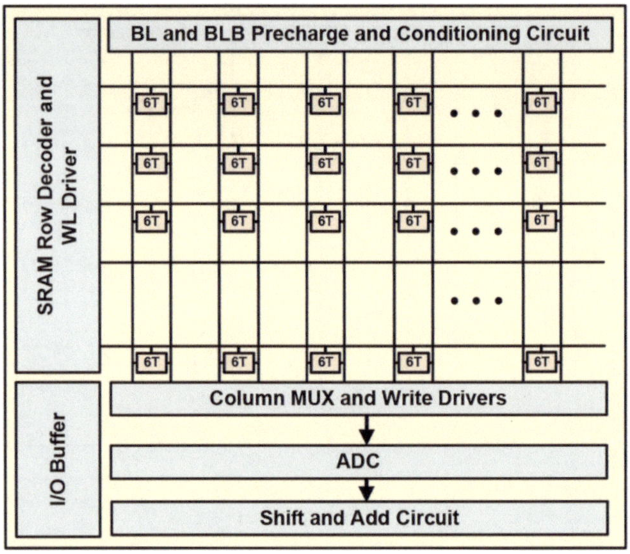

Figure 5.
Block diagram of a SRAM-based IMC crossbar array. An array of SRAM cells 6 T or 6 T + additional circuit) form the IMC crossbar array. Peripheral circuits such as bitline (BL) and bitline bar (BLB) precharge and conditioning circuits, row decoder and WL driver, column multiplexers, buffers, write drivers, ADC, and shift and add circuit complete the SRAM-based IMC.

6 T bit-cell with a parallel computation [72, 73]. The parallel computation was achieved by turning on all the WL together to perform the MAC operations. The WL is driven by the input vector where a one means it turns on that cell, while a zero means the cell is turned off. Next, a 6 T bit-cell with a local compute structure is utilized where a special compute engine is designed to perform the MAC operation [70]. Here, the MAC operation is performed row-by-row, similar to the serial read-out in RRAM-based IMC. Finally, in addition to the 6 T cell, extra transistors are added in each bit-cell to perform parallel compute [51, 68, 69]. In addition to the bit-cell structure, peripheral circuits such as precharge circuits, ADCs, write drivers, column multiplexers, row decoders, and row drivers are used.

2.1.3 Challenges with IMC Architectures

IMC architectures are known for their improved energy efficiency and throughput, but they have some drawbacks. One such drawback is the limited precision of the IMC crossbar array, particularly in the memory cell and ADC, which can affect the accuracy of DNN inference [74, 75]. Additionally, noise within analog computation can also harm DNN inference accuracy. The challenges associated with an RRAM-based IMC architecture are discussed next. RRAM devices suffer from several non-idealities, including limited resistance levels, device-to-device write variations, stuck-at-faults, and limited R_{off}/R_{on} ratio, which make it difficult to design reliable RRAM-based IMC architectures [60, 61, 76–84]. These non-idealities can cause programmed weight values (resistance value) to deviate, significantly reducing post-mapping accuracy for DNNs. Moreover, the limited array size of the IMC crossbar structure necessitates the splitting of large convolution (conv) or FC layers into partial operations, which can introduce additional errors due to the limited precision of the peripheral circuits (ADC and shift and add) of the RRAM-based IMC crossbar.

Previous research has proposed several methods to address the post-mapping accuracy loss associated with RRAM-based in-memory acceleration of DNNs. Two methods that have been proposed are Closed-Loop-on-Device (CLD) and Open-Loop-off-Device (OLD), which involve iterative read-verify-write (R-V-W) operations at the RRAM device until the resistance converges to the desired value [85, 86]. Other methods, such as in References [78, 87], utilize variation-aware-training (VAT) based on known device variation (σ) characterized from RRAM devices. In [76], VAT is combined with dynamic precision quantization to mitigate the post-mapping accuracy loss. Another approach proposed in [75] involves injecting RRAM macro measurement results that include variability and noise during the DNN training process to improve the DNN accuracy of the RRAM IMC hardware. Mohanty et al. [88] proposes post-mapping training by selecting a random subset of weights and mapping them to an on-chip memory to recover the accuracy. Charan et al. [79] utilizes knowledge distillation and online adaptation for accuracy mitigation, using an SRAM-based IMC as the parallel network, while [88] proposes to use a register file and a randomization circuit. Finally, [77, 80] propose a custom unary mapping scheme by mapping the most significant bit (MSB) and least significant bit (LSB) of the weights to RRAM devices based on individual cell variations and bit significance.

Next, we discuss the challenges associated with SRAM-based IMC architectures. A compromise between parallelism and reliability is employed for best performance. In a conventional 6 T SRAM IMC architecture, parallel computation is achieved by turning on all or multiple rows. The higher parallelism raises the critical issue of read disturbance, resulting in the WL voltage being driven with a lower voltage [72, 73, 89]. To mitigate this, reduced parallelism is employed by exploiting the local compute engine [70]. The reduced parallelism results in reduced throughput for DNN inference. Yin et al. [51, 68, 69] proposes to utilize additional transistors that isolate the bit-cell and employ parallel computation. Such a solution comes at the cost of increased area overhead, thus limiting the density of the SRAM-based IMC architecture. The additional transistor solution is typically implemented using a resistance or a capacitance. The resistive IMC method implements a multi-bit MAC operation by utilizing a resistive pull-up/down using transistors [51, 68, 69]. The pull-up/down characteristics of the transistors exhibit a non-linear behavior for the read bitline (RBL) transfer curve across different voltage ranges, thus reducing reliability. At the same time, the capacitive SRAM-based IMC utilizes a capacitor per bit-cell and charge sharing and capacitive coupling to perform the MAC operations [52]. The capacitive SRAM IMC exhibits a more linear transfer characteristic on the RBL but at the cost of a capacitor per bit-cell. Finally, the limited precision of the ADC and the noise on the bitline (BL) requires careful algorithm design to achieve the best inference accuracy [89].

2.2 2.5D/Chiplet-based AI accelerators

The area of monolithic hardware accelerators increases with the increasing number of parameters of AI algorithms. The higher silicon area of a single monolithic system reduces the yield, increasing fabrication cost [46]. The chiplet-based system solves the issue of higher fabrication cost by integrating multiple small chips (known as chiplets) on a single die. Since the area of each chiplet in the system is considerably lower than a monolithic chip (for the same AI algorithm), the yield of the chiplet-based system increases, which reduces the fabrication cost. The communication between chiplets is performed through network-on-package (NoP), as shown in **Figure 6**. Several works

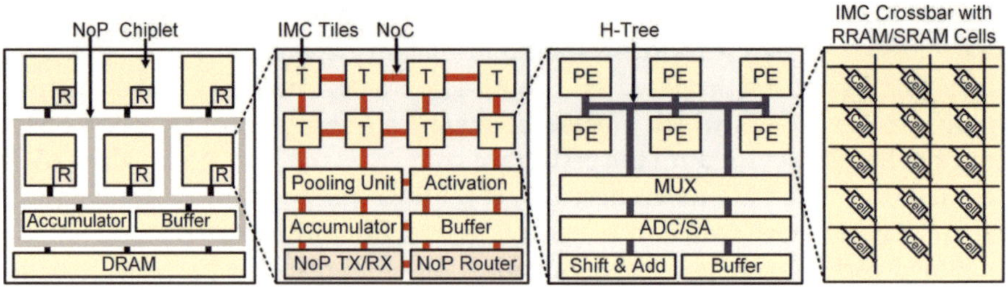

Figure 6.
Chiplet-based IMC architecture [46, 64] that includes an NoP for on-package communication, NoC for on-chip communication within each chiplet, and a point-to-point network like H-Tree for within tile communication.

in the literature propose NoP for chiplet-based systems considering different performance objectives (e.g., latency, energy) [42, 46, 90–92].

Kite is a family of NoP proposed in [90], mainly targeted for general-purpose processors. In this work, three topologies are proposed—Kite-Small, Kite-Medium, and Kite-Large. First, an objective function is constructed with a combination of the average delay between the source and destination, diameter, and bisection bandwidth of the NoP. Experimental evaluations on synthetic traffic show that the proposed Kite topologies reduce latency by 7% and improve the peak throughput by 17% with respect to other well-known interconnect topologies. A chiplet-based system with a 96-core processor, INTACT, is proposed in [91]. The chiplets are connected through a generic chiplet-interposer interface (called 3D-plugs in the paper). 3D-plugs consist of micro-bump arrays. However, both Kite and INTACT are not specific to AI workloads.

Shao et al. designed and fabricated a 36-chiplet system called SIMBA for deep learning inference [92]. The chiplets in the system are connected through a mesh NoP. Ground-referenced signaling (GRS) is used for intra-package communication. The NoP follows a hybrid wormhole/cut-through flow control. The NoP bandwidth is 100 GBps/chiplet, and the latency for one hop is 20 ns. Extensive evaluation on the fabricated chip shows up to 16% speed up compared to baseline layer mapping for ResNet-50. A simulator for the chiplet-based systems, SIAM, is proposed in [46], targeting AI workloads. In this simulator, a mesh topology is considered for NoP. It is shown that up to 85% of the total system area is contributed by NoP. In this work, multiple studies were performed by varying NoP parameters. For example, it is shown that increasing the NoP channel width increases the energy-delay product of the NoP for ResNet-110. This phenomenon is demonstrated for systems with 25 and 36 chiplets. However, none of the prior works considered any workload-aware optimization for the NoP. Therefore, there is ample opportunity for future research considering NoP optimization for AI accelerators.

3. Benchmarking simulators

3.1 Monolithic AI benchmarking simulators

Numerous researchers have put forth assessment frameworks for evaluating the performance of IMC-based AI accelerators. One such framework is NeuroSim [93], the first simulator designed to evaluate the performance of IMC-based AI accelerators. It includes various performance evaluation metrics such as area, latency, and power consumption of an IMC system under a given workload of DNN. NeuroSim is highly flexible and allows users to assess the performance of IMC-based AI accelerators under different system specifications, considering both traditional CMOS-based memory technology (such as SRAM) and emerging nonvolatile memory technologies (like ReRAM and STTMRAM) for the in-memory compute elements. NeuroSim assumes a tile-based architecture [44], consisting of multiple tiles where processing elements (PEs) and IMC-based crossbar arrays reside. Predictive Technology Model (PTM) [94] is used to simulate lower-level components, such as buffers, ADC, and multiplexers, and verified against circuit simulation (e.g., SPICE), reaching more than 90% accuracy. The interface between NeuroSim and popular ML frameworks such as PyTorch and TensorFlow has also been created to make it more user-friendly [95]. However, one major drawback of NeuroSim is that it assumes H-Tree based bus interconnect for inter-tile communication, which can consume up to 90% of the total energy consumption of DNN inference [96]. To overcome this issue, Krishnan et al. [97] (**Figure** 7) proposed an evaluation framework for IMC-based AI accelerators that considers cycle-accurate network-on-chip (NoC) simulation [98]. Similarly, MNSIM [99] also evaluates the performance of IMC-based systems to execute AI applications like NeuroSim. In addition, MNSIM integrates software-hardware co-design techniques in the evaluation framework. GeneiX, proposed by Chakraborty et al. [78], is another evaluation framework for crossbar-based IMC accelerators considering the non-idealities in the memory elements.

While it is essential to evaluate hardware performance under AI workloads, it is also important to assess the accuracy of the AI workload when implemented on-chip. Memory imperfections can lower the accuracy of DNNs. RxNN [100] and SpikeSim [66] are frameworks that evaluate the accuracy of DNN workloads in the presence of memory imperfections. These techniques focus on evaluating the performance of IMC systems executing DNN inference. However, emerging edge devices require online learning, which involves training the DNN. Therefore, evaluating the performance of AI accelerators while executing DNN inference alone is insufficient. An evaluation framework for IMC-based AI accelerators with on-chip training is introduced in [101]. The authors incorporate non-linearity, asymmetry, device-to-device, and cycle-to-cycle variation of weight update into the Python wrapper and peripheral circuits for error/weight gradient computation in NeuroSim core for a given AI

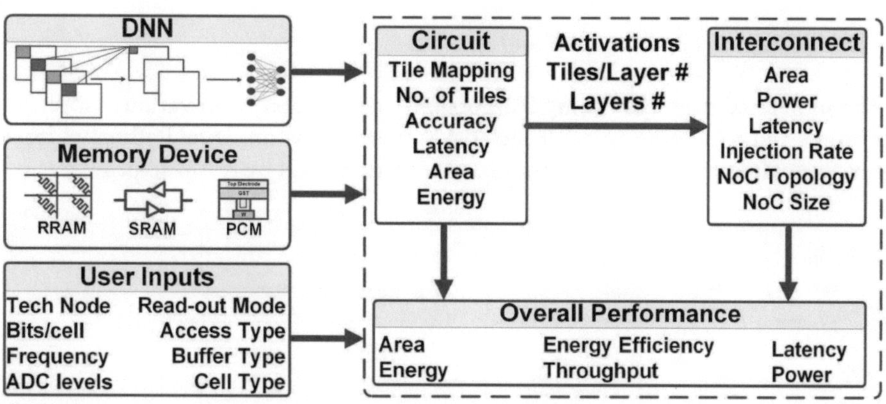

Figure 7.
Block diagram on an IMC benchmarking simulator proposed in [97]. The simulator consists of a circuit part and an interconnect part that perform system-level benchmarking of IMC architectures.

workload. The training framework is based on the authors' prior work [102], where they proposed an SRAM-based transposable function. Since SRAM-based arrays can perform write operations quickly while consuming minimal energy, the weight-gradient computation function is implemented through SRAM-based arrays rather than other nonvolatile memory technologies.

3.2 SIAM: Chiplet-based AI benchmarking simulator

This section introduces SIAM [46], a benchmarking simulator for IMC architectures based on chiplets. SIAM supports both generic or homogeneous and custom chiplet-based IMC architectures. A homogeneous architecture has a fixed number of chiplets that the user determines. On the other hand, a custom architecture comprises a specific number of chiplets required to map the DNN under consideration. In both cases, the chiplet structure contains a fixed number of user-defined IMC crossbar arrays. **Figure 6** illustrates a homogeneous chiplet-based IMC architecture SIAM uses.

Figure 8 illustrates how the comprehensive framework provided by SIAM can be used to benchmark the performance of chiplet-based in-memory compute (IMC) architectures. SIAM generates a chiplet-based IMC architecture based on user-defined inputs. It assesses the corresponding hardware performance by computing various performance metrics, including area, energy, latency, energy efficiency, power, leakage energy, and IMC utilization. The SIAM framework is developed using Python and C++ programming languages and has a top-level Python wrapper that integrates the different components of the simulator. Additionally, SIAM is designed to interface with widely used deep learning frameworks, such as PyTorch and TensorFlow, and supports a range of network structures in current literature. Therefore, SIAM can be used to explore neural architecture search (NAS) techniques. **Table 1** describes the user inputs and their associated parameters for the SIAM benchmarking tool. SIAM consists of four engines:

- Partition and mapping engine (Python)
- Circuit and NoC engine (C++)
- NoP engine (Python and C++)
- DRAM engine (Python and C++)

The individual engines within SIAM operate independently on various subsets of user inputs and communicate with each other through the top-level Python wrapper. **Figure 8** provides an overview of the simulation flow used by SIAM to provide a better

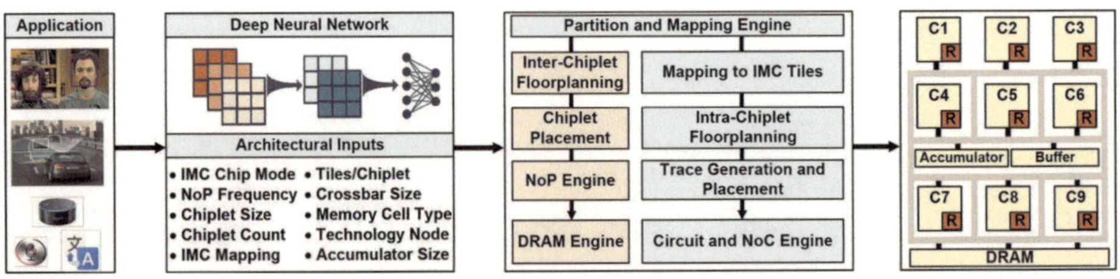

Figure 8.
Block diagram of the chiplet-based IMC architecture simulator SIAM [46].

User input	Description	User input	Description
Intra-chiplet architecture		Inter-chiplet architecture	
ADC resolution	Bit-precision of flash ADC	Chiplet size	Number of IMC tiles within each chiplet
Read-out method	Sequential or parallel	Total chiplet count	Fixed count or DNN specific custom count
Crossbar size	IMC crossbar array size	Chip mode	Monolithic or chiplet-based IMC architecture
Buffer type	SRAM or register file	Chiplet structure	Homogeneous or custom chiplet structure
NoC width	Number of channels in the NoC	NoP frequency	Frequency of the NoP driver and interconnect
NoC topology	Mesh or tree	Global accumulator size	Size of global accumulator
Frequency	Frequency of operation	NoP channel width	Number of parallel links for TX and RX
DNN algorithm		Device and technology	
Data precision	Weights and activation precision	Memory cell	RRAM or SRAM
Network structure	DNN network structure	Tech node	Technology node for fabrication
Sparsity	DNN layer-wise sparsity	Bits/cell	Number of levels in RRAM

Table 1.
Definition of the user inputs to SIAM.

understanding of the framework. Firstly, the partition and mapping engine is used to perform layer partition and mapping onto the chiplets and IMC crossbars, which generates the IMC architecture structure, the number of required chiplets and IMC tiles per layer, the IMC architecture utilization, the volume of intra-chiplet and inter-chiplet data movement, and the number of global accumulator accesses. Next, the circuit and NoC engine evaluate the intra-chiplet and global circuit performance, respectively, providing hardware performance metrics such as area, energy, and latency. Meanwhile, the NoP engine evaluates the cost of chiplet-to-chiplet data movement. Lastly, the DRAM engine assesses the memory access cost, providing energy and latency performance metrics. Except for the partition and mapping engine, all engines operate concurrently, resulting in shorter simulation times. Additionally, SIAM can be used to benchmark traditional monolithic IMC architectures. In the following sections, we provide detailed information on the four engines that comprise SIAM's core functionality.

3.3 Partition and mapping engine

The operational steps of the partition and mapping engine are explained in Algorithm 1. The engine is responsible for partitioning the DNN layers and mapping them onto the IMC chiplets and crossbar arrays. The partitioning and mapping are carried out layer by layer for the entire DNN. The engine takes various user inputs such as DNN structure, the precision of DNN weights, the mapping scheme for IMC chiplets, the size of IMC chiplets, and the size of IMC crossbars.

To begin with, we will describe the IMC mapping approach that is employed in SIAM. Suppose we have a layer i with weight matrix W_i represented by $Kx_i \times Ky_i \times Nif_i \times Nof_i$, where Kx and Ky indicate the kernel size, Nif refers to the number of input features, and Nof denotes the number of output features. SIAM uses the same mapping scheme presented in [44, 45]:

$$N_i^r = \left\lceil \frac{Kx_i \times Ky_i \times Nif_i}{(PE_x)} \right\rceil ; N_i^c = \left\lceil \frac{Nof_i \times N_{bits}}{(PE_y)} \right\rceil \quad (1)$$

The equation given above calculates the required number of rows and columns for the IMC crossbars to map layer i of the DNN. N_i^r and N_i^c represent the number of rows and columns required, while N_{bits}, PE_x, and PE_y denote the DNN weight precision and the number of rows and columns in the IMC crossbar array. To find the total number of required IMC crossbar arrays to map layer i of the DNN, one can multiply N_i^r and N_i^c together, which gives N_i^{Total} (line 7 of Algorithm 1).

SIAM can create homogeneous and custom chiplet-based IMC architectures using two types of chiplet partitions. The partition and mapping engine generates architectures based on the assumption that each DNN layer cannot be divided across multiple chiplets and that each chiplet can support multiple layers for optimal chiplet utilization. To map an entire layer of the DNN, multiple chiplets with IMC crossbar arrays are required due to numerous multi-bit weights in each layer. Dividing a layer across multiple chiplets increases overhead in control logic required for routing inputs, higher chiplet-to-chiplet communication energy and latency, and greater inter-chiplet data communication volume. The engine uniformly divides the layer across multiple chiplets during partitioning to prevent workload imbalance issues. The engine determines the number of chiplets required to map layer i of the DNN based on the total number of required IMC crossbar arrays, N_i^{Total}, using the equation $N_i^{Chiplet} = \left\lceil \frac{N_i^{Total}}{S} \right\rceil$, where S represents the total number of IMC crossbar arrays within a chiplet (the chiplet size). The resulting architectures generated by the partition and mapping engine can be seen in **Figure 9**.

After determining the number of required chiplets to map a layer of the DNN, the next step is to determine the total number of chiplets in the architecture. This is done at line 9 of Algorithm 1. In the *homogeneous chiplet partition* scheme, the user inputs a fixed number of chiplets to map the DNN. The engine compares the total number of

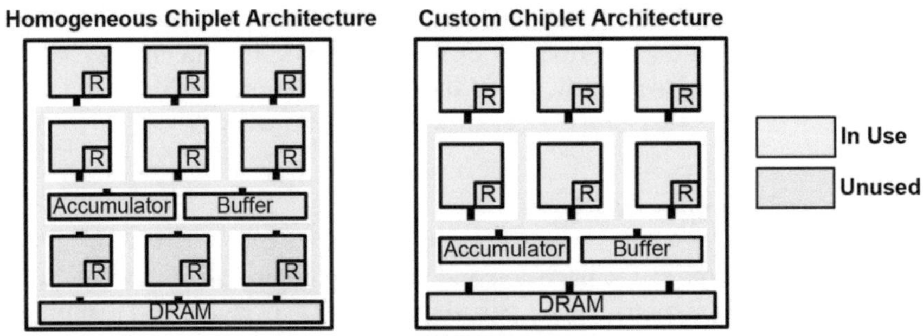

Figure 9.
The figure illustrates two chiplet-based IMC architectures, namely homogeneous (left) and custom (right), generated by SIAM for the same DNN. The homogeneous architecture is a generic architecture, whereas the custom architecture is tailored for the specific DNN. The NoP router is denoted by R in both architectures.

chiplets required to map the DNN $\left(N^{Chiplet}\right)$ with the maximum available chiplets in the architecture (C). If the number of required chiplets exceeds the maximum available chiplets, the engine throws an error and requests an increase in the number of available chiplets. On the other hand, if the number of required chiplets is less than or equal to the maximum available chiplets, the engine continues with the partition and mapping process for the subsequent layers in the DNN.

The custom partition scheme in SIAM creates a chiplet-based IMC architecture tailored to the specific DNN being considered without any upper limit on the number of chiplets used. Each chiplet in this scheme has a consistent structure with a fixed number of IMC tiles containing IMC crossbar arrays and peripheral circuitry. On the other hand, the homogeneous partition scheme uses a fixed number of chiplets (user input) to map the DNN in a generic manner. SIAM allows for the comparison of both architectures on a single platform. After partitioning and mapping the layers onto the IMC chiplets, the engine calculates the total amount of data communicated within and across the chiplets, taking into account instances where a layer is partitioned across multiple chiplets. The global accumulator is used to generate the layer output in such cases. The engine also determines the number of additions performed by the global accumulator and the number of global buffer accesses required. The engine output includes the layer partition across chiplets, the necessary number of chiplets and IMC crossbars, IMC crossbar utilization, intra- and inter-chiplet data movement volume, and the number of the global accumulator and buffer accesses. This information is then used by the circuit, NoC, and NoP engines to evaluate the performance of the chiplet-based IMC architecture.

3.4 Circuit and NoC engine

After the partitioning and mapping of the DNN, the next step in SIAM is to arrange the inter- and intra-chiplet components in a floorplan and place them. This process leads to the final design of the chiplet-based IMC architecture. Once the architecture is determined, the circuit and NoC engine estimate the performance of the hardware, as illustrated in **Figure 10**. The engine uses a model-based estimator to evaluate the circuit aspect, while it employs a trace-based estimator for the interconnect portion.

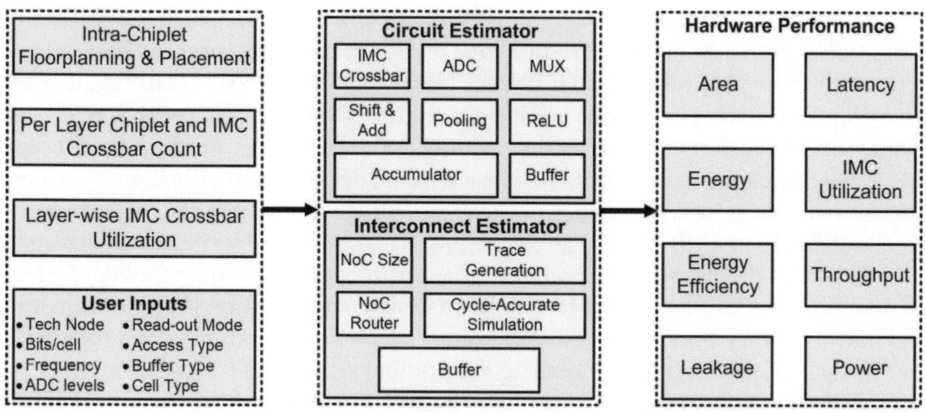

Figure 10.
Block diagram of the circuit and NoC engine within SIAM. The engine utilizes a separate circuit and NoC simulators that perform the overall hardware performance estimation.

3.4.1 Circuit estimator

The circuit estimator evaluates the hardware performance of the chiplets, global accumulator, and global buffer in the entire chiplet-based IMC architecture. To perform this evaluation, the engine considers a range of inputs such as the placement of components within and across chiplets, the number of chiplets and IMC crossbars per layer, the IMC utilization per layer, the technology node, the operating frequency, the type of IMC cell, the number of bits per cell, and the ADC precision. The intra-chiplet circuits include the IMC crossbar array, buffer, accumulator, activation unit, and pooling unit. In contrast, the peripheral circuits consist of the ADC, multiplexer circuit, shift and add circuit, and decoders. The circuit estimator is calibrated using NeuroSim [95].

The circuit estimator evaluates the performance of the entire chiplet-based IMC architecture, where each layer of the DNN is considered separately, and each chiplet is responsible for computations for a subset of layers. The partition and mapping engine provides the chiplet count, IMC crossbar count, and IMC utilization values for each layer. The estimator estimates the area, energy, and latency from the device level to the circuit and architecture levels, using user inputs such as technology node, IMC cell type, IMC crossbar size, ADC precision, and read-out mode. For each IMC crossbar within the chiplet, the cost of a single crossbar and its peripheral circuits are evaluated to obtain the total area, energy, and latency of the IMC chiplet, which includes the buffer cost, shift and adder circuitry, accumulator, pooling, and activation units. The global accumulator and global buffer accumulate the partial sum of a layer across chiplets at the chiplet-level. The estimator utilizes the number of additions performed, data volume from each chiplet, and the accumulator size to determine the global accumulator and buffer's area, energy, and latency. Finally, the estimator repeats the estimation for all chiplets required for a given layer of the DNN to determine the overall hardware performance.

3.4.2 NoC Estimator

Effective communication is essential for achieving optimal hardware performance in DNN accelerators [63]. In [103], communication-intensive DNN accelerators are extensively discussed. Every layer of a DNN transmits a significant amount of data to other layers. Studies have demonstrated that communication can account for up to 90% of the total inference latency in DNNs [45]. Therefore, it is vital to develop an efficient communication protocol for DNNs. To achieve this goal, we incorporate the cost of communication between multiple layers within a chiplet. As NoC is the standard interconnect fabric used in the SoC-domain [104], we consider using an NoC for intra-chiplet communication. We customize a cycle-accurate NoC simulator, BookSim [98], to evaluate NoC performance. First, we generate a trace file for each chiplet following Algorithm 2. The algorithm considers the number of tiles, input activations, chiplets, layer-to-chiplet mapping, quantization bit-precision, and bus width. We then determine the source and destination tile information for each layer in each chiplet. Next, we calculate the number of packets for each source-destination pair and generate a trace file as a tuple consisting of the source tile ID, destination tile ID, and timestamp. Finally, we simulate the trace file using BookSim to obtain the area, energy, and latency for on-chip communication within each chiplet.

3.5 NoP engine

Using specialized signaling techniques and driver circuits, the NoP handles on-chip data movement between different chiplets. This is achieved through a silicon interposer or organic substrate, as shown in [105, 106]. **Figure 11** (left) shows the cross-sectional image of a 2.5D integration with chiplets and an interposer. However, modeling the NoP's performance can be challenging due to its complex interconnect structure, specialized driver architectures, and corresponding signaling techniques. To accurately estimate performance, the NoP engine models each component of the NoP. **Figure 11** (right) shows different NoP implementations with the corresponding energy-per-bit (E_{bit}) proposed in prior works. The NoP performance evaluation comprises two main components: (1) NoP latency estimation and (2) NoP area and power estimation. To estimate NoP latency, the engine uses a cycle-accurate simulator to evaluate the interconnect. It generates the NoP trace using Algorithm 2, similar to the one used for NoCs, based on the chiplet-to-chiplet data volume generated by the partition and mapping engine. The generated traces are simulated using a cycle-accurate simulator or the NoP estimator to determine the latency of the NoP interconnect. To estimate NoP area and power consumption, the engine obtains interconnect parameters such as wire length, pitch, width, and stack-up. The engine then determines the interconnect capacitance and resistance using the PTM interconnect models [107]. Based on these parameters, it generates timing parameters for the interconnect and compares them to the target bandwidth. If the timing parameters do not meet the bandwidth requirements, the NoP engine chooses the maximum allowable bandwidth.

The NoP engine evaluates the NoP transmitter/receiver (TX/RX) circuits, which includes clocking circuitry, based on the energy per bit (E*bit*), number of TX/RX channels, bandwidth, chiplet-to-chiplet data volume, and operating frequency to estimate the energy and latency cost of the TX/RX circuits. The energy calculation for the NoP driver is provided in Algorithm 3. The NoP engine first calculates the number of bits being transferred between chiplets. It then retrieves the energy per bit (E_{bit}) from previous research, which is illustrated in **Figure 11** (right). The total energy for the TX/RX channel is computed by multiplying the number of bits and the energy per bit, as indicated in line 9 of Algorithm 3. Afterward, the NoP engine determines the NoP driver area using the NoP driver area cost from previous implementations (**Figure 11**). Finally, the engine integrates the interconnect and driver performance metrics to determine the overall NoP performance.

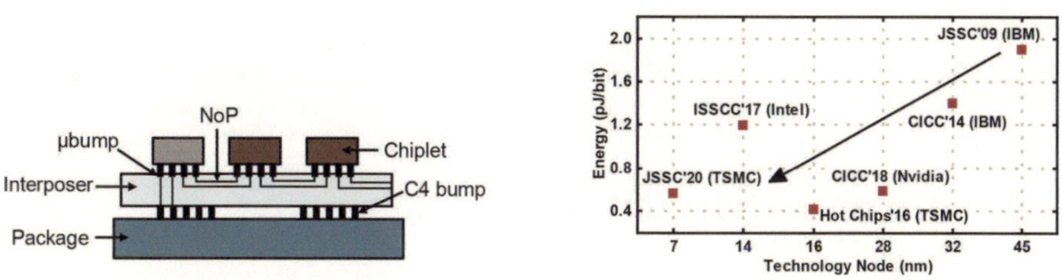

Figure 11.
(Left) Cross-sectional image of the NoP interconnect. The NoP is routed within the interposer connecting different chiplets across the architecture. M bumps connect the chiplets to the interposer, (Right) Energy per bit for different NoP driver circuit and signaling techniques proposed in prior works.

The functional flow of the NoP engine is summarized below:

- NoP trace generation based on the chiplet placement, inter-chiplet data transfer volume, and inter-chiplet layer partition,
- NoP interconnect evaluation using a cycle-accurate simulator to generate energy, latency, and area metrics,
- NoP router modeling and TX/RX driver based on real measurements. Finally, the NoP engine combines the NoP driver and the interconnect metrics to generate the overall NoP performance.

3.6 DRAM engine

The architecture of the IMC based on chiplets includes a DRAM chiplet that serves as the external memory for the IMC chiplets. The DRAM engine estimates the external memory access required for this architecture. The assumption is that the DRAM will transfer the entire set of weights to the chiplet only once before performing the inference task. This assumption remains constant across different architectural configurations and inference runs for a specific DNN model. The DRAM engine includes a DRAM request generator, RAMULATOR [108] for estimating the latency for DRAM transactions, and VAMPIRE [109] to estimate the DRAM transaction power. The DRAM choice depends on user input and currently supports DDR3 and DDR4 [110, 111]. For a given DNN model, the DRAM engine generates the required traces and memory requests with timestamps, which include the location within the DRAM memory and the operation. SIAM uses a customized version of the cycle-accurate simulator RAMULATOR and the model-based power analysis tool VAMPIRE. To reduce simulation time, the DRAM engine estimates smaller sets of instructions, which are then multiplied by the total number of sets required to represent all the weights in the DNN. An experiment was performed to calibrate the method (**Figure 12(a)** and **(b)**), which showed that a reduction of 50% of DRAM instructions to the engine results in less than 2% EDP accuracy degradation than that at 100% instructions. The overall EDP for different networks across different datasets for DDR4 shows an exponential increase in EDP with the increase in the model size of the

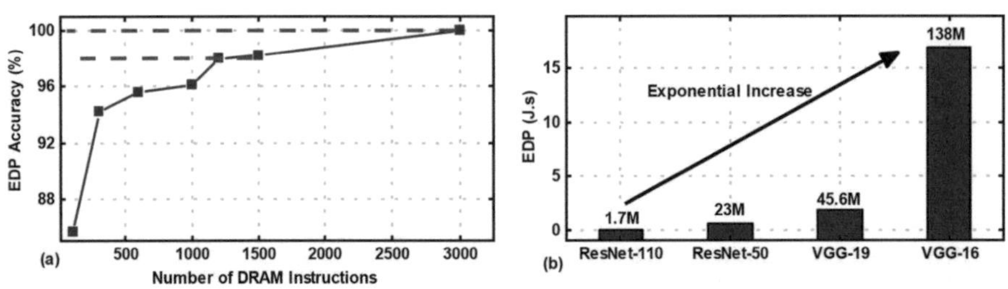

Figure 12.
(a) The accuracy of EDP prediction for different numbers of instructions processed to represent 3000 DRAM instructions. Reduction in the number of instructions to half results in less than 2% EDP accuracy degradation for half the simulation time. (b) EDP of DRAM transactions (DDR4) for different DNNs. There is an exponential increase in DRAM cost with an increase in DNN model size.

DNN. The DRAM engine provides a fast and accurate estimation of external memory access for the entire range of DNNs through this method. To summarize, the DRAM engine's execution involves the following essential steps:

- Generate DRAM requests based on the data precision and DNN model size,
- Calculate the latency cost of DRAM transactions using a customized version of RAMULATOR, and calculate the power consumption by utilizing a customized version of VAMPIRE,
- Combine the results to produce the total cost of DRAM access.

4. SIAM dataflow

This section outlines the default dataflow of the SIAM chiplet-based IMC architecture. The computation dataflow is illustrated in **Figure 13**. Before starting the inference task, the weights are obtained from the DRAM and allocated to the IMC chiplets according to the partition and mapping engine described in Section III-C. There are two possible scenarios based on this: either no layer is spread across several chiplets, or a layer is distributed across multiple chiplets. Let us consider the case where layer N of the DNN is assigned to the first chiplet, as depicted in **Figure 13(a)**. During computation, the entire layer is processed within one chiplet, generating the computed output activations for layer N. The global accumulator and buffer are not utilized during this process and are turned off. When the computation is done, the output activations are sent to the chiplets that execute layer N + 1. If two chiplets are required to map the weights for layer N + 1, the NoP transfers the output activation from layer

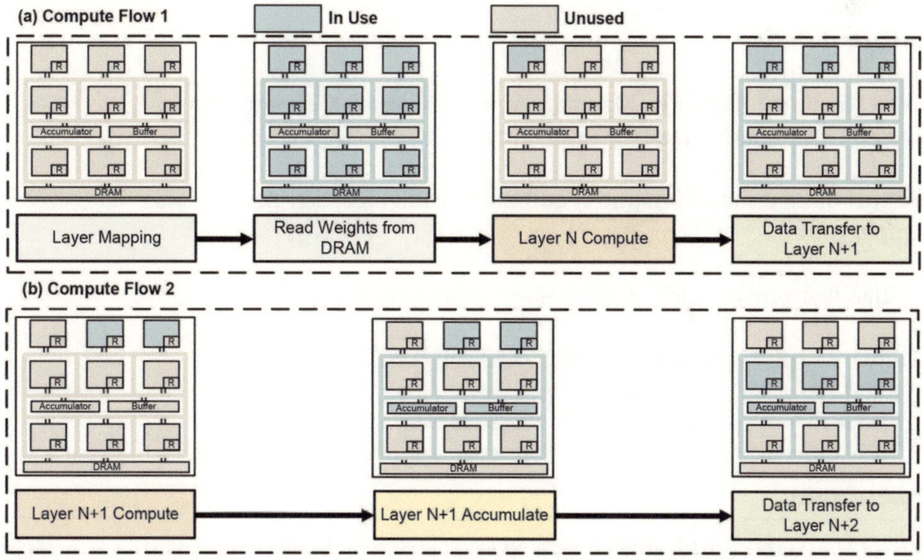

Figure 13.
Computation dataflow within the chiplet-based IMC architecture in SIAM. Two cases arise: (a) no layer is partitioned across two or more chiplets and (b) a layer is partitioned across two or more chiplets.

N to both chiplets, as shown in **Figure 13(a)**. **Figure 13(b)** presents the computation flow for layer N + 1, where both chiplets execute computations in parallel. The mapping ensures that an equal number of weights are assigned to each chiplet, avoiding any workload imbalances. After the computation is completed, the generated partial sums are aggregated using the global accumulator and buffer. Then, the accumulated outputs from layer N + 1 are transferred to the chiplets that hold the weights for layer N + 2. This process is repeated until all layers are processed and the final output is generated. The algorithmic implementation of the dataflow used in the SIAM IMC chiplet architecture is explained in Algorithm 4.

5. Conclusion

In this chapter, we discussed benchmarking chiplet-based IMC-based AI accelerators. We discuss various IMC architectures proposed in the literature. CMOS- (e.g., SRAM) and memristor- (e.g., RRAM) based IMC architectures are discussed. Although IMC improves the energy efficiency of computing elements, it increases on-chip communication volume. To address this, we discuss chiplet-based in-memory architectures. We also discuss different benchmarking simulators for monolithic and chiplet-based IMC architectures in detail. Finally, we dive deeply into SIAM, a chiplet-based IMC benchmarking simulator. SIAM provides a unified framework for performance benchmarking of chiplet-based IMC architectures. SIAM supports both *homogeneous* (*generic*) and custom chiplet-based IMC architectures. Finally, SIAM interfaces with popular deep learning frameworks such as PyTorch and TensorFlow and can be integrated with modern NAS techniques.

Author details

Gokul Krishnan[1], Sumit K. Mandal[2], Alper A. Goksoy[3], Zhenyu Wang[1], Chaitali Chakrabarti[1], Jae-sun Seo[1], Umit Y. Ogras[3] and Yu Cao[1*]

1 School of Electrical, Computer, and Energy Engineering, Arizona State University, Tempe, USA

2 Indian Institute of Science, Bangalore, India

3 University of Wisconsin-Madison, Madison, USA

*Address all correspondence to: yu.cao@asu.edu

References

[1] Krizhevsky A, Sutskever I, Hinton GE. Imagenet classification with deep convolutional neural networks. Advances in Neural Information Processing Systems. 2012;**1**:1097-1105

[2] Deng L, Hinton G, Kingsbury B. New types of deep neural network learning for speech recognition and related applications: An overview. In: 2013 IEEE International Conference on Acoustics, Speech and Signal Processing. Vancouver, Canada: IEEE; 2013. pp. 8599-8603

[3] Litjens G, Kooi T, Bejnordi BE, Setio AAA, Ciompi F, Ghafoorian M, et al. A survey on deep learning in medical image analysis. Medical Image Analysis. 2017;**42**:60-88

[4] Lin T-Y, Maire M, Belongie S, Hays J, Perona P, Ramanan D, et al. Microsoft coco: Common objects in context. In: European Conference on Computer Vision. Springer; 2014. pp. 740-755

[5] Hamilton W, Ying Z, Leskovec J. Inductive representation learning on large graphs. Advances in Neural Information Processing Systems. 2017; **30**:1025-1035

[6] Liu B, Chen Y, Liu S, Kim H-S. Deep learning in latent space for video prediction and compression. Proceedings of the IEEE/CVF Conference on Computer Vision and Pattern Recognition. 2021:701-710

[7] Rubinstein R, Bruckstein AM, Elad M. Dictionaries for sparse representation modeling. Proceedings of the IEEE. 2010;**98**(6):1045-1057

[8] Gagniuc PA. Markov Chains: From Theory to Implementation and Experimentation. John Wiley & Sons; 2017

[9] Kotsiantis SB. Decision trees: A recent overview. Artificial Intelligence Review. John Wiley & Sons Publisher; 2013; **39**(4):261-283

[10] Pisner DA, Schnyer DM. Support vector machine. In: Machine Learning. Elsevier; 2020. pp. 101-121

[11] Goodfellow I, Bengio Y, Courville A. Deep Learning. MIT Press; 2016

[12] Howard AG, Zhu M, Chen B, Kalenichenko D, Wang W, Weyand T, et al. Mobilenets: Efficient convolutional neural networks for mobile vision applications. arXiv Preprint arXiv: 1704.04861. 2017

[13] Szegedy C, Liu W, Jia Y, Sermanet P, Reed S, Anguelov D, et al. Going deeper with convolutions. In: Proceedings of the IEEE Conference on Computer Vision and Pattern Recognition, Boston. 2015. pp. 1-9

[14] He K, Zhang X, Ren S, Sun J. Deep residual learning for image recognition. In: Proceedings of the IEEE Conference on Computer Vision and Pattern Recognition, Las Vegas. 2016. pp. 770-778

[15] Huang G, Liu Z, Van Der Maaten L, Weinberger KQ. Densely connected convolutional networks. In: Proceedings of the IEEE Conference on Computer Vision and Pattern Recognition, Hawaii. 2017. pp. 4700-4708

[16] Iandola FN, Han S, Moskewicz MW, Ashraf K, Dally WJ, Keutzer K. Squeezenet: Alexnet-level accuracy with 50x fewer parameters and¡ 0.5 mb model

size. arXiv Preprint arXiv:1602.07360. 2016

[17] Krishnan G, Ma Y, Cao Y. Small-world-based structural pruning for efficient FPGA inference of deep neural networks. In: 2020 IEEE 15th International Conference on Solid-State & Integrated Circuit Technology (ICSICT). IEEE; 2020. pp. 1-5

[18] Krishnan G, Du X, Cao Y. Structural pruning in deep neural networks: A small-world approach. arXiv Preprint arXiv:1911.04453. 2019

[19] Du X, Krishnan G, Mohanty A, Li Z, Charan G, Cao Y. Towards efficient neural networks on-a-chip: Joint hardware-algorithm approaches. In: 2019 China Semiconductor Technology International Conference (CSTIC). Shanghai, China: IEEE; 2019. pp. 1-5

[20] Zoph B, Vasudevan V, Shlens J, Le QV. Learning transferable architectures for scalable image recognition. In: Proceedings of the IEEE Conference on Computer Vision and Pattern Recognition, Salt Lake City, Utah. 2018. pp. 8697-8710

[21] Wu B, Dai X, Zhang P, Wang Y, Sun F, Wu Y, et al. Fbnet: Hardware-aware efficient convnet design via differentiable neural architecture search. In: Proceedings of the IEEE/CVF Conference on Computer Vision and Pattern Recognition, Long Beach, California. 2019. pp. 10734-10742

[22] Real E, Aggarwal A, Huang Y, Le QV. Regularized evolution for image classifier architecture search. Proceedings of the AAAI Conference on Artificial Intelligence. 2019;**33**(01): 4780-4789

[23] Liu C, Zoph B, Neumann M, Shlens J, Hua W, Li L-J, et al. Progressive neural architecture search. In: Proceedings of the European Conference on Computer Vision (ECCV), Munich, Germany. 2018. pp. 19-34

[24] Zhou D, Zhou X, Zhang W, Loy CC, Yi S, Zhang X, et al. Econas: Finding proxies for economical neural architecture search. In: Proceedings of the IEEE/CVF Conference on Computer Vision and Pattern Recognition. 2020. pp. 11396-11404

[25] Tan M, Chen B, Pang R, Vasudevan V, Sandler M, Howard A, et al. Mnasnet: Platform-aware neural architecture search for mobile. In: Proceedings of the IEEE/CVF Conference on Computer Vision and Pattern Recognition, Long Beach, California. 2019. pp. 2820-2828

[26] Jordan MI. Serial order: A parallel distributed processing approach. Advances in Psychology. 1997;**121**: 471-495

[27] Sutskever I, Vinyals O, Le QV. Sequence to sequence learning with neural networks. Advances in Neural Information Processing Systems. 2014;**2**: 3104-3112

[28] Lipton ZC, Berkowitz J, Elkan C. A critical review of recurrent neural networks for sequence learning. arXiv Preprint arXiv:1506.00019. 2015

[29] Hochreiter S, Schmidhuber J. Long short-term memory. Neural Computation. 1997;**9**(8):1735-1780

[30] Schuster M, Paliwal KK. Bidirectional recurrent neural networks. IEEE Transactions on Signal Processing. 1997;**45**(11):2673-2681

[31] M. Gori, G. Monfardini, and F. Scarselli, A new model for learning in

graph domains. In: Proceedings 2005 IEEE International Joint Conference on Neural Networks. Vol. 2. Montreal, Canada: IEEE; 2005. pp. 729–734

[32] Scarselli F, Gori M, Tsoi AC, Hagenbuchner M, Monfardini G. The graph neural network model. IEEE Transactions on Neural Networks. 2008; **20**(1):61-80

[33] Gallicchio C, Micheli A. Graph echo state networks. In: The 2010 International Joint Conference on Neural Networks (IJCNN). Barcelona, Spain: IEEE; 2010. pp. 1-8

[34] Liu Z, Chen C, Li L, Zhou J, Li X, Song L, et al. Geniepath: Graph neural networks with adaptive receptive paths. Proceedings of the AAAI Conference on Artificial Intelligence. 2019;**33**(01): 4424-4431

[35] Xu K, Hu W, Leskovec J, Jegelka S. How powerful are graph neural networks? arXiv Preprint arXiv: 1810.00826. 2018

[36] Chiang W-L, Liu X, Si S, Li Y, Bengio S, Hsieh C-J. Cluster-gcn: An efficient algorithm for training deep and large graph convolutional networks. In: Proceedings of the 25th ACM SIGKDD International Conference on Knowledge Discovery & Data Mining, Anchorage, Alaska. 2019. pp. 257-266

[37] Simonovsky M, Komodakis N. Graphvae: Towards generation of small graphs using variational autoencoders. In: International Conference on Artificial Neural Networks. Springer; 2018. pp. 412-422

[38] Ma T, Chen J, Xiao C. Constrained generation of semantically valid graphs via regularizing variational autoencoders. arXiv Preprint arXiv: 1809.02630. 2018

[39] De Cao N, Kipf T. Molgan: An implicit generative model for small molecular graphs. arXiv Preprint arXiv: 1805.11973. 2018

[40] Horowitz M. Computing's energy problem (and what we can do about it). IEEE ISSCC. 2014:10-14

[41] Gholami A, Kim S, Dong Z, Yao Z, Mahoney MW, Keutzer K. A survey of quantization methods for efficient neural network inference. arXiv Preprint arXiv:2103.13630. 2021

[42] Krishnan G, Goksoy AA, Mandal SK, Wang Z, Chakrabarti C, Seo J-s, et al. Big-little chiplets for in-memory acceleration of DNNS: A scalable heterogeneous architecture. In: Proceedings of the 41st IEEE/ACM International Conference on Computer-Aided Design, San Diego, California. 2022. pp. 1-9

[43] Wang Z, Nair GR, Krishnan G, Mandal SK, Cherian N, Seo J-s, et al. AI computing in light of 2.5 d interconnect roadmap: Big-little chiplets for in-memory acceleration. In: 2022 International Electron Devices Meeting (IEDM). San Francisco, California: IEEE; 2022. pp. 23-26

[44] Shafiee A et al. ISAAC: A convolutional neural network accelerator with in-situ analog arithmetic in crossbars. ACM SIGARCH Computer Architecture News. 2016; **44**(3):14-26

[45] Krishnan G, Mandal SK, Chakrabarti C, Seo J-s, Ogras UY, Cao Y. Interconnect-aware area and energy optimization for in-memory acceleration of DNNS. IEEE Design & Test. 2020; **37**(6):79-87

[46] Krishnan G, Mandal SK, Pannala M, Chakrabarti C, Seo J-s, Ogras UY, et al.

SIAM: Chiplet-based scalable in-memory acceleration with mesh for deep neural networks. ACM Transactions on Embedded Computing Systems (TECS). 2021;**20**(5s):1-24

[47] Song L, Qian X, Li H, Chen Y. Pipelayer: A pipelined ReRAM-based accelerator for deep learning. In: 2017 IEEE International Symposium on High Performance Computer Architecture (HPCA), Austin, Texas. 2017. pp. 541-552

[48] Valavi H, Ramadge PJ, Nestler E, Verma N. A 64-tile 2.4-mb in-memory-computing CNN accelerator employing charge-domain compute. IEEE Journal of Solid-State Circuits. 2019;**54**(6): 1789-1799

[49] Yin S, Zhang B, Kim M, Saikia J, Kwon S, Myung S, et al. Pimca: A 3.4-MB programmable in-memory computing accelerator in 28 nm for on-chip DNN inference. In: 2021 Symposium on VLSI Technology. Kyoto, Japan: IEEE; 2021. pp. 1-2

[50] Yin S, Jiang Z, Kim M, Gupta T, Seok M, Seo J-s. Vesti: Energy-efficient in-memory computing accelerator for deep neural networks. IEEE Transactions on Very Large Scale Integration (VLSI) Systems. 2019;**28**(1):48-61

[51] Yin S, Jiang Z, Seo J-s, Seok M. XNOR-SRAM: In-memory computing SRAM macro for binary/ternary deep neural networks. IEEE Journal of Solid-State Circuits. 2020;**55**(6):1733-1743

[52] Jiang Z, Yin S, Seo J-s, Seok M. C3SRAM: An in-memory-computing SRAM macro based on robust capacitive coupling computing mechanism. IEEE Journal of Solid-State Circuits. 2020; **55**(7):1888-1897

[53] Chih Y-D, Lee P-H, Fujiwara H, Shih Y-C, Lee C-F, Naous R, et al. An 89tops/w and 16.3 tops/mm 2 all-digital SRAM-based full-precision compute-in memory macro in 22nm for machine-learning edge applications. In: 2021 IEEE International Solid-State Circuits Conference (ISSCC). Vol. 64. San Francisco, California: IEEE; 2021. pp. 252-254

[54] Kim H, Yoo T, Kim TT-H, Kim B. Colonnade: A reconfigurable SRAM-based digital bit-serial compute-in-memory macro for processing neural networks. IEEE Journal of Solid-State Circuits. 2021;**56**(7):2221-2233

[55] Yue J, Liu Y, Yuan Z, Feng X, He Y, Sun W, et al. Sticker-IM: A 65 nm computing-in-memory NN processor using block-wise sparsity optimization and inter/intra-macro data reuse. IEEE Journal of Solid-State Circuits. 2022;**57** (8):2560-2573

[56] Fujiwara H, Mori H, Zhao W-C, Chuang M-C, Naous R, Chuang C-K, et al. A 5-nm 254-tops/w 221-tops/mm 2 fully-digital computing-in-memory macro supporting wide-range dynamic-voltage-frequency scaling and simultaneous mac and write operations. In: 2022 IEEE International Solid-State Circuits Conference (ISSCC). Vol. 65. San Francisco, California: IEEE; 2022. pp. 1-3

[57] Spetalnick SD, Chang M, Crafton B, Khwa W-S, Chih Y-D, Chang M-F, et al. A 40nm 64kb 26.56 tops/w 2.37 mb/mm 2 rram binary/compute-in-memory macro with 4.23 x improvement in density and 75% use of sensing dynamic range. In: 2022 IEEE International Solid-State Circuits Conference (ISSCC). Vol. 65. San Francisco, California: IEEE; 2022. pp. 1-3

[58] Mao M et al. MAX2: An ReRAM-based neural network accelerator that maximizes data reuse and area

utilization. IEEE Journal on Emerging and Selected Topics in Circuits and Systems. 2019;**9**(2):398-410

[59] Liehr M, Hazra J, Beckmann K, Rafiq S, Cady N. Impact of switching variability of 65nm CMOS integrated hafnium dioxide-based ReRAM devices on distinct level operations. In: IIRW. IEEE; 2020. pp. 1-4

[60] Krishnan G, Sun J, Hazra J, Du X, Liehr M, Li Z, et al. Robust RRAM-based in-memory computing in light of model stability. In: IRPS. IEEE; 2021. pp. 1-5

[61] Krishnan G, Yang L, Sun J, Hazra J, Du X, Liehr M, et al. Exploring model stability of deep neural networks for reliable RRAM-based in-memory acceleration. IEEE Transactions on Computers. 2022;**71**(11):2740-2752

[62] He W, Yin S, Kim Y, Sun X, Kim J-J, Yu S, et al. 2-bit-per-cell RRAM-based in-memory computing for area-/energy-efficient deep learning. IEEE Solid-State Circuits Letters. 2020;**3**: 194-197

[63] Mandal SK, Krishnan G, Chakrabarti C, Seo J-s, Cao Y, Ogras UY. A latency-optimized reconfigurable NOC for in-memory acceleration of DNNS. IEEE Journal on Emerging and Selected Topics in Circuits and Systems. 2020;**10**(3):362-375

[64] Krishnan G, Wang Z, Yang L, Yeo I, Meng J, Joshi RV, et al. IMC architecture for robust DNN acceleration. In: 2022 IEEE 16th International Conference on Solid-State & Integrated Circuit Technology (ICSICT). IEEE; 2022. pp. 1-4

[65] Krishnan G, Wang Z, Yeo I, Yang L, Meng J, Liehr M, et al. Hybrid RRAM/SRAM in-memory computing for robust DNN acceleration. IEEE Transactions on Computer-Aided Design of Integrated Circuits and Systems. 2022;**41**(11): 4241-4252

[66] Moitra A, Bhattacharjee A, Kuang R, Krishnan G, Cao Y, Panda P. Spikesim: An end-to-end compute-in-memory hardware evaluation tool for benchmarking spiking neural networks. arXiv Preprint arXiv:2210.12899. 2022

[67] Krishnan G. Energy-Efficient In-Memory Acceleration of Deep Neural Networks Through a Hardware-Software Co-Design Approach [Technical Report]. Arizona State University; 2022

[68] Si X, Chen J-J, Tu Y-N, Huang W-H, Wang J-H, Chiu Y-C, et al. 24.5 a twin-8t SRAM computation-in-memory macro for multiple-bit CNN-based machine learning. In: 2019 IEEE International Solid-State Circuits Conference-(ISSCC). San Francisco, California: IEEE; 2019. pp. 396-398

[69] Dong Q, Sinangil ME, Erbagci B, Sun D, Khwa W-S, Liao H-J, et al. 15.3 a 351tops/w and 372.4 GOPS compute-in-memory SRAM macro in 7 nm finfet CMOS for machine-learning applications. In: 2020 IEEE International Solid-State Circuits Conference-(ISSCC). San Francisco, California: IEEE; 2020. pp. 242-244

[70] Su J-W, Si X, Chou Y-C, Chang T-W, Huang W-H, Tu Y-N, et al. 15.2 a 28nm 64kb inference-training two-way transpose multibit 6t sram compute-in-memory macro for ai edge chips. In: 2020 IEEE International Solid-State Circuits Conference-(ISSCC). San Francisco, California: IEEE; 2020. pp. 240-242

[71] Seo JS, Saikia J, Meng J, He W, Suh HS, Liao Y, et al. Digital Versus Analog Artificial Intelligence Accelerators:

Advances, trends, and emerging designs. IEEE Solid-State Circuits Magazine. 2022;**14**(3):65-79

[72] Kang M, Kim Y, Patil AD, Shanbhag NR. Deep in-memory architectures for machine learning–accuracy versus efficiency trade-offs. IEEE Transactions on Circuits and Systems I: Regular Papers. 2020;**67**(5): 1627-1639

[73] Zhang J, Wang Z, Verma N. In-memory computation of a machine-learning classifier in a standard 6t sram array. IEEE Journal of Solid-State Circuits. 2017;**52**(4):915-924

[74] Krishnan G, Hazra J, Liehr M, Du X, Beckmann K, Joshi RV, et al. Design limits of in-memory computing: Beyond the crossbar. In: 2021 5th IEEE Electron Devices Technology & Manufacturing Conference (EDTM). Chengdu, China: IEEE; 2021. pp. 1-3

[75] Cherupally SK, Meng J, Rakin AS, Yin S, Yeo I, Yu S, et al. Improving the accuracy and robustness of rram-based in-memory computing against rram hardware noise and adversarial attacks. Semiconductor Science and Technology. 2022;**37**(3):034001

[76] Long Y, She X, Mukhopadhyay S. Design of reliable DNN accelerator with un-reliable ReRAM. In: DATE. Grenoble France: IEEE; 2019. pp. 1769-1774

[77] Ma C et al. Go unary: A novel synapse coding and mapping scheme for reliable Reram-based neuromorphic computing. In: DATE. Grenoble France: IEEE; 2020. pp.1432-1437

[78] Chakraborty I, Ali MF, Kim DE, Ankit A, Roy K. Geniex: A generalized approach to emulating non-ideality in memristive Xbars using neural networks. In: 2020 57th ACM/IEEE Design Automation Conference (DAC), San Francisco, California. 2020. pp. 1-6

[79] Charan G et al. Accurate inference with inaccurate RRAM devices: Statistical data, model transfer, and on-line adaptation. In: DAC. San Francisco, California: IEEE; 2020. pp. 1-6

[80] Sun Y et al. Unary coding and variation-aware optimal mapping scheme for reliable ReRAM-based neuromorphic computing. TCAD. 2021;**40**(12):2495-2507

[81] Zhou C, Kadambi P, Mattina M, Whatmough PN. Noisy machines: Understanding noisy neural networks and enhancing robustness to analog hardware errors using distillation. arXiv Preprint arXiv:2001.04974. 2020

[82] Yang X et al. Multi-objective optimization of ReRAM crossbars for robust DNN inferencing under stochastic noise. In: ICCAD. IEEE/ACM; 2021. pp. 1-9

[83] Joshi V et al. Accurate deep neural network inference using computational phase-change memory. Nature Communications. 2020;**11**(1):2473

[84] Charan G, Mohanty A, Du X, Krishnan G, Joshi RV, Cao Y. Accurate inference with inaccurate RRAM devices: A joint algorithm-design solution. IEEE Journal on Exploratory Solid-State Computational Devices and Circuits. 2020;**6**(1):27-35

[85] Hu M, Li H, Chen Y, Wu Q, Rose GS. BSB training scheme implementation on memristor-based circuit. In: IEEE CISDA. Singapore: IEEE; 2013. pp. 80-87

[86] Liu B et al. Reduction and IR-drop compensations techniques for reliable

neuromorphic computing systems. In: ICCAD. San Jose, CA: IEEE; 2014. pp. 63-70

[87] Chen L et al. Accelerator-friendly neural-network training: Learning variations and defects in RRAM crossbar. In: DATE. Lausanne, Switzerland: IEEE; 2017. pp. 19-24

[88] Mohanty A et al. Random sparse adaptation for accurate inference with inaccurate multi-level RRAM arrays. In: IEDM. San Francisco: IEEE; 2017. pp. 3-6

[89] Saikia J, Yin S, Cherupally SK, Zhang B, Meng J, Seok M, et al. Modeling and optimization of sram-based in-memory computing hardware design. In: 2021 Design, Automation & Test in Europe Conference & Exhibition (DATE). IEEE; 2021. pp. 942-947

[90] Bharadwaj S, Yin J, Beckmann B, Krishna T. Kite: A family of heterogeneous interposer topologies enabled via accurate interconnect modeling. In: 2020 57th ACM/IEEE Design Automation Conference (DAC). San Francisco, California: IEEE; 2020. pp. 1-6

[91] Vivet P, Guthmuller E, Thonnart Y, Pillonnet G, Fuguet C, Miro-Panades I, et al. IntAct: A 96-core processor with six chiplets 3D-stacked on an active interposer with distributed interconnects and integrated power management. IEEE Journal of Solid-State Circuits. 2020;**56**(1):79-97

[92] Shao YS, Clemons J, Venkatesan R, Zimmer B, Fojtik M, Jiang N, et al. Simba: Scaling deep-learning inference with multi-chip-module-based architecture. In: Proceedings of the 52nd Annual IEEE/ACM International Symposium on Microarchitecture, Columbus, Ohio. 2019. pp. 14-27

[93] Chen P-Y, Peng X, Yu S. Neurosim: A circuit-level macro model for benchmarking neuro-inspired architectures in online learning. IEEE Transactions on Computer-Aided Design of Integrated Circuits and Systems. 2018; **37**(12):3067-3080

[94] Zhao W, Cao Y. New generation of predictive technology model for sub-45 nm early design exploration. IEEE Transactions on Electron Devices. 2006; **53**(11):2816-2823

[95] Peng X, Huang S, Luo Y, Sun X, Yu S. DNN+ NeuroSim: An end-to-end benchmarking framework for compute-in-memory accelerators with versatile device technologies. In: 2019 IEEE International Electron Devices Meeting (IEDM), San Francisco, California. 2019. pp. 32-35

[96] Krishnan G, Mandal SK, Chakrabarti C, Seo J-s, Ogras UY, Cao Y. Impact of on-chip interconnect on in-memory acceleration of deep neural networks. ACM Journal on Emerging Technologies in Computing Systems (JETC). 2021;**18**(2):1-22

[97] Krishnan G, Mandal SK, Chakrabarti C, Seo J-s, Ogras UY, Cao Y. Interconnect-centric benchmarking of in-memory acceleration for DNNS. In: 2021 China Semiconductor Technology International Conference (CSTIC). Shanghai, China: IEEE; 2021. pp. 1-4

[98] Jiang N et al. A detailed and flexible cycle-accurate network-on-chip simulator. In: 2013 IEEE International Symposium on Performance Analysis of Systems and Software (ISPASS). Austin, Texas: IEEE; 2013. pp. 86-96

[99] Zhu Z, Sun H, Qiu K, Xia L, Krishnan G, Dai G, et al. MNSIM 2.0: A behavior-level modeling tool for memristor-based neuromorphic

computing systems. In: Proceedings of the 2020 on Great Lakes Symposium on VLSI, Beijing, China. 2020. pp. 83-88

[100] Jain S, Sengupta A, Roy K, Raghunathan A. RxNN: A framework for evaluating deep neural networks on resistive crossbars. IEEE Transactions on Computer-Aided Design of Integrated Circuits and Systems. 2020;**40**(2):326-338

[101] Peng X, Huang S, Jiang H, Lu A, Yu S. DNN+ NeuroSim V2. 0: An end-to-end benchmarking framework for compute-in-memory accelerators for on-chip training. IEEE Transactions on Computer-Aided Design of Integrated Circuits and Systems. 2020;**40**(11): 2306-2319

[102] Jiang H, Huang S, Peng X, Su J-W, Chou Y-C, Huang W-H, et al. A two-way SRAM array based accelerator for deep neural network on-chip training. In: 2020 57th ACM/IEEE Design Automation Conference (DAC), San Francisco, California. 2020. pp. 1-6

[103] Nabavinejad SM, Baharloo M, Chen K-C, Palesi M, Kogel T, Ebrahimi M. An overview of efficient interconnection networks for deep neural network accelerators. IEEE Journal on Emerging and Selected Topics in Circuits and Systems. 2020;**10**(3):268-282

[104] Jeffers J et al. Intel Xeon Phi Processor High Performance Programming. Knights Landing Edition; 2016

[105] Poulton JW et al. A 0.54 pJ/b 20Gb/s ground-referenced single-ended short-haul serial link in 28nm CMOS for advanced packaging applications. In: 2013 IEEE ISSCC. San Francisco, California: IEEE; 2013. pp. 404-405

[106] Lin M-S et al. A 7-nm 4-GHz Arm[1]-core-based CoWoS[1] chiplet design for high-performance computing. IEEE Journal of Solid-State Circuits. 2020; **55**(4):956-966

[107] Sinha S, Yeric G, Chandra V, Cline B, Cao Y. Exploring sub-20nm FinFET design with predictive technology models. In: DAC 2012. San Francisco, California: IEEE; 2012. pp. 283-288

[108] Kim Y, Yang W, Mutlu O. RAMULATOR: A fast and extensible DRAM simulator. IEEE Computer Architecture Letters. 2015;**15**(1):45-49

[109] Ghose S et al. What your DRAM power models are not telling you: Lessons from a detailed experimental study. Proceedings of the ACM on Measurement and Analysis of Computing Systems. 2018; **2**(3):1-41

[110] MICRON, Datasheet for DDR3 model, 2011. Available at: https://media-www.micron.com/-/media/client/global/documents/products/data-sheet/dram/ddr3/2gb_ddr3l-rs.pdf?rev=f43686e89394458caff410138d9d2152 (Accessed March 29, 2021).

[111] MICRON, Datasheet for DDR4 model. 2014. Available at: https://www.micron.com/-/media/client/global/documents/products/data-sheet/dram/ddr4/4gb_ddr4_dram_2e0d.pdf [Accessed March 29, 2021].

Chapter 2

Implementation of Associative Memory Learning in Mobile Robots Using Neuromorphic Computing

Noah Zins, Yan Zhang and Hongyu An

Abstract

Fear conditioning is a behavioral paradigm of learning to predict aversive events. It is a form of associative learning that memorizes an undesirable stimulus (e.g., an electrical shock) and a neutral stimulus (e.g., a tone), resulting in a fear response (such as running away) to the originally neutral stimulus. The association of concurrent events is implemented by strengthening the synaptic connection between the neurons. In this paper, with an analogous methodology, we reproduce the classic fear conditioning experiment of rats using mobile robots and a neuromorphic system. In our design, the acceleration from a vibration platform substitutes the undesirable stimulus in rats. Meanwhile, the brightness of light (dark vs. light) is used for a neutral stimulus, which is analogous to the neutral sound in fear conditioning experiments in rats. The brightness of the light is processed with sparse coding in the Intel Loihi chip. The simulation and experimental results demonstrate that our neuromorphic robot successfully, for the first time, reproduces the fear conditioning experiment of rats with a mobile robot. The work exhibits a potential online learning paradigm with no labeled data required. The mobile robot directly memorizes the events by interacting with its surroundings, essentially different from data-driven methods.

Keywords: neuromorphic computing, associative memory learning, mobile robot, Hebbian learning, fear conditioning

1. Introduction

Associative memory learning is a ubiquitous online learning paradigm in animals [1–3]. Unlike the data-driven learning schemes of current Artificial Intelligence (AI), animals have the capability of memorizing the events that occur at the same time or within a certain time interval. The underlying memorization mechanism in the nervous system is the synaptic connection that becomes strengthened under the stimulus of the firing neurons evoked by concurrent events. The strengthened synaptic connection enables the response neurons at the conditional pathway to receive a larger amount of the synaptic transmitter. As a result, the response neuron in the conditional signal pathway will fire, even though it originally did not become active. In other words, the memorization of the relationship between concurrent events is achieved by signal pathway modification rather than backpropagation. The signal pathway modification is accomplished by synaptic plasticity. An AI system with associative memory potentially provides an alternative way of active self-learning by constantly interacting with environments. The signal pathway modification can be accomplished with a few training processes, leading to less dependence on large-size datasets. For data-driven deep learning, for example, Deep Neural Networks (DNNs), the large datasets prolong training time and increase energy demands. Consequently, the application of deep learning is highly reliant on bulky supercomputers that are not feasible and applicable to scenarios that require Size, Weight, and Power (SWaP) constraints [4, 5]. In addition, massive and labeled data are costly to build or even not practical to collect, such as the Lunar and Martian terrain data [5].

Numerous studies have implemented associative memory with neuromorphic systems [2, 6–13]. However, these studies merely complete a small-scale association with a few neurons in simulation environments. It is far away from the capability of associative memory learning to enable animals to self-learn and explore independently in an unknown environment. In addition, pretraining processes with labeled datasets are still required for these studies [9–13]. In order to resolve these limitations of studies on associative learning, we have designed several experiments of associative memory in real-world scenarios using a mobile robot and neuromorphic chips. Our system of associative memory is validated by reproducing one of the classic associative memory learning in rats: *fear conditioning*. In fear conditioning experiments, the rats learn to associate a particular neutral Conditional Stimulus (CS), for example, tone, with an aversive Unconditional Stimulus (US), such as an electrical foot shock, and show a fear response, freezing or running away. The rats learn fear conditioning after several training sessions and exhibit long-lasting behavioral changes. Several brain regions have been proven to be involved in the learning process, including frontotemporal amygdala, hippocampus, and so on. The process of fear conditioning cannot be reproduced by other state-of-the-art associative memory models [2, 6–13] due to their limited neural network sizes. The simple neural network models cannot process informative signals, such as visual signals. These informative signals are processed with large-scale neural assemblies rather than simply a few neurons in the brains [14–19]. To resolve these limitations, in our design, we use large-scale biological plausible neurons to process the visual signals. Specifically, in our system and experimental designs, the mobile robot with sensors serves as the substitute for the rats in fear conditioning experiments. The neuromorphic chip (Intel Loihi) provides a computational platform for the associative memory learning operation. In our experiment, the brightness of a light emulates the visual stimulus, and the vibration signals from the accelerometer mimic the shock signals to the rats. Thus, the vibration signals are the unpleasant stimulus, and light is the neutral stimulus. The movement of the mobile robot emulates the fear response. The perception of the light and the vibration are separately processed within two different neural assemblies. Two neural assemblies connect to the response neuron, which stimulates the movement of the robot, with two signal pathways. One signal pathway with a weak synaptic connection serves as the conditional signal pathway, while another one with a stronger synaptic connection is the unconditional signal pathway. Thanks to the mobile robot providing a platform directly interacting with the environment, we for the first time, to our best knowledge, implement associative memory as real-time online learning with no pretrained procedure. The contributions of this paper are summarized as follows:

1. Compared to other state-of-the-art works [9–13], we have implemented associative memory learning with a mobile robot for an online learning scenario using an Intel Loihi chip.

2. The work reproduces the classic fear conditioning of rats with solid biological rationales from a cellular level (Hebbian learning) to the behavior level (neural assembly).

3. Aiming signal pathway modification as the neural network training purpose, which is a novel learning paradigm of associative memory learning.

4. No labeled datasets are required.

2. Research background

This work aims to develop a novel self-learning paradigm by emulating associative memory learning of animals. Thus, the study is built upon the reverse engineering of brain function. In specific, the large-scale computational models of associative memory learning are implemented by a neuromorphic system. In this section, we first introduce the state-of-the-art development of neuromorphic computing and systems. Then, the mechanism of associative memory learning at both macroscopic and microscopic levels is analyzed.

2.1 Neuromorphic system

A neuromorphic system emulates nervous systems, such as human brains, aiming at implementing Artificial Intelligence [20–25]. Human brains have the capability of executing sophisticated missions in unbelievably ultra-low energy. The average power of human brains is as small as ~20 watts [1]. In addition, unlike the training process required for Artificial Neural Networks (ANNs) using big data, the nervous systems can adjust their responses by constantly interacting with their surroundings. This learning process is referred to as associative memory learning [1]. These incredible capabilities of nervous systems are attributed to their parallelization, high degree of connectivity, adjustable network topology, the colocation of data memory computation, and spike-based information representation.

Human brains consist of billions of neurons and trillions of synapses forming a high-degree and three-dimensional neural network. Through this extraordinarily complex network, an individual neuron can communicate with more than ten thousand other neurons simultaneously. Within this complex neural network, neurons are mainly signal-processing units, and the synapse between neurons connects organs. As computing units, the neurons integrate the received spiking signals in their cell body and send another sequence of spiking signals to other neurons through synapses. The signal strength received by other neurons is depended on the connection strength of the synapses. The connection strength among neurons can be adjusted. This feature is named as synaptic plasticity [1, 26, 27]. In specific, the connection strength among neurons becomes strong if the presynaptic neuron and postsynaptic neuron fire together. This synaptic connection strength change inspired a learning paradigm known as Hebbian learning [28–31].

In addition, the computational units (neurons) and the memory units (synapses) are located in close proximity. This structure eliminates one of the biggest inefficiencies in the von Neumann architecture that separates computing units and memory at different locations. The physical separation leads to data needing to be constantly transferred back and forth between memory and central processing units (CPUs). Furthermore, neuromorphic systems use sparse and event-based computation, meaning that only a small percentage of the available computing resources are active for a given task, and they are only activated and consuming power as needed in response to present events. Neuromorphic computing attempts to exploit these useful properties by modeling the architecture, neuron and synaptic cells, and the way of learning observed in the brain, enabling a new era of computers and AI [32].

Neuromorphic systems utilize specialized neuromorphic chips with artificial neurons. These chips are generally used to operate spiking neural networks (SNNs), which encode the information with a sequence of spikes just like nervous systems. In an SNN, neurons communicate with each other with discrete "spike" signals. There are various types of neuromorphic chips, such as Intel's Loihi [33, 34]. Unlike traditional GPUs and CPUs built upon the von Neumann architecture operating on digital information, Loihi chips are specifically designed for neuromorphic computing and asynchronous SNNs. To date, two generations of Loihi chips have been released. The first generation of Loihi chip was revealed in 2017 [33, 34]. Loihi-1 chips consist of 130,000 electronic neurons and 130 million synapses at 128 neuromorphic cores. The advanced 14 nm process of Intel renders the area of the Loihi-1 chip as small as 60 mm^2. Loihi-1 chips implement the digital leaky-and-fire neurons located on 128 cores. At each core, the communication among neurons is organized in a mesh configuration. The synapses in Loihi-1 chips are fully configurable and further support weight-sharing and compression features. The plasticity of synapses can be manipulated with various biologically plausible learning rules, such as Hebbian rules, STDP, and reward-modulated rules [33, 34]. The firing behavior of neurons in Loihi chips is implemented when received spikes accumulate to a threshold value in a certain time; the neurons will fire off their own spikes to their connected neurons.

Loihi-1 chips are offered with several neuromorphic platforms providing distinct interfaces for integrating the Loihi-1 chip with other computer systems or Field-Programmable Gate Array (FPGA) devices. Kapoho bay includes 1–2 Loihi chips with a USB interface. Nahuku is a 32-chip Loihi board with a standard FPGA Mezzanine Card (FMC) connector. The FMC connector allows the Nahuku system to communicate with the Arria FPGA development board. Pohoiki Spring is a large-scale Loihi chip with 100 million neurons equipped as a server for remote access. The second generation of the Loihi chips, namely Loihi-2, was introduced in late 2021 [35]. Loihi-2 is fabricated in Intel 4 process, previously referred to as 7 nm technology. Powered by this advanced technology, the area of the Loihi-2 has been reduced to 31 mm^2 from 60 mm^2 of the first generation Loihi chips. Unlike the rigid neuron models in the last generation of Loihi chips, Loihi-2 realizes fully programmable neuron models. In Loihi-2, the specific behavior of the neurons can be programmed with microcode instructions. The microcode instructions support basic bitwise and math operations that can be used to specify custom neuron models. The Loihi-2 chip is dedicatedly designed for neuromorphic computing and edge devices with parallel computations, achieving high computational and energy efficiency. The comparison between two generations of Loihi chips is summarized in **Table 1**.

Feature	Loihi-1 [33, 34].	Loihi-2 [35]
Technology	Intel 14 nm	Intel 4 (7 nm)
Die Area	60 mm^2	31 mm^2
Max # Neurons/Chip	128,000	1 million
Max # Synapses/Chip	128 million	120 million
Neuron Model	Generalized digital LIF	Fully programmable

Table 1.
Introduction to Loihi and Loihi 2 chips.

2.2 Associative memory learning

Animals have the capability of memorizing different events if they occur at the same time or with a small time lag. The capability is referred to as associative memory [1]. Associative memory learning was first studied by Ivan Pavlov in the 1890s when he was studying salivation reflex actions in dogs [1]. During Pavlov's experiments, the dogs originally had a salivation reflex to the presence of food, instead of the sound of whistles. However, if these two signals were presented together several times, the dogs salivated even if they only listened to the sound of a whistle with no food provided. This means the dogs can memorize the sound of whistles as a sign of food [1, 6, 36] through a learning/memorizing process. Through a series of experiments, Pavlov concluded that dogs have the capability of associating two originally irrelevant signals together through a training process, which is referred to as associative memory learning later. In general, two types of stimuli exist in associative memory learning: unconditional stimuli (US) and conditional stimuli (CS). The unconditional stimuli evoke the response with no training required. On the contrary, conditional stimuli demand an associative learning process to acquire corresponding reactions. For instance, in Pavlov's experiments, the presence of food was the unconditional stimulus, and the sound of whistles was a conditional stimulus (CS). After dogs, further studies have demonstrated that associative memory learning is a self-learning paradigm of a large variety of animals such as rats, bats, and sea slugs [1].

The studies in neuroscience exhibit that signal pathway modification and synaptic plasticity are highly related to associative memory learning [1, 22]. In a nervous system, the shapes of the spiking signals are almost identical (spikes), whether the signals come from the sensation of light or hearing. Thus, neuroscientists hypothesize that brains distinguish these signals by the signal pathways they are traveling to rather than their shapes. This hypothesis is much more straightforward in invertebrates that have simple nervous systems. **Figure 1** illustrates part of the nervous system of Aplysia that has two signal pathways from siphon to the gill and from tail to the gill, separately.

With these two signal pathways, Aplysia can accomplish a simple version of associative memory learning by memorizing the touch on the tail and stimulus from the siphon. When the tail of an Aplysia is touched, its gill shrinks, demonstrating an unconditional signal pathway. On the contrary, the gill does not shrink if the siphon is cut, exhibiting a conditional signal pathway. By applying a touch to the tail and stimulus on the siphon at the same time several times, the gill motor neuron becomes more responsive to the touch on the siphon alone. At the cellular level, the concurrent stimuli on the siphon and tail lead to a spiking signal overlapping when the stimuli are

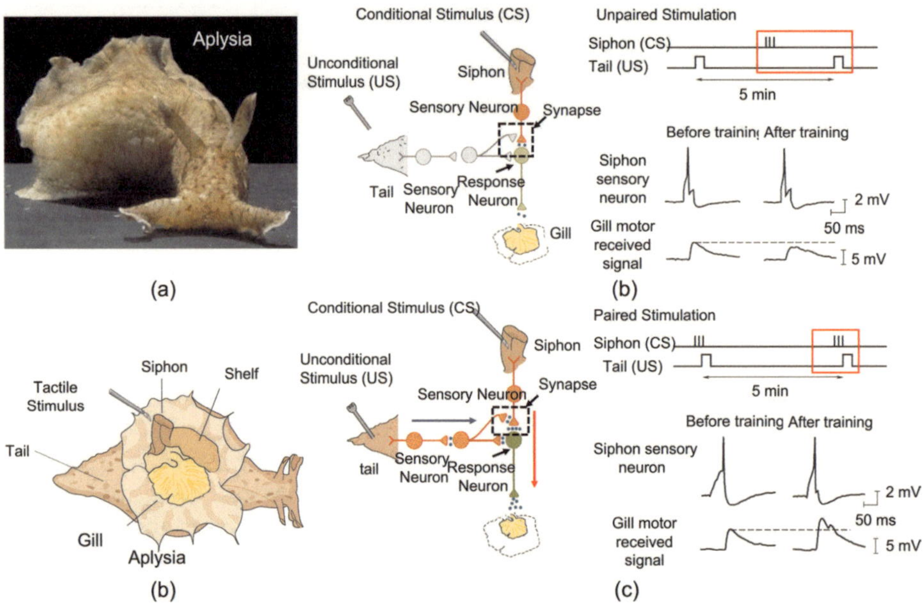

Figure 1.
Illustration of associative memory learning of Aplysia.

applied at the same time, shown in **Figure 1**. As a result, the synaptic connection among neurons, from the siphon to the gill, becomes stronger than the original state. This means the signal pathway from the siphon to the gill becomes unimpeded from blocked. These experiments on Aplysia demonstrate two critical factors for associative memory learning: (1) signal pathway modification and (2) synaptic plasticity.

For more complicated animals, such as rats, the sensation signals are processed not in individual neurons but in a group of neurons. These groups of neurons are referred to as neural assemblies [29, 37–39]. For example, fear conditioning experiments in rats involve two types of stimuli: electric shock on the food and a sound as neutral stimuli. These two types of signals are processed at different neural regions: auditory thalamus and somatosensory thalamus. The experimental goal is to let the rats associate the neutral sound with undesired electric shock by applying these two stimuli at the same time. Thus, it is one type of associative memory learning scheme. The studies have strong experimental evidence showing that signal pathway modification potentially occurs in lateral nucleus because the output signals from the auditory thalamus and somatosensory thalamus converge at the lateral nucleus [1]. This hypothesizes that associative memory learning in higher animals is accomplished via the association of two, or several, neural assemblies together rather than individual neurons.

3. Reproducing fear conditioning using mobile robots

In this section, we introduce our experimental design for reproducing fear conditioning using mobile robots. In specific, we select visual signals, the brightness of light as the conditional stimulus, and vibration signals as the unconditional stimulus. Vibration signals emulate electric shock applied on the foot of rats in fear conditioning experiments, while light signals serve as a neutral stimulus. We select the Leaky

Integrate and Fire (LIF) neuron model to build neural assemblies for both UC and CS signal pathways due to its simplicity. In our experiments, the Nengo simulator has been used to implement the system [40]. Our mobile robot is controlled with the Robot Operating System (ROS) [41].

3.1 Neuron model

In our experiment, a mobile robot is placed on a vibration platform and unconditionally responds to the acceleration signals, which will be detected with an *inertial measurement unit (IMU)*. The acceleration signals emulate the unpleasant stimulus in the fear conditioning experiment in rats. The conditional stimulus is mimicked with the brightness of a light. Two neural assemblies will be built for processing these UC and CS signals. At last, the movement of a mobile robot is controlled by the motion neurons. We design several specific neurons for precepting the brightness of lights, detecting vibrations, and controlling mobile robots. All these neurons are customized from classic Leaky Integrate and Fire (LIF) neurons, which are characterized using the following eqs. [42]:

$$C_m \frac{dV_m}{dt} = G_L(E_L - V_m) + A * I_{app},$$
$$if V_m > V_{th} then V_m = V_{reset}, \tag{1}$$
$$\tau_{RC} = C_m / G_L, \tag{2}$$

where C_m is the membrane capacitance, V_m is the membrane potential, $G_L$is the leak conductance, A is the input signal gain, $E_L$is the leak potential, I_{app} is the input current, and τ_{RC} is the RC time constant. The specific values of these parameters in our design are summarized in **Table 2**.

For all the neurons in **Table 2**, the membrane potential is fixed at 1 V, and input gain is modified instead. They all use the Nengo LIF model's default τ_{RC} of 0.02 seconds because it is sufficient for our desired functionality. The other two parameters are calculated and optimized based on our experimental setups so that they can produce the desired responses for their respective uses. In specific, for the vibration detection neuron, gain (A) and bias (V_{reset}) are empirically derived so that it fires with vibration stimulus input instead of a small sudden move. The movement response neuron is a typical LIF configured to spike whenever it receives any sustained input spikes, either from vibration neurons or from brightness neurons. The brightness neuron is the Layer 3 neuron in **Table 3** and is a LIF neuron with empirically derived gain and bias so that it fires only when the Light Feature neurons have a high enough collective output. More details regarding how these neurons work will be introduced in subsequent sections.

Neuron types	τ_{RC}	A	V_{reset} (V)	$V_{th}(V)$
Vibration neuron	0.02	1.3	0.6	1.0
Brightness neuron	0.02	0.3	−1.0	1.0
Movement neuron	0.02	1.0	0.01	1.0

Table 2.
LIF neuron parameters.

LCA neuron layer	τ_{RC}(s)	A	V_{reset} (V)	V_{th} (V)
Layer 2	∞	1.0	-λ = 0.85	1.0
Layer 3	0.02	0.3	−1.0	1.0

Table 3
LCA neuron parameters in layers 2 and 3.

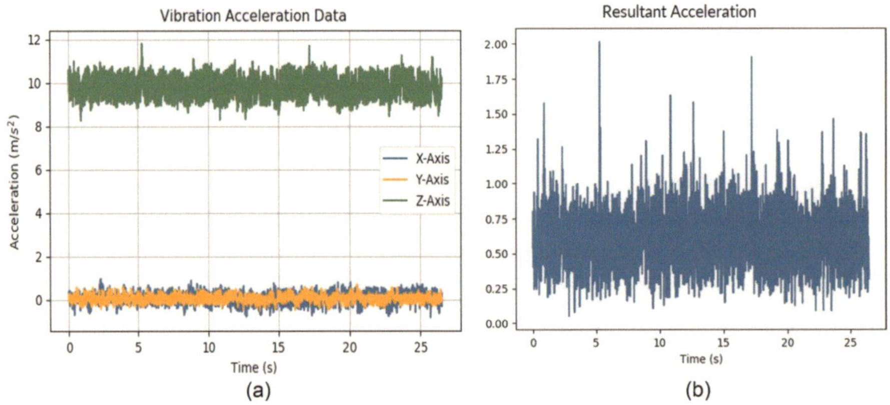

Figure 2.
(a) Acceleration data of IMU for three dimensions denoted as X-axis (blue), Y-axis (orange), and Z-axis (green). (b) Resultant acceleration.

3.2 Vibration perception

The accelerometer within the onboard IMU of the mobile robot is used to measure the acceleration. In our experiment design, acceleration is used for evaluating the degree of vibration. Our vibration platform generates vibration signals with 1.2 mm amplitude at 25 Hz. In general, the IMU measures three-dimensional accelerations from x, y, and z directions as shown in **Figure 2a**. In **Figure 2a**, the acceleration in the z-axis (vertical) has the largest magnitude as it includes the intrinsic gravity of the earth (9.8 m/s^2). Because the mobile robot should only count the acceleration from the vibration platform, the gravity effect is removed from z-direction by subtracting the standard gravity (9.8 m/s^2). Thus, the resultant acceleration, which is used for evaluating the vibration states, is calculated by the equation:

$$a_{res} = \sqrt{a_x{}^2 + a_y{}^2 + (a_z - 9.8)^2}, \tag{3}$$

where a_{res} is the resultant acceleration and a_x, a_y, and a_z are the accelerations in X-axis, Y-axis, and Z-axis, respectively. The resultant acceleration measured by our mobile robot is depicted in **Figure 2b**. The resultant acceleration is imported into the vibration detection neuron. A vibration detection neuron, implemented with LIF, is connected to the accelerometer, and it fires if the detected acceleration is larger than its threshold (shown in **Figure 3**). At last, a movement neuron is specifically designed to control our mobile robot moving away from the vibration platform with specific direction and speed. The motion neuron is also implemented by a LIF neuron. The parameters of motion neuron are listed in **Table 2**. The active motion neuron will trigger a specific movement (escape) response that commands the mobile robot to move away from the vibration platform at a speed of 0.3 m/s.

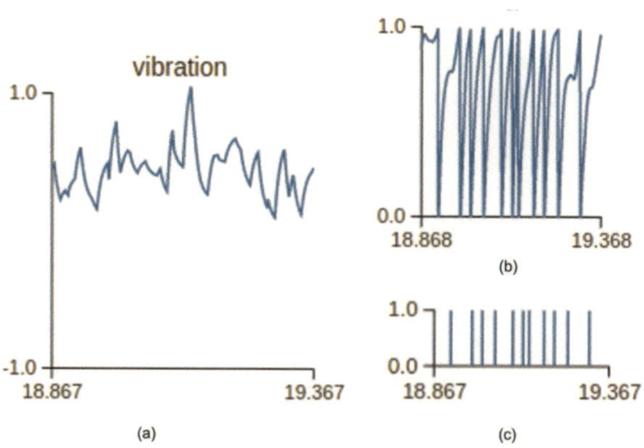

Figure 3.
Vibration detection neuron response to the acceleration input: (a) input vibration signals; (b) membrane potential of the vibration detection neuron; (c) output spiking signals of the vibration detection neuron.

3.3 Visual perception with sparse coding and locally competitive algorithm

In order to process visual stimuli, we designed an artificial neural network (ANN) to activate the output neuron if the light is on. **Figure 4** illustrates the brightness of the light captured by the stereo camera equipped with our mobile robot. The stream of visual signals captured by the camera is sent to the computer via ROS. As the images arrive, their resolution is 24x48 pixels, and the pixel brightness is normalized to the range between −1 and 1.

An ANN model based on 2D sparse coding is used for detecting the brightness of lights to further determine whether the light is on or off. The goal of sparse coding is to represent an input vector with a linear combination of features from a dictionary. This can be modeled by the LASSO optimization function:

$$E = \frac{1}{2}\|x - \Phi \cdot a\|_2^2 + \lambda \cdot \|a\|_1, \tag{4}$$

$$a^* = \underset{a}{\text{argmin}} E(a), \tag{5}$$

where the features Φ_i are columns of the dictionary matrix Φ, and the sparse code a^* is the set of coefficients a_i for which the reconstruction $(\Phi \cdot a)$ of the input x minimizes the cost (E). The sparsity penalty λ reduces the amount of non-zero terms in a by penalizing its 1-norm.

Typically, the sparse code a^* is used to reconstruct x or as the input to a classifier. In 2D image sparse coding, we divided the image into patches the size of the features

Figure 4.
Sights of the light off and on.

Φ_j. Thus, each patch is a sparse-coding optimization problem. As shown in **Figure 5**, the light image is divided into 9 sub-regions, which are further partitioned into patches. The Spiking Locally Competitive Algorithm (LCA) model [43] is used to solve Eq. (6) and Eq. (7):

$$\dot{u} = \frac{1}{\tau}\left(\Phi^T x - u - \left(\Phi^T \phi - I\right) \cdot a\right), a = T_\lambda(u) \tag{6}$$

$$T_\lambda(u) = 0 \; if \; u \leq \lambda, else \; T_\lambda(u) = u - \lambda \tag{7}$$

where a_i is the firing rate of neuron i, u_i is the average soma current, τ is the discrete time step, and T_λ is the thresholding function that determines if neuron i will fire.

Figure 6 illustrates a spiking LCA network for solving one image patch (**Figure 5**). The model consists of the first layer for the input x and the second layer for the sparse code a^*. The firing rates of each neuron are the coefficients a_i. The neurons in the second layer are referred to as feature neurons as each of them is associated with one feature Φ_i. Typically, an overcomplete dictionary is used in sparse coding, resulting in a having a larger size than x. Our model represents each patch with only two features, light and dark. Thus, the dictionary in **Figure 6** only contains two features. The dictionary could be made complete by having more feature neurons than the size of x,

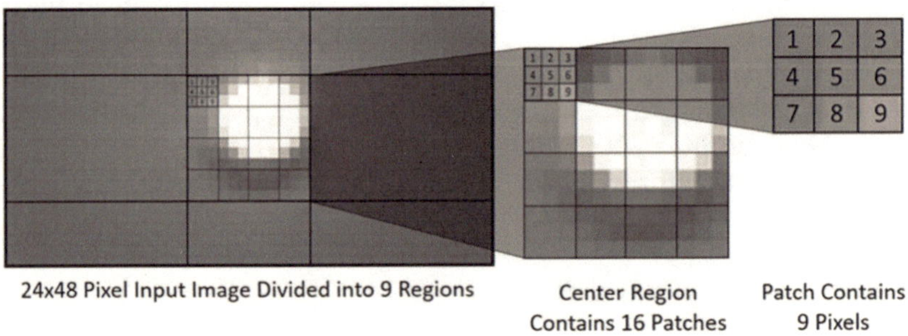

Figure 5.
Image region layout and patch structure.

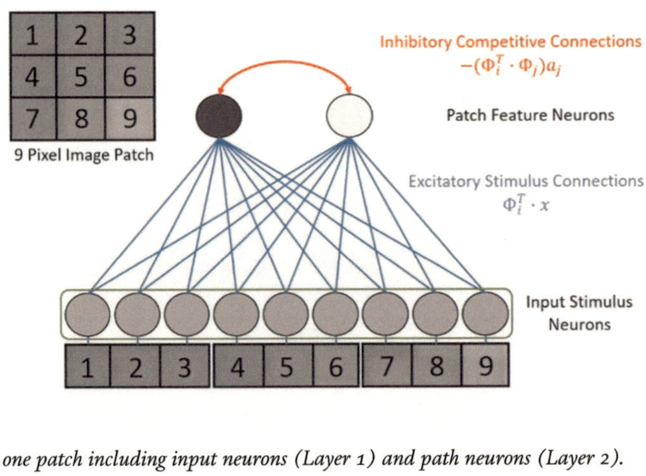

Figure 6.
LCA network for one patch including input neurons (Layer 1) and path neurons (Layer 2).

9 in this case. There could be copies or variations of light and dark features to make the dictionary overcomplete.

The convolutional stride between the image patches is equal to the width of one patch (3 pixels), resulting in no overlap between them. This simplifies the LCA model by removing connections between feature neurons in overlapping patches. The neural network contains a third layer with one neuron for each of the 9 regions in the image as illustrated in **Figure 7**. The third layer has one neuron as an output neuron integrating the light feature neurons of every patch. In **Figure 7**, each neuron at layer 1 simply has a firing rate proportional to the pixel intensity it represents, serving as a spike generator. The neurons at layer 2 and layer 3 are Integrate and Fire neurons with the parameters listed in **Table 3**. The layer 2 neurons are the LCA neurons in a single-layer LCA configuration. The parameters of the Integrate and Fire LCA feature neuron in layer 2 are implemented via the V_{reset} parameter, which was empirically adjusted until the desired response was achieved from the feature neurons. The other parameters are assigned according to the LCA model [43]. The neurons in layer 3 have the parameters derived for brightness neurons from **Table 2**.

When the high brightness signals of the light (**Figure 4a**) enter the network, the light feature neurons in the center patches start to fire. The light-off image (**Figure 4b**) subsequently reduces the activity in the neurons. These images are alternately presented as inputs, shown in **Figure 8**. This causes the neurons in layer 3 to fire for the center region from the time periods 0 s to 2 s and 4 s to 6 s, as shown in **Figure 9**. Those are the times the light is on in our experiment.

The sparse coding network operates at Intel Loihi neuromorphic chip for power and energy estimation. Our associative memory neural networks are implemented with Nengo simulator and further deployed into Loihi chips by Intel's NxSDK as a backend [44]. The NxSDK contains a built-in LCA network implementation, which can be connected to the rest of the SNN. The parameters from **Table 3** were used to create the same LCA network from Nengo for deployment on Loihi. In the Loihi chip, the synaptic weights only have a 4-bit resolution, instead of a 24-bit resolution in typical

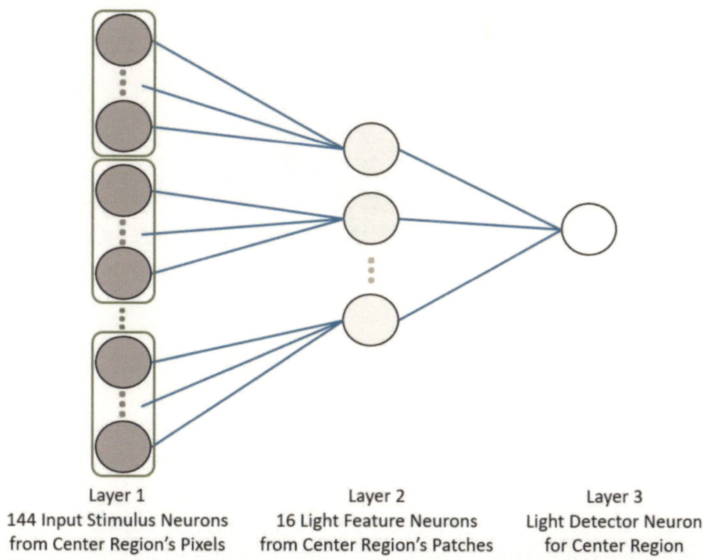

Figure 7.
Neural network for light detection in the center region. Note: The dark feature neurons in Layer 2 are not shown.

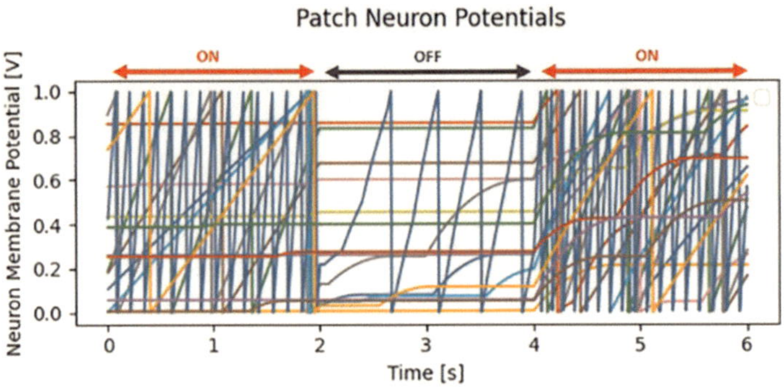

Figure 8.
Membrane potential of light feature neurons (Layer 2) in center region image patches.

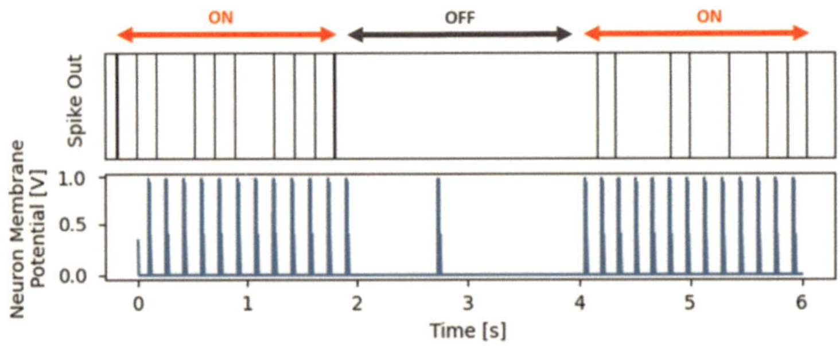

Figure 9.
Spiking signals of Layer 3 neuron.

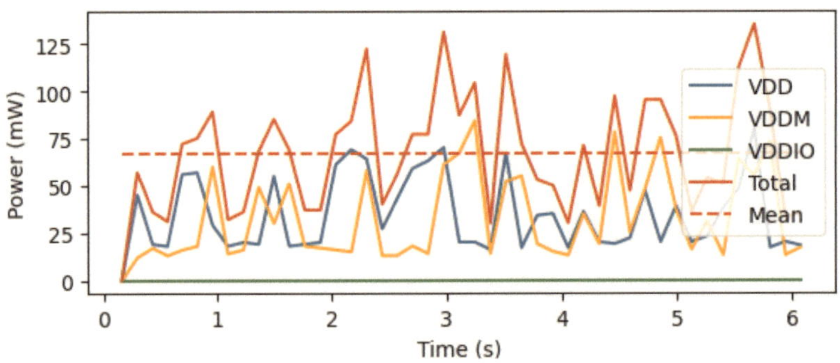

Figure 10.
Power consumption for sparse coding network with the Loihi chip. The VDD represents the compute logic; VDDM is the SRAMs, and VDDIO is the IO interface.

computational platforms. Thus, the time of one simulation is reduced from 1 ms to 20 ms at the Loihi platform. The measured powers are shown in **Figure 10**. The average VDD power is 30 mW, and the average VDDM power is 29 mW. In **Figure 10**, the VDDIO has a relatively negligible contribution to the total power, while VDD and VDDM have effectively equal contributions. The power is measured and reported with the average consumption across every 8 timesteps while the experiment is under operation.

3.4 Associative memory learning experiment and results

Figure 11 illustrates the comparison of signal pathway modification between our SLN system and fear conditioning of rats. In our design, the brightness signal from light serves as a conditional stimulus, and the acceleration from the vibration platform is the unconditional stimulus. These two signals emulate the electric shock and sound signals in fear conditioning experiments. The movement of our mobile robot away from the vibration platform emulates the fear response of rats. The red arrows (**Figure 11**) represent the weak synaptic connections in the conditional signal pathway, while the blue arrows represent the unconditional pathway with weaker synaptic connections. During associative memory learning, the synaptic connection (red arrow) will be strengthened by Hebbian learning.

In the brains of rats, shock signals and tone signals are processed at different neural regions: auditory thalamus and somatosensory thalamus. The output signals from these two regions converge at lateral nucleus [1]. Originally, the rat has no fear reaction to a neutral tone. However, when the tone is presented with a foot shock associated with tone signals, the rat starts memorizing the relationship between the tone and the shock. After multiple times, the tone alone will stimulate a fear response, indicating the accomplishment of the associative memory learning [1].

Our experimental setup to emulate fear conditioning using our mobile robot is shown in **Figure 12**. The mobile robot is placed on a testing platform, which is constructed with 9 wooden boxes. The dimension of each box is 23 in (L) × 23 in (W) × 8 in (H). A vibration plate is installed underneath the center platform, which is marked in red square in **Figure 12a**. The vibration plate can provide vibration signals (15–40 Hz) emulating an unpleasant stimulus (the electrical shock) in fear learning on the rats. The other eight platforms with no installed vibration plate are marked in green-dashed squares in **Figure 12**.

For the fear response, the robot is expected to move away from the vibration platform to the neutral location if unpleasant stimuli, such as vibration, are applied.

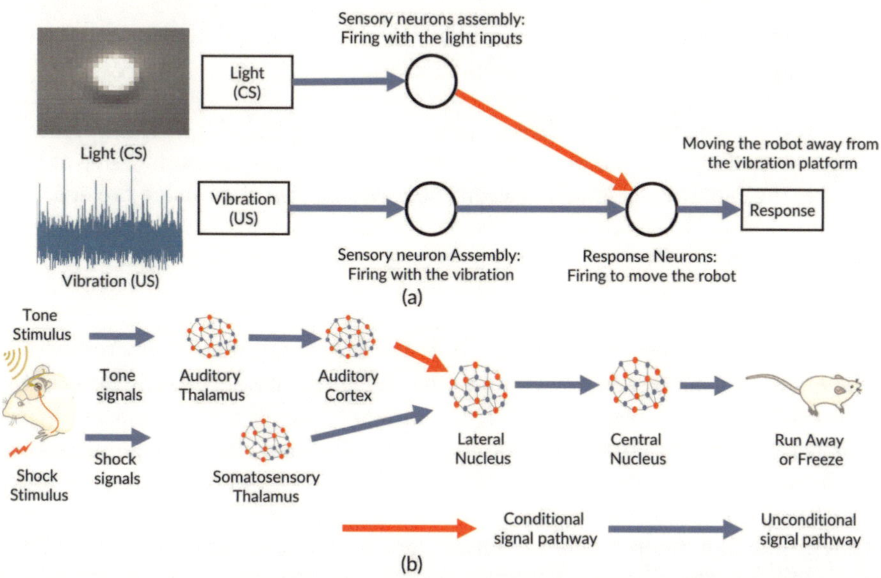

Figure 11.
(a) Our SLN system for associative memory learning implementation. (b) Fear conditioning in rats.

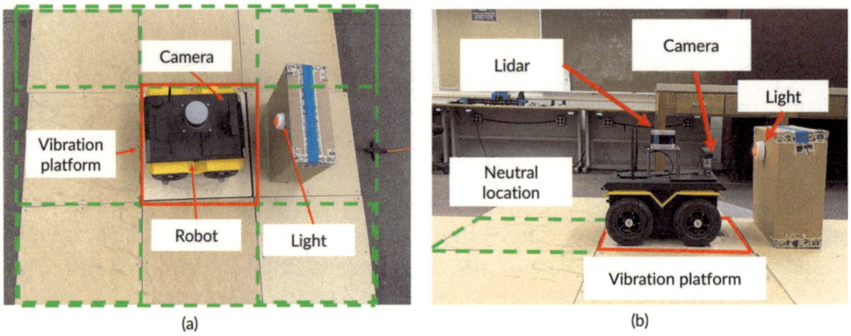

Figure 12.
Fear conditioning imitation experimental setup with the mobile robot: (a) top view; (b) side view.

The specific experimental procedures aiming to reproduce the experimental process of rats in fear conditioning are:

1. Turn on the light, expecting no response (testing conditional signal pathway).

2. Turn on the vibration platform, and the robot moves (testing the unconditional signal pathway).

3. Reset the robot position.

4. Turn on the light, and then turn on the vibration platform; the robot moves (conducting the associative memory learning process such that the conditional and unconditional stimuli are provided at the same time).

5. Reset the robot position.

6. Turn on the light alone, and the robot moves, indicating that associative memory learning is accomplished successfully.

The synaptic weights are modified based on Hebbian learning [42, 45]. Hebbian learning states that when the pre- and postsynaptic neurons are both active at the same time, the synaptic weights between them will be modified using the eqs. [42, 45]:

$$w = \eta r_i r_j, \tag{8}$$

where the r_i and r_j are firing rates of pre- and postsynaptic neurons, respectively, and the η is the learning rate, determining the changing rate of synaptic weight. In our experiment, the learning rate η is 2×10^{-4}. **Figure 13** shows that the initial synaptic weights between the brightness detection neuron (CS) and the movement neuron are small. Consequently, the brightness stimulus of the light cannot be delivered to the movement neuron to stimulate it to fire. As a result, the synaptic weights between the brightness detection neuron and movement neuron stay constant. When the vibration (US) is applied to the vibration detection neuron, the movement neuron starts to fire. When the vibration and light are applied to the system, both the brightness detection neuron and the movement neuron are active, resulting in the synaptic weight's

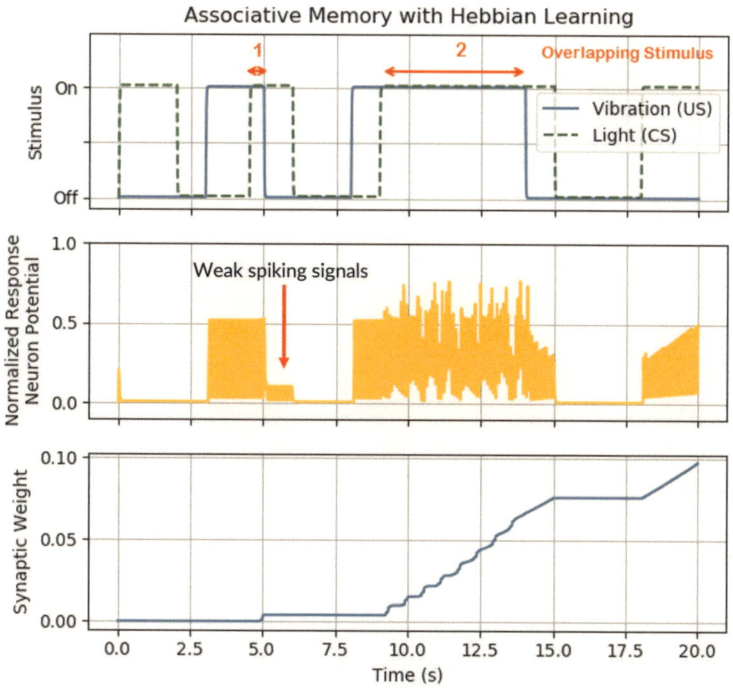

Figure 13.
Change of synaptic weight during associative memory learning.

increase. **Figure 13** illustrates that the synaptic weights increase when the vibration and light stimuli are both applied. However, the first overlapping time frame is not long enough to establish a significant synaptic weight modification. Thus, a sequence of weak spiking signals is observed from the response neuron after the vibration is removed, which is marked in **Figure 13**. In contrast, the second overlapping period is longer than the first one, which leads to a larger increase in synaptic weights. Thereby, after learning, the response neuron (movement detection neuron) will fire with a visual stimulus (light) even with no vibration stimulus. This demonstrates an accomplishment of associative memory learning.

In the experiments, the synaptic weights between the brightness detection neuron and the movement neuron are modified during the training process. As a result, after associative memory learning, the mobile robot will move away from the vibration platform under the stimulus of light, even with no vibration signal presented, demonstrating successful online learning in real-time. Compared to other state-of-the-art associative memory works listed in **Table 4**, we reproduce the classic fear conditioning experiments of rats using a mobile robot and the Loihi chip rather than simply simulation. In addition, the scales of our neural networks outperform other works.

4. Conclusion

In this paper, we implement a classic self-learning paradigm in rats: associative memory learning (fear conditioning) using a mobile robot and a neuromorphic system

	Neuron	Synapse	Dataset	Learning scheme	Biology scenarios
[9]	6	3	N/A	Simulation	N/A
[10]	3	1	N/A	Simulation	N/A
[11]	5	6	N/A	Simulation	N/A
[12]	3	1	N/A	Simulation	N/A
[46]	3	1	N/A	Simulation	N/A
[47]	3	2	N/A	Simulation	N/A
[48]	3	2	N/A	Simulation	Cellular Association in Aplysia
[6]	20	100	Pretrained with datasets	Simulation	N/A
This work	1419	1420	No dataset for pretraining	Experiment	Fear conditioning of rats

Table 4.
Comparison of scale and association capability with other state-of-the-art works.

(Loihi chip) in an online learning scenario. In specific, we use a mobile robot as the substitute for the rats in fear conditioning experiments. Two signal pathways are assigned for conditional and unconditional stimuli. In our experiments, vibration signals emulate unconditional stimuli, while brightness of lights is assigned as conditional stimuli. Originally, the mobile robot only moves when it detects vibration signals. After providing these two signals at the same time several times, the robot performs a movement when light signals are present alone. The detections of lights and vibrations are implemented with Integrate and Fire Neurons. In addition, the movement of the robot is controlled by the specific-designed response neurons. The signal pathway modification during associative memory learning is implemented with Hebbian learning. Compared to other state-of-the-art works, our work successfully reproduces the fear conditioning of rats in a real-world scenario with no labeled data and pretraining process.

Acknowledgement

As Intel Neuromorphic Research Community (INRC) members, the authors would like to thank Intel Labs for providing the neuromorphic Loihi chips for our online learning studies.

Author details

Noah Zins[1], Yan Zhang[2] and Hongyu An[1*]

1 Department of Electrical and Computer Engineering, Michigan Tech, Houghton, Michigan, USA

2 Department of Biological Sciences, Michigan Tech, Houghton, Michigan, USA

*Address all correspondence to: hongyua@mtu.edu

References

[1] Kandel ER, Schwartz JH, Jessell TM, Siegelbaum SA, Hudspeth A. Principles of Neural Science. New York: McGraw-Hill; 2000

[2] Sun J, Han G, Zeng Z, Wang Y. Memristor-based neural network circuit of full-function pavlov associative memory with time delay and variable learning rate. In: IEEE Transactions on Cybernetics. New York, USA: IEEE; 2019

[3] Kohonen T. Self-Organization and Associative Memory. New York, USA: Springer Science & Business Media; 2012

[4] Goodfellow I, Yoshua B, Aaron C. Deep learning. Learning. 2016;**2016**:785. DOI: 10.1016/B978-0-12-391420-0.09987-X

[5] Devlin J, Chang M.-W., Lee K, Toutanova K. Bert: Pre-training of deep bidirectional transformers for language understanding. *arXiv preprint arXiv:1810.04805.* 2018

[6] An H, An Q, Yi Y. Realizing behavior level associative memory learning through three-dimensional memristor-based neuromorphic circuits. In: IEEE Transactions on Emerging Topics in Computational Intelligence. New York, USA: IEEE; 2019

[7] Hu SG et al. Associative memory realized by a reconfigurable memristive Hopfield neural network. Nature Communications. 2015;**6**:1-5. DOI: 10.1038/ncomms8522

[8] Moon K et al. Hardware implementation of associative memory characteristics with analogue-type resistive-switching device. Nanotechnology. 2014;**25**:(49). DOI: 10.1088/0957-4484/25/49/495204

[9] Yang J, Wang L, Wang Y, Guo T. A novel memristive Hopfield neural network with application in associative memory. Neurocomputing. 2017;**227**:142-148. DOI: 10.1016/j.neucom.2016.07.065

[10] Liu X, Zeng Z, Wen S. Implementation of memristive neural network with full-function pavlov associative memory. IEEE Transactions on Circuits and Systems I: Regular Papers. 2016;**63**(9):1454-1463

[11] Hu X, Duan S, Chen G, Chen L. Modeling affections with memristor-based associative memory neural networks. Neurocomputing. 2017;**223**:129-137. DOI: 10.1016/j.neucom.2016.10.028

[12] Moon K et al. Hardware implementation of associative memory characteristics with analogue-type resistive-switching device. Nanotechnology. 2014;**25**(49):495204

[13] Eryilmaz SB et al. Brain-like associative learning using a nanoscale non-volatile phase change synaptic device array. Frontiers in Neuroscience Switzerland. 2014;**8**. Available from: https://www.frontiersin.org/articles/10.3389/fnins.2014.00205/full

[14] Roy DS et al. Brain-wide mapping reveals that engrams for a single memory are distributed across multiple brain regions. Nature Communications. 2022;**13**(1):1-16

[15] Josselyn SA, Tonegawa S. Memory engrams: Recalling the past and imagining the future. Science. 2020;**367**(6473):eaaw4325

[16] Nomura H, Teshirogi C, Nakayama D, Minami M, Ikegaya Y.

Prior observation of fear learning enhances subsequent self-experienced fear learning with an overlapping neuronal ensemble in the dorsal hippocampus. Molecular Brain. 2019; **12**(1):1-8

[17] DeNardo LA et al. Temporal evolution of cortical ensembles promoting remote memory retrieval. Nature Neuroscience. 2019;**22**(3): 460-469

[18] Khalaf O, Resch S, Dixsaut L, Gorden V, Glauser L, Gräff J. Reactivation of recall-induced neurons contributes to remote fear memory attenuation. Science. 2018;**360**(6394): 1239-1242

[19] Kitamura T et al. Engrams and circuits crucial for systems consolidation of a memory. Science. 2017;**356**(6333): 73-78

[20] Mead C. Neuromorphic electronic systems. Proceedings of IEEE. 1990; **78**(10):1629-1636

[21] Roy K, Jaiswal A, Panda P. Towards spike-based machine intelligence with neuromorphic computing. Nature. 2019; **575**(7784):607-617

[22] Zins N, Zhang Y, Yu C, An H. Neuromorphic computing: A path to artificial intelligence through emulating human brains. In: Frontiers of Quality Electronic Design. New York, USA: Springer; 2023. pp. 259-296

[23] An H. Powering Next-Generation Artificial Intelligence by Designing Three-Dimensional High-Performance Neuromorphic Computing System with Memristors. DriveBlacksburg, VA: Virginia Tech; 2020

[24] Bai K, Yi Y. Opening the "Black Box" of Silicon Chip Design in Neuromorphic Computing in Bio-Inspired Technology. London, UK: IntechOpen; 2019

[25] Bai K, Yi Y. DFR: An energy-efficient Analog delay feedback reservoir computing system for brain-inspired computing. ACM Journal on Emerging Technologies in Computing Systems (JETC). 2018;**14**(4):45

[26] Baird E, Srinivasan MV, Zhang S, Cowling A. Visual control of flight speed in honeybees. Journal of Experimental Biology. 2005;**208**(20): 3895-3905

[27] Kern R, Boeddeker N, Dittmar L, Egelhaaf M. Blowfly flight characteristics are shaped by environmental features and controlled by optic flow information. Journal of Experimental Biology. 2012;**215**(14):2501-2514

[28] Kempter R, Gerstner W, Van Hemmen JL. Hebbian learning and spiking neurons. Physical Review E. 1999;**59**:4498-4514. DOI: 10.1103/PhysRevE.59.4498

[29] Levy N, Horn D, Meilijson I, Ruppin E. Distributed synchrony of spiking neurons in a Hebbian cell assembly. Advances in Neural Information Processing Systems. 2000; **14**:129-135

[30] Van Rossum MC, Bi GQ, Turrigiano GG. Stable Hebbian learning from spike timing-dependent plasticity. Journal of Neuroscience. 2000;**20**(23): 8812-8821

[31] Caporale N, Dan Y. Spike timing–dependent plasticity: A Hebbian learning rule. Annual Review of Neuroscience. 2008;**31**:25-46

[32] Bai K, An Q, Yi Y. Deep-DFR: A memristive deep delayed feedback reservoir computing system with hybrid

neural network topology. In: Proceedings of the 56th Annual Design Automation Conference 2019. New York, USA: ACM; 2019. p. 54

[33] Davies M et al. Advancing neuromorphic computing with Loihi: A survey of results and outlook. Proceedings of IEEE. 2021;**109**(5): 911-934

[34] Davies M et al. Loihi: A neuromorphic manycore processor with on-chip learning. IEEE Micro. 2018; **38**(1):82-99

[35] Orchard G et al. Efficient neuromorphic signal processing with loihi 2. In: 2021 IEEE Workshop on Signal Processing Systems (SiPS). New York, USA: IEEE; 2021. pp. 254-259

[36] Pavlov PI. Conditioned reflexes: An investigation of the physiological activity of the cerebral cortex. Annals of Neurosciences. 2010;**17**(3):136. DOI: 10.5214/ans.0972-7531.1017309

[37] Lansner A. Associative memory models: From the cell-assembly theory to biophysically detailed cortex simulations. Trends in Neurosciences. 2009;**32**(3):178-186

[38] Snider GS. Self-organized computation with unreliable, memristive nanodevices. Nanotechnology. 2007;**18**. DOI: 10.1088/0957-4484/18/36/365202

[39] Greig WJ. Integrated circuit packaging, assembly and interconnections. Integrated Circuit Packaging, Assembly and Interconnections. 2007;**2007**:1-296. DOI: 10.1007/0-387-33913-2

[40] Bekolay T et al. Nengo: A Python tool for building large-scale functional brain models. Frontiers in Neuroinformatics. 2014;**7**:48

[41] DiLuoffo V, Michalson WR, Sunar B. Robot operating system 2: The need for a holistic security approach to robotic architectures. International Journal of Advanced Robotic Systems. 2018;**15**(3):172

[42] Miller P. An Introductory Course in Computational Neuroscience. Cambridge, MA: MIT Press; 2018

[43] Tang PTP, Lin T.-H, Davies M. Sparse coding by spiking neural networks: Convergence theory and computational results. *arXiv preprint arXiv:1705.05475*. 2017

[44] Rueckauer B, Bybee C, Goettsche R, Singh Y, Mishra J, Wild A. NxTF: An API and compiler for deep spiking neural networks on Intel Loihi. ACM Journal on Emerging Technologies in Computing Systems (JETC). 2022;**18**(3):1-22

[45] Dayan P, Abbott LF. Theoretical neuroscience: computational and mathematical modeling of neural systems. Computational Neuroscience Series. 2001;**2001**:281-328

[46] Ziegler M et al. An electronic version of Pavlov's dog. Advanced Functional Materials. 2012;**22**(13):2744-2749. DOI: 10.1002/adfm.201200244

[47] Pershin YV, Di Ventra M. Experimental demonstration of associative memory with memristive neural networks. Neural Networks. 2010;**23**(7):881-886

[48] An H, Zhou Z, Yi Y. Memristor-based 3D neuromorphic computing system and its application to associative memory learning. In: 2017 IEEE 17th International Conference on Nanotechnology, NANO 2017. New York, USA: IEEE; 2017. pp. 555-560. DOI: 10.1109/NANO.2017.8117459

Section 2

Emerging Devices for Neuromorphic Computing

Chapter 3

Study of RRAM-Based Binarized Neural Networks Inference Accelerators Using an RRAM Physics-Based Compact Model

Tommaso Zanotti, Paolo Pavan and Francesco Maria Puglisi

Abstract

In-memory computing hardware accelerators for binarized neural networks based on resistive RAM (RRAM) memory technologies represent a promising solution for enabling the execution of deep neural network algorithms on resource-constrained devices at the edge of the network. However, the intrinsic stochasticity and nonidealities of RRAM devices can easily lead to unreliable circuit operations if not appropriately considered during the design phase. In this chapter, analysis and design methodologies enabled by RRAM physics-based compact models of LIM and mixed-signal BNN inference accelerators are discussed. As a use case example, the UNIMORE RRAM physics-based compact model calibrated on an RRAM technology from the literature, is used to determine the performance vs. reliability trade-offs of different in-memory computing accelerators: i) a logic-in-memory accelerator based on the material implication logic, ii) a mixed-signal BNN accelerator, and iii) a hybrid accelerator enabling both computing paradigms on the same array. Finally, the performance of the three accelerators on a BNN inference task is compared and benchmarked with the state of the art.

Keywords: RRAM, logic-in-memory, binarized neural networks, compact modeling, in-memory computing

1. Introduction

The increasing adoption of devices for the internet of things (IoT) combined with the diffusion of data-driven computing approaches, such as artificial neural networks [1], is promoting innovation in multiple sectors. Thus, to sustain the diffusion of these technologies, a constant research effort is being directed to the development of ultra-low power computing devices and architectures that will enable the execution of complex computations directly on low-power devices at the edge of the network, reducing the burden on the cloud computing infrastructure. In-memory computing (IMC) architectures based on new nanoelectronics devices, are considered a key enabling technology. In these architectures, computations are executed directly inside or in the proximity of the memory array, removing the main source of inefficiencies, i.e., the von Neumann bottleneck (VNB), which is caused by the time and energy hungry data transfer between the memory and the processing unit. Among novel nonvolatile memory devices, resistive random access memory (RRAM) technologies are one of the most mature and are considered as the frontrunners for enabling the diffused adoption of IMC accelerators, thanks to their low-cost, back end of line compatibility, high-density, high switching speed, and relatively low programming voltages. Still, RRAM technologies present several nonideal effects (e.g., cycle-to-cycle (C2C) and device-to-device variations, and random telegraph noise (RTN)) which can negatively impact the reliability of RRAM-based circuits and limit the number of bits that can be reliably stored in a single device.

Thus, to address these nonideal effects appropriate design methodologies need to be followed, and IMC accelerators in which RRAM devices are used as binary elements preferred. Among these IMC accelerators, logic-in-memory (LIM) [2–5] and mixed-signal binarized neural networks (BNN) inference accelerators [6–8] are promising solutions for resource-constrained edge devices. LIM accelerators enable the in-memory computation of logic operations. BNNs [9] are neural networks in which neurons' weights and activations are encoded using a single bit, and 2D RRAM arrays can be used to compute vector-matrix multiplication (VMM) operations (i.e., the most executed computation in neural networks) in the analog domain, achieving high energy efficiency and performance.

In this chapter, analysis and design methodologies enabled by an RRAM physics-based compact model, are discussed and applied to LIM and mixed-signal BNN inference accelerators. As a use case example, the UNIMORE RRAM physics-based compact model [10, 11] calibrated on an RRAM technology from the literature [12], is used to extract performance vs. reliability trade-offs of different IMC accelerators: (i) a LIM accelerator based on the material implication logic, (ii) a mixed-signal BNN accelerator, and (iii) a hybrid accelerator enabling the coexistence of both computing paradigms on the same array. Finally, the performance of the three accelerators on a BNN inference task are compared and benchmarked with a state-of-the-art solution.

2. Resistive RAM memory technologies and compact modeling

RRAM technologies are considered a promising nonvolatile memory technology for next-generation embedded systems applications, thanks to their low fabrications cost, small feature size, back end of line compatibility, and performance. Specifically, RRAMs typically provide fast switching speed (i.e., < 10 ns [13]), low programming energy (i.e., < 1pJ [13]), long retention (i.e., > 10 years [13]), high endurance (i.e., 10^6–10^{12} [13]), and a relatively large memory window (i.e., > 10^2 [13]) when used as binary memory elements. These devices can be electrically switched between nonvolatile a low and a high resistive state (LRS and HRS, respectively) that can be used to encode a logic 1 and a logic 0, respectively. Depending on the materials used for their fabrication, different switching mechanisms can occur [14, 15]. In this chapter, bipolar metal-oxide-based RRAM cells that exhibit filamentary switching are considered, but the methods described apply also to other resistive switching memory technologies.

A bipolar metal-oxide-based RRAM is programmed into a LRS by biasing it with a sufficiently large positive voltage, which causes the bond breakage and drift toward the top electrode (TE) of oxygen ions in the metal oxide layer, resulting in the formation of an oxygen deficient conductive filament (CF) assisting ohmic-like drift conduction [14, 16]. During a device set, a current compliance (I_C) must be provided to prevent dielectric breakdown. The device is reset into a HRS by biasing it with a sufficiently negative voltage (i.e., V_{RESET}) which induces drift and recombination of oxygen ions with oxygen vacancies in the dielectric which partially dissolve the CF. The current conduction in HRS is associated with trap-assisted tunneling in the dielectric [14, 16]. Due to the physical mechanisms behind the device's operation, these devices present intrinsic nonidealities. Specifically, programming C2C variations of the CF morphology and the dielectric composition [17], result in stochastically distributed resistive states which can influence the available memory window and circuit reliability. Also, random telegraph noise (RTN), which is caused by the effect of interstitial oxygen ions and vacancies [11, 18], is typically observed in low voltage reads.

2.1 The UNIMORE RRAM physic-based compact model

When designing novel RRAM-based circuits, RRAM devices nonideal effects and complex switching mechanism, result in non-trivial design constraints that need appropriate analysis tools to be studied. Thus, compact models of RRAM devices have been developed to enable circuit simulations [19]. Different compact models typically adopt different approximations, and two main categories can be identified, i.e., general purpose and physics-based compact models. The formers, typically adopt simpler equations that enable to simulate larger circuits, however at the expense of lower accuracy when simulating the device operation outside the operating range considered for parameter calibration, and a lack of a clear mapping between model and technology related parameters. Conversely, in physics-based compact models, parameters and equations are linked to the device physics, enabling to reproduce the device characteristic in multiple operating conditions [11] and simplifying the parameter extraction [20]. Thus, as a first step when studying the performance and reliability tradeoffs of novel RRAM-based circuits, physics-based compact models should be employed [21].

For the analysis and use-case examples reported in this chapter, the UNIMORE RRAM physics-based compact model available in [10, 11], is used. Other physics-based compact models could also be used to perform the analysis [21–23], however for other compact models, clear parameter extraction procedures are currently not available [20]. The UNIMORE RRAM compact model is implemented in Verilog-A and reproduces the effects of self-heating, C2C variability, multi-level RTN, and is completed by an automated parameter extraction procedure [20]. The RRAM device approximated by the compact model is shown in **Figure 1a** and **b** for a device in LRS and HRS, respectively. The compact model considers a CF and a dielectric barrier whose thickness x is modulated by means of a system of differential equations, which take into account thermal effects, the field-driven oxygen ions drift and recombination during the device reset, and the field-accelerated bond breaking during the device set [10, 11]. Internally, the compact model includes two sub modules that reproduce the effects of C2C variability and RTN, as shown in **Figure 1b**. A detailed description of the modeling of C2C variability and RTN is available in [11, 18]. The compact model was calibrated in [11] on 4 RRAM metal-oxide-based bipolar RRAM technologies. In the rest of the chapter, data and results of simulations are reported considering the model calibrated on the *TiN/HfO$_x$/AlO$_x$/Pt* RRAM technology from [12], which has an I_C of 100 μA. The calibrated compact model can well reproduce the experimental data in different operating conditions, as shown in **Figure 1c** and **d**, the

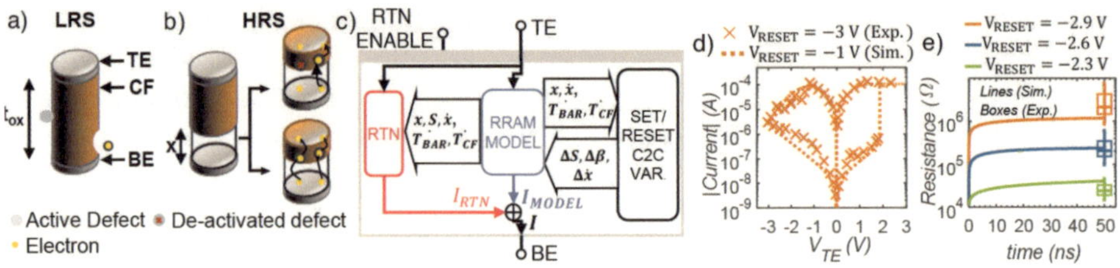

Figure 1.
Representation of an RRAM device as approximated in the compact model for a device in (a) LRS and (b) HRS, respectively. The defects causing RTN are also shown. (b) Functional block diagram of the compact model. The internal models for the device nonidealities are shown. Experimental and simulated, (c) quasi-static IV, and (d) pulsed reset characteristics of a TiN/HfOx/AlOx/Pt device. Data from [12].

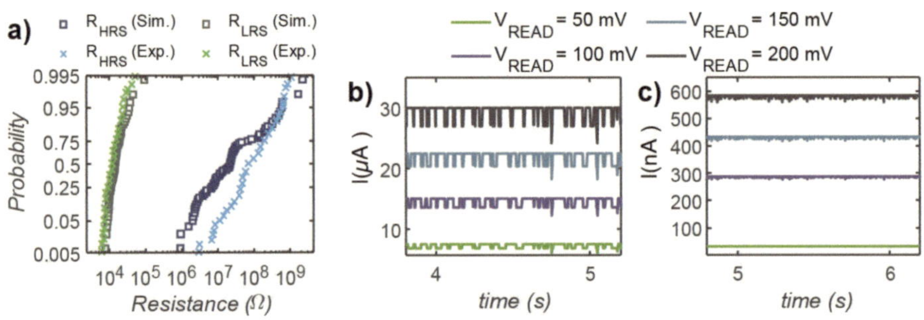

Figure 2.
Experimental and simulated cycle-to-cycle variability at different reset voltages. (b) and (c) simulated RTN at different read voltages for a device in (b) LRS, and (c) HRS.

experimental probability distributions of the resistive states, see **Figure 2a**, and multi-level RTN signals, as shown in **Figure 2b** and **c**. Simulations are performed with Cadence Virtuoso®.

2.2 Using compact models to study RRAM-based IMC architectures

As discussed in the following sections, IMC accelerators can be quite complex, and composed of a large number of RRAM devices and components. As the size of the accelerators increases, circuit simulation can quickly require excessive computing resources that are often not available. A possible approach is to study the problem at different abstraction layers. Specifically, the three steps approach shown in **Figure 3** can be followed. The first step of this methodology requires the electrical characterization of novel emerging nonvolatile memory devices, and the application of parameter extraction strategies to calibrate compact models on the specific technology. In the second step, the compact model is used to perform circuit simulations of smaller portion of the entire architecture. Specifically, the functionality of the core of operations of the in-memory computing framework of interest is first simulated using the compact model without including nonideal effects. The results of these simulations can provide indications regarding valid circuit operating points and the existing design tradeoffs. Parametric simulations including the device nonideal effects are then used to identify appropriate operating points that satisfy the desired reliability vs. performance tradeoffs, and to estimate the performance in the worst-case (WC in **Figure 3**) corners of the circuit. Finally, the results from the previous analysis can be used to estimate the performance of a more complex design based on the core

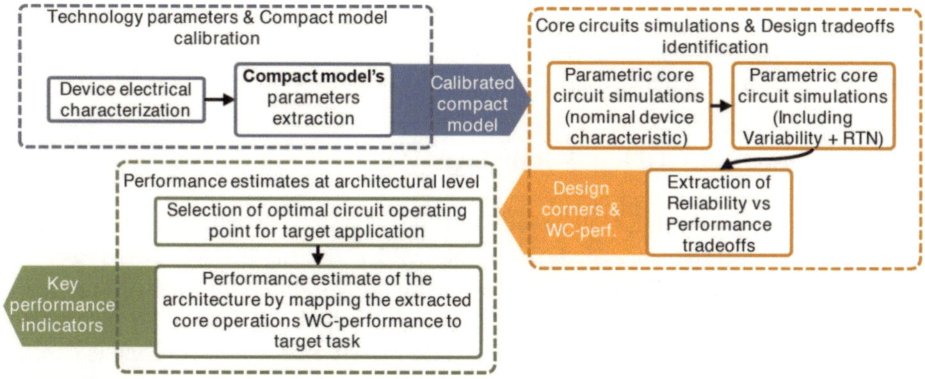

Figure 3.
Flow-chart of the methodology enabled by the RRAM physics-based compact model that can be used to design reliable RRAM-based in-memory computing architectures and to estimate their performance.

operations studied with the compact model. Specifically, the worst-case performance estimates for the core operations are mapped to each operation executed at the architectural level. Although this step introduces several approximations, the use of the worst-case performance for each operation results in an underestimation of the energy efficiency therefore providing a sufficient headroom to account, to first order, for additional energy overheads possibly existing in the complete architecture.

A use case example of this methodology is discussed in the following sections of this chapter. In Section 3, the compact model is used to study the performance and reliability of the core operations of a LIM architecture, while in Section 5 the results of analysis enabled by the compact model are used to estimate the performance, at a higher abstraction level, of an inference accelerator based on BNNs.

3. Logic-in-memory with RRAM devices

In recent years, several IMC hardware accelerators based on resistive memory technologies have been proposed in the literature as a solution for improving computation efficiency, especially for data intensive tasks. Among these accelerators, LIM circuits enable the computation of logic operations on arrays of RRAM devices. Two different approaches can be commonly distinguished, i.e., stateful and non-stateful computing paradigms. Stateful LIM frameworks include the material implication [4, 24] and the memristor-aided logics (MAGIC) [2]. In these approaches, RRAM devices are treated at the same time as computing and storing elements. When performing an operation, inputs are stored as the resistive state of RRAM devices, and by using specific voltage pulses, the resistive state of an output device is either changed or preserved depending on the input combination. Non-stateful LIM approaches such as the scouting logic [25] and the smart material implication logic (SIMPLY) [26], instead, exploit the peripheral circuit to perform part of the computations. Inputs are encoded in the nonvolatile resistance of RRAM devices which are read in parallel with small voltages (i.e., ≈100 mV). The voltage at a specific circuit node changes depending on the input combination, and by sensing this voltage it is possible to implement different logic operations.

In the following subsections, the compact model is used to study the performance and reliability of a stateful and a non-stateful LIM frameworks based on the material implication logic. In addition, in Section 3.3 the performance and characteristics of different LIM frameworks are compared considering the computation of a 1-bit full addition operation as a benchmark.

3.1 RRAM-based material implication logic

The material implication logic is type of stateful logic that is functionally complete [27] and is based on two core operations, the IMPLY and the FALSE. These two operations were demonstrated in [4] to be easily implemented with RRAM devices and a circuit such as the one reported in **Figure 4a**, which consists of RRAM devices, that store the inputs and outputs of IMPLY and FALSE operations and act as computing elements, a control logic and analog tri-state drivers that deliver appropriate voltages to RRAM devices, and a resistor R_G. The IMPLY is a two inputs one output operation and its truth table is reported in **Figure 4b**. To execute an IMPLY operation, the inputs are encoded in the resistive state of devices P and Q. Then, the control logic delivers two voltage pulses with amplitudes V_{COND} and V_{SET}, on P and Q, respectively. The voltage pulse amplitudes are dimensioned so that the state of device P never changes, while the state of device Q changes according to the truth table in **Figure 4b**. The FALSE operation is a 1 input 1 output logic operation which always results in logic 0 output and is implemented by delivering a negative voltage V_{FALSE} to an RRAM device.

However, the use of high voltage pulses to conditionally program the output device in the IMPLY operation, leads to intrinsic reliability challenges that need to be studied with accurate compact models, including the effects of temperature and variability. Using the UNIMORE RRAM physics-based compact model, simulations of the IMPLY operation performed in [5] for different pairs of V_{SET} and V_{COND} voltages show that only a very narrow design space exists. As shown in **Figure 4c**, voltage variation of tens of mV, easily introduced by voltage drivers and line parasitic resistances, can lead to a malfunction of the circuit [5]. Also, the use of high voltage pulses induces the problem of logic state degradation [3, 5, 26] on devices that should preserve a HRS after the operation execution. These are devices P and Q in the first and third cases of the truth table, respectively. In these cases, although the voltage drop across these RRAM devices cannot fully switch a device, it can induce a drop of their resistance. Repeatedly executing IMPLY operations on these devices eventually leads to a bit

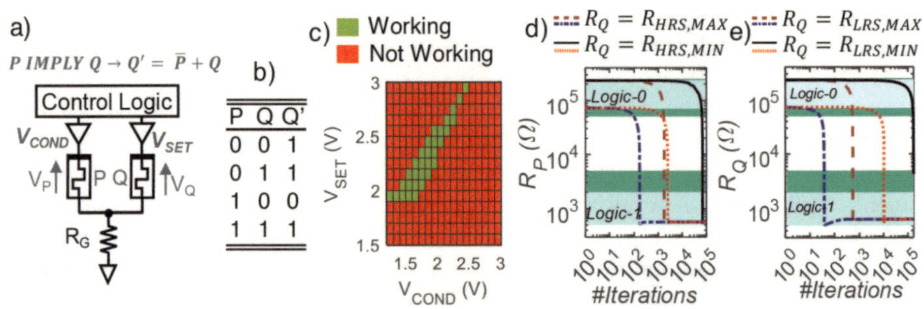

Figure 4.
(a) RRAM-based material implication logic core gate. (b) IMPLY operation truth table. Q' is the state of Q after the operation execution. (c) Map of the V_{SET} and V_{COND} pairs leading to a correct IMPLY gate operation. The effect of C2C variations is considered. Data from [5]. d), and e) effect of the resistive state degradation on device P and Q, due to the repeated execution of IMPLY operations on the same input device when both inputs are zero and when P = 1 and Q = 0, respectively.

corruption. This effect cannot be prevented completely, due to existence of opposite requirements on V_{COND} for the two cases of the truth table. The number of cycles before a corruption depends on the V_{SET} and V_{COND} pair, but also on the initial device resistance, as shown in **Figure 4d** and **e**.

Thus, although promising, the analysis performed with the compact model highlighted that this stateful LIM framework is intrinsically unreliable and other approaches should be preferred.

3.2 The smart material implication logic

The SIMPLY LIM framework was proposed in [26] as a more reliable and efficient solution to implement the material implication logic on RRAM devices. The circuit used to implement an IMPLY operation is similar to the one used in the stateful approach with the addition of a comparator which is feedback to the control logic, see **Figure 5a**. Since, during an IMPLY operation, the state of the output device Q changes only when both P and Q are in HRS, in the SIMPLY framework the operation is split into a read step, which detects this input configuration, followed by a conditional programming step, as sketched in **Figure 5b**. During a read step small (i.e., ≈ 100 mV) voltage pulses are delivered to the input RRAM devices. As a result, the voltage across R_G (i.e., V_N) when both inputs are in HRS is lower than in all the other cases. Thus, a sufficient read margin exists to discriminate the first case of the truth table from all the others by using a comparator with an appropriate threshold V_{TH}. In the conditional programming step, the control logic delivers V_{SET} to Q only when the comparator detects the first case of the truth table, while it keeps the analog drivers in high impedance otherwise. Since only a sufficient read margin needs to be ensured, this approach is intrinsically more reliable than the stateful one. In fact, circuit simulations performed in [26, 28, 29] with the compact model, including the effect of variability and RTN showed that a sufficiently large read margin is easily obtained by tuning the read voltage and V_{FALSE} (i.e., higher HRS leads to larger memory windows). Also, simulations confirmed that by delivering only small read voltages to the RRAM devices, the effect of variability is virtually solved. In addition, V_{COND} is no more

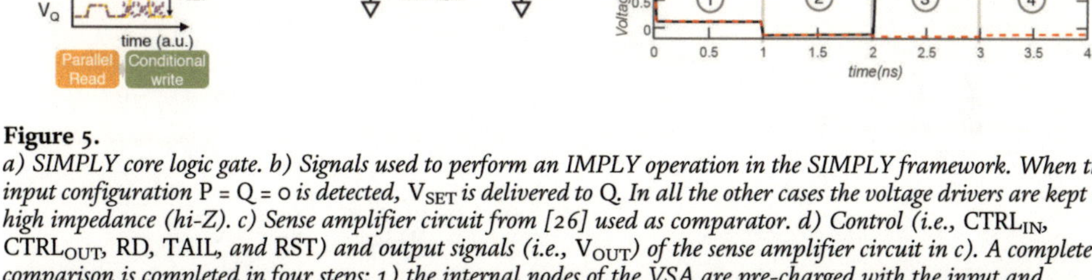

Figure 5.
a) SIMPLY core logic gate. b) Signals used to perform an IMPLY operation in the SIMPLY framework. When the input configuration P = Q = 0 *is detected,* V_{SET} *is delivered to Q. In all the other cases the voltage drivers are kept in high impedance (hi-Z). c) Sense amplifier circuit from [26] used as comparator. d) Control (i.e.,* $CTRL_{IN}$, $CTRL_{OUT}$, RD, TAIL, *and* RST*) and output signals (i.e.,* V_{OUT}*) of the sense amplifier circuit in c). A complete comparison is completed in four steps: 1) the internal nodes of the VSA are pre-charged with the input and threshold voltage (i.e.,* V_{TH}*); 2) the comparison is performed by turning on the* TAIL *transistor; 3) the result of the comparison is passed to the output node; 4) the control logic delivers the conditional* V_{SET} *pulse and resets the internal nodes of the VSA.*

required, and since V_{SET} is applied only when P and Q are zero, in three out of four cases of the truth table the energy consumption is greatly reduced with respect to the stateful implementation, as discussed in [28]. Additional energy can be saved using the same approach on the FALSE operation, thus splitting the operation into a read step followed by a conditional V_{FALSE} pulse. When the device is already in the HRS, delivering V_{FALSE} to the device would waste additional energy.

Also, the multi-input IMPLY operation was proposed in [30] to speed up computations, and its feasibility up to four inputs operations was demonstrated by means of circuit simulations on SIMPLY-based architectures in [31]. In the SIMPLY framework the multi-input IMPLY operation (hereafter called n-IMPLY, where n indicates the number of inputs) is executed in the same way as the two inputs IMPLY operation, with the only difference that all the input devices are read in parallel. V_{SET} is then delivered to the output device only when all the inputs are in HRS. The logic operation is equivalent to $Q' = Q + \overline{P + S + \ldots}$, where the output is written on Q while P and S are other input devices. Although the read margin at the input of the comparator decreases for increasing number of inputs, the same voltage threshold V_{TH} can be used. The SIMPLY architecture enabling the execution of n-IMPLY operations is referred to as n-SIMPLY in this chapter.

To correctly evaluate the performance of SIMPLY-based architectures it is important to also evaluate the performance of the comparator. The latter was implemented as a voltage sense amplifier (VSA) designed in [26] with a 45 nm technology from [32]. The circuit of the VSA and the timing of the control signals are reported in **Figure 5c** and **d**. The proposed design consumes less than 8 fJ per comparison when V_{DD} is 2 V and the temperature is in the range 0 to 85°C.

3.2.1 Full adder implementation on a n-SIMPLY-based architecture

To implement more complex logic functions, sequences of IMPLY and FALSE operations and a sufficient number of devices are required. To highlight the remarkable energy efficiency of the n-SIMPLY architecture here we consider the implementation of a full adder (FA). As shown in **Figure 6a**, to implement a 1-bit FA an array with at least 8 RRAM devices is needed. These devices are used to store the inputs (i.e., IN_1, IN_2, and C_{IN}), the outputs (i.e., S, C_{OUT}), and partial results (i.e., M_1, M_2, and M_3) of the operation. To compute the result of the 1-bit addition, the 15 computing steps of n-IMPLY and FALSE operations reported in **Figure 6b** need to be executed sequentially [31]. The circuit in **Figure 6a** was simulated in [31] using the compact model and considering a clock frequency of 500 MHz, to estimate its performance. As reported in [31] the computation of a 1-bit FA consumes a worst-case

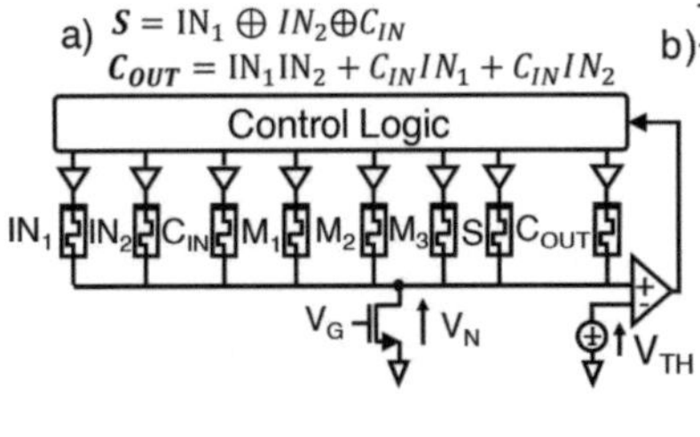

b)

#Step	Operation	Result
1-5	FALSE(M_1,M_2,M_3,S,C_{OUT})	$M_1 = M_2 = M_3 = S = C_{OUT} = 0$
6	2-IMPLY(IN_1, M_1)	$M_1 = \overline{IN_1}$
7	2-IMPLY(IN_2, M_2)	$M_2 = \overline{IN_2}$
8	2-IMPLY(C_{IN}, M_3)	$M_3 = \overline{C_{IN}}$
9	3-IMPLY(M_1,M_2, C_{OUT})	$C_{OUT} = C_{OUT} + \overline{M_1 + M_2}$
10	3-IMPLY(M_2, M_3, C_{OUT})	$C_{OUT} = C_{OUT} + \overline{M_2 + M_3}$
11	3-IMPLY(M_1, M_3, C_{OUT})	$\boldsymbol{C_{OUT} = C_{OUT} + \overline{M_1 + M_3}}$
12	4-IMPLY(M_1, IN_2, C_{IN}, S)	$S = S + \overline{M_1 + IN_2 + C_{IN}}$
13	4-IMPLY(IN_1, IN_2, M_3, S)	$S = S + \overline{IN_1 + IN_2 + M_3}$
14	4-IMPLY(M_1, M_2, M_3, S)	$S = S + \overline{M_1 + M_2 + M_3}$
15	4-IMPLY(IN_1, M_2, C_{IN}, S)	$\boldsymbol{S = S + \overline{IN_1 + M_2 + C_{IN}}}$

Figure 6.
a) Implementation of a 1-bit FA on the n-SIMPLY architecture. b) Sequence of computing steps from [31] required to compute the output for a 1-bit FA when exploiting the n-SIMPLY framework.

energy of 4.2 pJ and it is computed in 60 ns. Although these numbers are orders of magnitude higher than those achievable with full CMOS implementations, n-SIMPLY-, and LIM-based architectures in general, provide considerable advantages when data intensive tasks are considered. Thus, in [31] the parallel execution on conventional CMOS circuits and on the n-SIMPLY architecture of 512 32-bit FA operations was compared. When including the VNB overhead, the n-SIMPLY architecture was demonstrated to achieve a $> 10^6$ energy delay product (EDP) improvement with respect to the CMOS implementation.

3.3 Comparison with other RRAM-based logic circuits

Although in this chapter material implication-based LIM frameworks are analyzed as a use case example, other approaches have been studied in the literature. Here follows a description and an analysis of the main characteristics and performance of the most common LIM frameworks in which RRAM devices are used both as computing and as memory elements, and hence can enable the implementation of non-von Neumann computing architectures.

3.3.1 Memristor ratioed logic

Although the Memristor Ratioed Logic (MRL) [33, 34] enables to fabricate logic gates with smaller footprints with respect to conventional CMOS logic circuits, it is not included in the analysis because in this framework RRAM devices are used only as computing elements, while inputs are encoded as voltages. Ali et al. demonstrated in [34] a possible realization on a crossbar array of a MRL-based FA design, however in this approach the computation parallelism is limited, and retrieving the input data would still incur in the VNB overhead.

3.3.2 Memristor-aided logic

The MAGIC LIM framework is similar to the stateful RRAM-based material implication logic one described in Section 3.1. The core logic operations in the MAGIC framework are the NOR and the NOT operations that can be implemented with RRAM devices using the circuit shown in **Figure 7a**. Differently from the stateful material implication logic gate, the resistor R_G is replaced with an RRAM device (i.e., *O* in **Figure 7a**) which stores the result at the end of an operation execution. Thus, before each operation *O* is set into LRS. The inputs are stored on the devices P and Q (see **Figure 7a**), and by delivering a negative voltage pulse with amplitude V_O to their TE, the resistive state of the device *O* changes depending on the resistive state of the inputs. As in the stateful material implication logic implementation, the conditional programming of an RRAM device reduces the circuit reliability, increasing the BER. As discussed in [35, 36], RRAM-based MAGIC implementations are affected by the small design space, and by logic state degradation, the latter affecting both the input and output devices. Also, as discussed in [35], the circuit reliability is strongly influenced by the RRAM technology characteristics. Indeed, larger $|V_{SET}/V_{RESET}|$and R_{OFF}/R_{ON} ratios can potentially improve the circuit reliability, provided that the effect of C2C variability is considered in the design phase. This further underlines the importance of accurate compact models for the implementation of device-circuit co-optimization strategies.

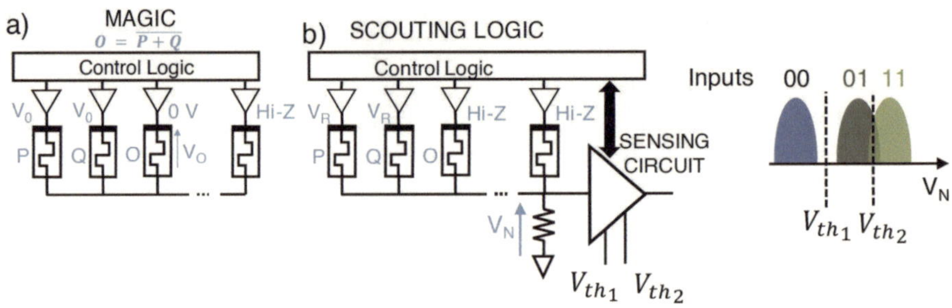

Figure 7.
a) NOR operation as implemented in the MAGIC LIM framework on a linear RRAM array. The resistive state of the output device O changes depending on the input configuration. b) Sketch of the scouting logic LIM framework and associated reliability issues. Similarly to SIMPLY, two devices are read in parallel and the voltage (V_N) at the input of a sensing circuit changes depending on the input configuration. The sensing circuit tries to detect how many devices are in LRS, but as shown in the sketched distribution of V_N for different input combinations, while a sufficient read margin exists to detect the 00 combination, the 01 and the 11 V_N distributions are partially overlapped, thus leading to high BER.

3.3.3 Scouting logic

The approach used in scouting logic is similar to the one employed in the SIMPLY framework. As shown in **Figure 7b**, also in this case RRAM devices encode the input bits of logic operations, and their state is read in parallel using current or voltage sense amplifiers. The equivalent parallel resistance of the input RRAM devices changes depending on the number of inputs in LRS, leading to different distributions at the input of the sense amplifier [25]. By using different thresholds, the sense amplifier can implement different logic operations. Specifically, by discriminating the 11 input combination from the others using the sense amplifier and V_{TH2} (see **Figure 7b**), the control logic in the array periphery can compute the AND and NAND logic operations. Conversely, the OR and the NOR operations can be implemented by discriminating the 00 input combination from the others using the V_{TH1} threshold (see **Figure 7b**). Also, in this framework the XOR and XNOR operations could be potentially implemented in only two read steps which discriminate the 01 input combination from the others. However, although as discussed in Section 3.2 for the SIMPLY framework, the 00 input combination can be reliably distinguished from the others, the distributions at the input of the sense amplifier for the 01 and 11 input combinations overlap due to the effects of variability and RTN, leading to very high bit error rates and a low circuit reliability when computing AND, NAND, XOR, or XNOR operations. Thus, in this framework only OR and NOR operations can be reliably computed [37]. Also, differently from SIMPLY, in which the output device is also an input of IMPLY operations, in scouting logic if more complex operations are implemented, partial results need to be stored on an additional device. Using the same device to accumulate multiple partial results (i.e., to OR together the partial results) may cause a set voltage pulse to be applied to a device already in LRS, thus causing high energy consumption.

3.3.4 Enhanced scouting logic

To improve the reliability of the scouting logic, Yu et al. introduced in [37] an enhanced version which is based on 2T1R memory arrays. As sketched in **Figure 8a** and **b**, the transistors in series with the RRAM devices can be used to either read the

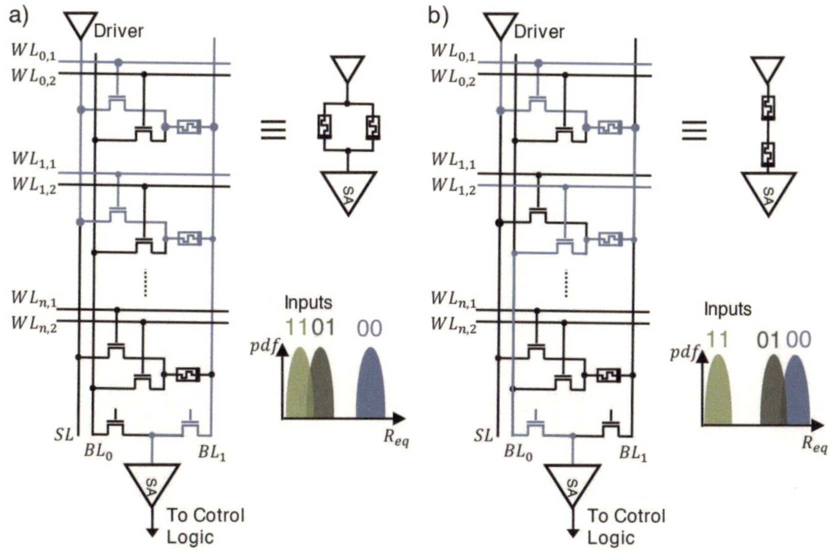

Figure 8.
Operation of the enhanced scouting logic implementation on 2T1R arrays from [37]. a) Execution of an OR/NOR operation. The WL and BL are controlled so that the parallel resistance between the input devices is connected to the sense amplifier. b) Execution of an AND/NAND operation. The WL and BL are controlled so that the series resistance between the input devices is connected to the sense amplifier.

equivalent parallel or series resistance between the two RRAM devices storing the input bits. As in the scouting logic, when reading the resistive state of the two devices in parallel the 00 input combination can be easily distinguished from the others to compute the OR or NOR operations. Instead, by reading the equivalent series resistance, the 11 input combination can be distinguished from the others thanks to a sufficiently large read margin, which enables the correct implementation of AND and NAND operations. Thus, by trading an increased chip area for reliability, in this framework both OR and AND operations can be reliably computed. Compared to the SIMPLY framework this approach can potentially reduce the number of computing steps required to compute more complex operations, however at the cost of a larger chip area that is required to accommodate the additional bit line and selector transistor, and a more complex sense amplifier circuit.

3.3.5 Performance comparison on a 1-bit FA implementation

The performance of LIM frameworks can be compared considering different metrics. As shown in **Table 1**, these include both metrics that can be directly associated with specific LIM frameworks (i.e., number of devices and of computing steps used to implement a specific logic function, and the feasibility in crossbar arrays), and other metrics that also depend on design choices and the specific technology employed to fabricate or to simulate the circuits, making the comparison between different LIM solutions non-trivial. To identify key differences between different LIM approaches the execution of a 1-bit FA operation is considered, since it represents a common simple benchmark for LIM frameworks.

Considering the stateful material implication-based implementations (i.e., "Stateful Mat. Imp." in **Table 1**) different approaches can be followed to optimize the number of required devices or the number of computing steps. As reported in the

Author(s)	Type of LIM (Sim/ Exp)	Physics-Based Model	Number of Devices	Feasible in RRAM Array	Number of Elementary Steps	Delay	Energy	Circuit parameters**	Retains input values	Endurance before Refresh*
Talati et al. [38]	MAGIC (Sim)	NO	5–19	YES	13–15	≈ 17–19.5 ns (Reported)	≈3–6pJ (Reported)	V_0 = 1 V V_{SET} = 2 V V_{RESET} = -1 V f_{op} = 0.77GHz $\overline{R_{LRS}} = 1k\Omega$ $\overline{R_{HRS}} = 300k\Omega$	NR (Tech. & Design Dependent)	NR
Yu [37]	Enhanced scouting logic	NR	9 RRAM	YES	17	68 ns (Estimated at 0.5 GHz)	NR	V_{SET} = NR V_{RESET} = NR V_{READ}=NR f_{op} = 0.33GHz $\overline{R_{LRS}} = 30k\Omega$ $\overline{R_{HRS}} = 16.6\ M\Omega$ R_G = 160kΩ	YES	NR (Expected to be high)
Siemon et al. [39]	n-bits Stateful Mat. Imp. (Sim)	YES	8 RRAM	YES	19	3.61 µs (estimated)	202 pJ (estimated)	V_{SET} = 1.45 V V_{COND} = 1.24 V f_{op} = 1 MHz $\overline{R_{LRS}}$ =NR $\overline{R_{HRS}}$ =NR R_G = 15kΩ	NO	NR
Lehtonen et al. [27]	Stateful Mat. Imp. (Sim)	NO	8 RRAM	YES	136	NR	NR	NR	YES	NR
Kvatinsky et al. [3]	Stateful Mat. Imp. (Sim)	NO	9 RRAM	YES	23	9.1 µs (estimated)	NR	V_{SET} = 1 V V_{COND} = 0.5 V f_{op} = 2.5 MHz $\overline{R_{LRS}} = 1k\Omega$ $\overline{R_{HRS}} = 100k\Omega$ R_G = 10kΩ	NO	≈ 300 up to 10^5 (trades with energy)
Kvatinsky et al. [3]	Stateful Mat. Imp. (Sim)	NO	6 RRAM	YES	29	11.5 µs (estimated)	NR	V_{SET} = 1 V V_{COND} = 0.5 V f_{op} = 2.5 MHz $\overline{R_{LRS}} = 1k\Omega$	NO	≈ 300 up to 10^5 (trades with energy)

Author(s)	Type of LIM (Sim/ Exp)	Physics-Based Model	Number of Devices	Feasible in RRAM Array	Number of Elementary Steps	Delay	Energy	Circuit parameters**	Retains input values	Endurance before Refresh*
								$\overline{R_{HRS}} = 100k\Omega$ $R_G = 10k\Omega$		
Cheng et al. [40]	Stateful Mat. Imp. (Exp)	NA	8 RRAM	YES	27	54 µs (reported)	19.5 pJ (reported)	V_{SET} = -2 V V_{COND} = -1 V V_{RESET} = 1.5 V f_{op} = 200 kHz $\overline{R_{LRS}} = 3k\Omega$ $\overline{R_{HRS}} = 300k\Omega$ $R_G = 20k\Omega$	NO	NR
Zanotti et al. [26]	Stateful Mat. Imp. (Sim)	YES	9 RRAM	YES	43	345 ns (Reported)	6.4 nJ (reported)	V_{SET} = 2.15 V V_{COND} = 1.7 V_{RESET} = -1.45 V f_{op} = 0.05GHz $\overline{R_{LRS}} = 2.8k\Omega$ $\overline{R_{HRS}} = 150k\Omega$ $R_G = 1k\Omega$	YES	67
Zanotti et al. [26]	Stateful Mat. Imp. (Sim)	YES	8 RRAM	YES	28	560 ns (Reported)	518 pJ	V_{SET} = 2.15 V V_{COND} = 1.7 V_{RESET} = -1.45 V f_{op} = 0.05GHz $\overline{R_{LRS}} = 2.8k\Omega$ $\overline{R_{HRS}} = 150k\Omega$ $R_G = 1k\Omega$	YES	≈ 30
Zanotti et al. [26]	SIMPLY (Sim)	YES	8 RRAM	YES	28	920 ns (Reported)	172 pJ	V_{SET} = 2.15 V V_{RESET} = -1.45 V V_{READ} = 0.2 V f_{op} = 25 MHz $\overline{R_{LRS}} = 2.8k\Omega$ $\overline{R_{HRS}} = 150k\Omega$ $R_G = 1k\Omega$	YES	$> 4.5 \cdot 10^6$ (no energy trade-off)

Author(s)	Type of LIM (Sim/Exp)	Physics-Based Model	Number of Devices	Feasible in RRAM Array	Number of Elementary Steps	Delay	Energy	Circuit parameters**	Retains input values	Endurance before Refresh*
Zanotti et al. [31]	n-SIMPLY (Sim)	YES	8 RRAM	YES	15	60 ns (Reported)	4.2 pJ	V_{SET} = 3 V V_{RESET} = -2.8 V V_{READ} = 0.2 V f_{op} = 0.25GHz $\overline{R_{LRS}} = 10.5k\Omega$ $\overline{R_{HRS}} = 550k\Omega$ R_G = 10kΩ	YES	> $4.5 \cdot 10^6$ (no energy trade-off)
Junsangsri et al. [41]	CMOS LIM (Sim)	YES (FET) NO (RRAM)	41 FET + 4 RRAM	NO	NA	52 ps (Reported-excludes RRAM delay)	2.2 fJ (reported-excludes RRAM energy)	V_{DD} = 3.3 V $\overline{R_{LRS}} = 30k\Omega$ $\overline{R_{HRS}} = 100\ M\Omega$	YES	NR (Limited by FET Reliability)

This estimate can be trusted only when a physics-based model was used. **Circuit parameters reported by the authors in their manuscripts. f_{op} is the frequency of single operation execution (e.g., 1/ t_{IMPLY}). NA stands for "not applicable." NR stands for not reported.*
In the table, reported indicates that the values were explicitly reported in the paper. Estimated indicates that the corresponding value was extrapolated from data in the paper since it was not explicitly reported.

Table 1.
Comparison of 1-bit FA implementations based on different LIM frameworks.

literature [3, 26, 27, 40], typically a 1-bit FA operation can be implemented using from 6 up to 9 RRAM devices, and sequences of IMPLY and FALSE operations which length vary from as low as 23 steps, when the inputs are overwritten, up to 136 steps when the sequence is not optimized. To further reduce the number of computing steps, the multi-input stateful material implication framework was proposed by Siemon et al. in [39], showing that by using three-inputs IMPLY operations the number of computing steps can be reduced down to 19 steps when using 8 RRAM devices. The n-SIMPLY framework discussed previously in Section 3.2, enables to reliably compute 4-inputs IMPLY operations, and thus it reduces the number of computing steps down to 15. In terms of number of computing steps, the different set of core operations used in the enhanced scouting logic and the MAGIC LIM frameworks provides additional advantages compared to most of the material implication-based solutions. Although, not explicitly reported in the literature, by combining AND, NAND, NOR, and OR operations in the enhanced scouting logic array implementation (see **Figure 8**), the 1-bit FA operation is executed in 15 steps, which could be further reduced as in the n-SIMPLY approach by increasing the parallelism of the OR/NOR operations (the parallelism of the AND/NAND operations cannot be increased in the architecture reported in [37]). MAGIC-based implementations in [38] can achieve the lowest number of computing steps (i.e., 13 steps) among the different LIM implementations reported in **Table 1**.

Although the number of computing steps is linked to the computing delay (i.e., $delay = \#steps \; t_{step}$) and can give an idea about the energy efficiency of the specific implementation (i.e., $E_{tot} = \sum_{i=1}^{\#steps} E_i$, where E_i is the energy dissipated in each step), the achievable time (t_{step}) and energy (E_i) per single computing step depend on the characteristics of the RRAM technology that is used on the adopted design choices. Also, the accuracy of the performance and reliability analysis depends on the type of compact model used in the circuit simulations. Indeed, compared to general purpose memristor models, the use of physics-based compact models can provide more accurate results when fast voltage pulses are used and enable to analyze important nonideal effects such as the logic state degradation. Thus, in **Table 1**, along with the energy and computing delay of different implementations from the literature, also the main circuit parameters and an indication regarding the type of compact model used in the simulations are reported. In general, to reduce the energy per computing step, increasing the LRS resistances can be a viable solution. However, larger voltages must be employed to program a device without increasing the programming pulse width, potentially conflicting with the desired circuit design space. In terms of energy efficiency, currently a hybrid FET-RRAM design from [41] shows the best performance. However, this implementation is not feasible in crossbar arrays, limiting the area efficiency, and simulations were performed with an extremely simplified RRAM model, suggesting the need for a more accurate performance analysis. Among the available data regarding crossbar-feasible implementations, MAGIC and n-SIMPLY achieve the best energy and delay performance. However, although MAGIC is slightly more efficient than n-SIMPLY, the 1-bit FA implementation does not retain the input states and is affected by the problem of logic state degradation, which can discourage its adoption in applications such as hardware accelerator for BNNs. In BNN the network parameters, stored in the nonvolatile resistance of RRAM devices, must be preserved through computations. Thus, the effect of logic state degradation introduces the need for frequent memory refresh cycles, causing high inefficiencies. Therefore, in the following sections, the design of a LIM-based accelerator for BNNs is discussed considering an n-SIMPLY implementation.

4. RRAM-based binarized neural networks accelerators

In BNNs [9] weights and activations of neural networks are represented with a single bit. Despite the considerable reduction in the required hardware resources, for some classification tasks, BNNs were demonstrated to achieve similar performance with respect to full precision neural network implementations, with only slight accuracy degradation [9]. The use of binary weights and activations, not only leads to a reduction of the size of the memory used to store the network parameters, but also reduces the complexity of the operations performed by each neuron of the network. As shown in **Figure 9a**, in full precision neural networks each neuron computes the dot product between its input activations (i.e., X) and the learned weights (i.e., W), adds a bias term (i.e., b) to the result, and finally applies a nonlinear activation function (i.e., *ReLu* in **Figure 9a**) to compute the output activation. In BNNs, these operations translate to simpler logic operations. Specifically, the dot product becomes a bitwise XNOR followed by an accumulation, or popcounting, of the results, and the activation function becomes the comparison with a learned threshold. Despite using less memory space to store the network parameters, BNNs are still characterized by a large number of parameters. Running inference tasks on conventional hardware would incur in the VNB, therefore degrading the performance. Thus, several IMC hardware accelerators based on RRAM technologies have been proposed in the literature [7, 8, 42], showing promising efficiency improvement without considerable accuracy degradation. In fact, binary storage is currently a more robust approach for state-of-the-arts RRAM technologies with respect to multibit storage on a single which is more affected by device nonidealities.

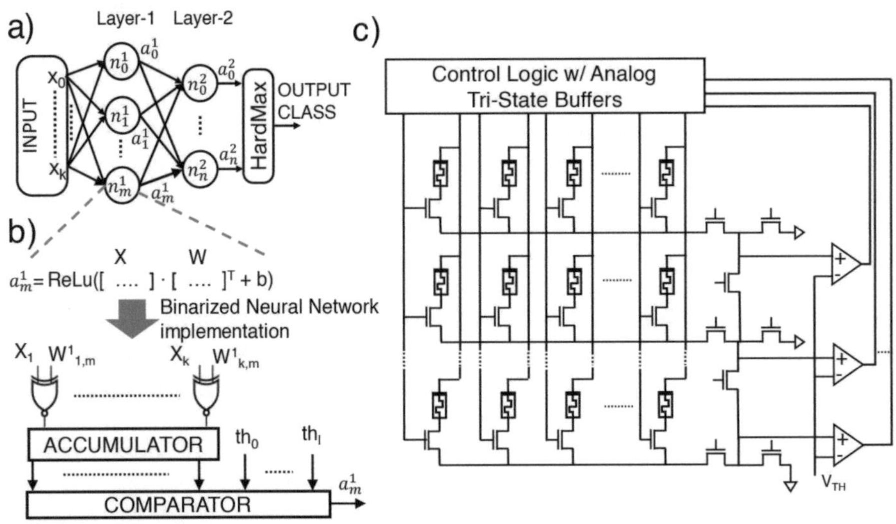

Figure 9.
(a) Sketch of a fully connected neural network, with k *inputs,* m *neurons in the hidden layer, and* n *neurons in the output layer. The operation performed by each neuron is shown. (b) Implementation of a neuron in BNNs. The vector-vector multiplication becomes the accumulation of the results from bitwise XNOR operations, and the activation function the comparison with a threshold. (c) 2D array of 1T1R devices that can be used to accelerate in hardware the operations required for classification tasks based on BNNs, in the SIMPLY framework. FETs control signals are managed by the control logic.*

In the following sections, three different BNN hardware accelerators based on 1T1R memory arrays are discussed and benchmarked. Specifically, Section 4.1 describes SIMPLY-based implementations, Section 4.2 discusses analog accelerators for VMM used in BNNs, and Section 4.3 describes a hybrid accelerator combining the previous two approaches on the same array.

4.1 SIMPLY-based BNN accelerators

Since the core operations of BNNs are logic operations, SIMPLY-based architectures can be used to realize energy efficient hardware accelerators for inference tasks. Such accelerators can be realized on memory arrays and a peripheral circuit such as the one shown in **Figure 9c**, enabling the implementation of single instruction multiple data architectures, which can exploit the intrinsic parallelism of BNN inference algorithms [28, 31, 43]. In fact, depending on the number of comparators placed in the array periphery, and the design of the drivers and control logic, multiple devices in different rows and in the same column can be read and programmed in parallel. For instance, to perform an IMPLY operation in the SIMPLY framework on multiple rows of the array, the columns of interest are biased with the read voltage and the corresponding select lines are biased to turn on the selector FET. At the same time, FET devices in the array periphery are biased to implement R_G and bring the voltage at the row of interest to the input of a VSA. Then the column corresponding to the output device is biased with V_{SET} and FET devices in the array periphery provide a low conductive path for the devices that need to be programmed.

In the rest of this section, the implementation of the different core operations of BNNs on the SIMPLY architecture is discussed.

4.1.1 XNOR operation

In BNNs, a dot product operation is equivalent to a bitwise XNOR operation between the input activations and the weights of the neuron, followed by an accumulation of positive results. As shown in **Figure 10**, for BNN applications, each XNOR operation can be implemented with 5 RRAM devices and 5 computing steps [31], provided that both the weight (i.e., W) and its complement (i.e., $\overline{W}$) are stored in the memory array and never overwritten. The other three devices store the input (i.e., *IN*), the output (i.e., *O*), and the partial result (i.e., N_1) of the XNOR. Since operations

a)

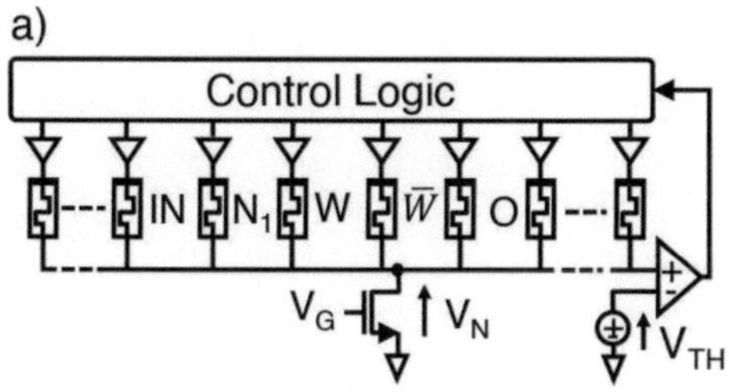

b) $O = \overline{IN \oplus W}$

#Step	Operation	Result
1-2	FALSE(O,N$_1$)	$O = N_1 = 0$
3	2-IMPLY(IN, N$_1$)	$N_1 = \overline{IN}$
4	3-IMPLY(W, IN, O)	$O = \overline{W IN}$
5	3-IMPLY($\overline{W}$, N$_1$, O)	$O = \overline{W + IN}$

Figure 10.
(a) XNOR gate implementation used for BNN inference tasks in the SIMPLY framework. To reduce the number of computing steps, the weight, and its complement (i.e., W, and $\overline{W}$, respectively) are stored in the array and preserved through the computations. (b) Sequence of computing steps in the n-SIMPLY framework.

on the same row of an array are sequentially executed, N_1 can be reused in the subsequent XNOR operation when computing a dot product.

4.1.2 Accumulation operation

To implement the accumulation operation, different approaches corresponding to different sequences of IMPLY and FALSE operations can be followed. As proposed in [31, 43], the accumulation operation can be implemented with a chain of half adders (HAs), see **Figure 11a** and **b**. Each HA receives in input its output from the previous step and a new input bit when it computes the least significant bit (LSB), or the output carry from the previous HA in the chain otherwise. In the SIMPLY implementation, when a new bit is accumulated, the result of each HA in the chain is computed sequentially, starting from the LSB. However, based on the position of an HA in the chain (i) its result is computed only after 2^i input bits have been accumulated, since when fewer bits have been accumulated the corresponding HA output would be zero. Although the number of HAs grows as the $log_2()$ of the number of input bits, the number of computing steps grows exponentially. As shown in **Figure 11e** and **f**, exploiting different maximum parallelism in the framework of n-SIMPLY to optimize the sequence of computing steps results in a > 15% reduction in the number of computing steps with respect to the 2-SIMPLY implementation [31], however, to further improve the efficiency different approaches can be followed.

Here we propose a SIMPLY-based implementation of a binary tree adder architecture as the one shown in **Figure 11d**. In each level of the tree, the results from the previous level are added two by two, until all the input bits are accumulated. Thus, the number of levels of the tree corresponds to the $log_2()$ of the number of input bits. From one level of the tree to the following, the size of the FA increases by 1 bit. Each adder is implemented as a ripple-carry adder, see **Figure 11c**, with a 1-bit HA requiring 9 computing steps, and a 1-bit FA requiring 15 steps (see **Figure 6b**). This

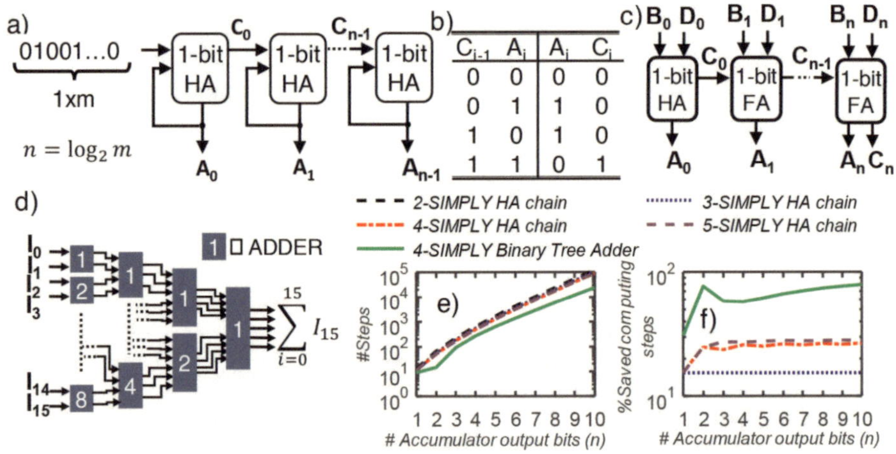

Figure 11.
(a) Implementation of the accumulation operation on m *input bits with a chain of* n *HAs, as in [31]. (b) Truth table of the 1-bit HA used in the HA chain. (c) n-bit ripple-carry adder implementation used as the core element of the binary tree adder-based accumulator. (d) Example of a binary tree adder used to accumulate 16 input bits. The number of levels in the binary tree grows as the* $\log_2(m)$*. (e) Comparison of the number of steps required to accumulate (*2^n*-1) bits for different accumulator implementations. (f) Percentage of saved computing steps with respect to the* 2-SIMPLY HA chain *implementation of the different accumulator implementations.*

approach provides a considerable performance improvement compared to the HA chain implementation as shown in **Figure 11e** and **f**.

4.1.3 Comparison operation

In the original BNN implementation form Courbariaux et al. [9], weights and activations are encoded with +1 and − 1 values, meaning that the result of the accumulation operation spans from negative to positive values. Thus, the activation operation is commonly implemented using the *sign()* function. Conversely, in the SIMPLY-based implementation, weights and activations are encoded with 0 and 1, and the result of the accumulation is a positive value, ranging from zero to the number of products performed in the neuron. Thus, the activation function in SIMPLY-based implementations corresponds to the comparison with a threshold (i.e., half the number of products performed in the corresponding neuron). The comparator outputs a logic 1 only when the result of the accumulation is above the threshold. The logic function implementing this operation in the sum of products form is reported in **Figure 12a**. As discussed in [31], increasing the parallelism in the read step of n-IMPLY operation, considerably reduces the required number of computing steps required for a comparison. In fact, when using only 2-IMPLY operations the number of computing step grows exponentially with the number of inputs, while using n-IMPLY operations with $n > 2$ leads to linear trends, as shown in **Figure 12b**. More details regarding the different sequence of computing steps implementing the comparison operation are available in [31].

4.1.4 Hard max operation

To determine the output class of an inference task the hard max function can be used. This function determines the index of the maximum value among a group of elements. A sketch of the approach used for the SIMPLY-based implementation is shown in **Figure 12c**. Similarly to the binary tree adder, also in this case a binary tree structure is used, and at each level of the tree pairs of elements are compared. To do so, for each pair of elements, the one stored in the upper position of the array is copied in the same row of the other element together with its ID. Then the two elements are

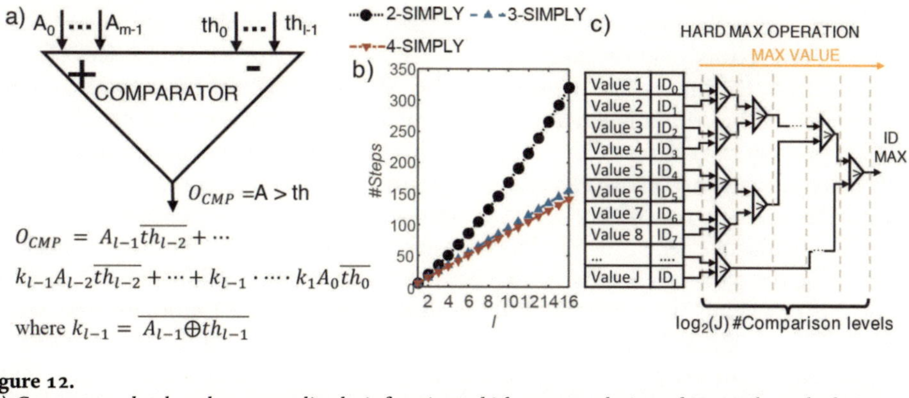

Figure 12.
(a) Comparator sketch and corresponding logic function, which outputs a logic 1 when a > th*, and a logic 0 otherwise. (b) Number of computing step required for different n-SIMPLY-based implementations and increasing number of compared bits (l). In the legend,* n *indicates the maximum degree of parallelism used in the parallel read step of IMPLY operations. (c) Sketch of the hard max activation function implemented in the n-SIMPLY framework. Inputs are compared pairwise using the comparator implementation in (a) and (b).*

compared using the same comparator implementation used for the neuron activation function. For each pair, the initial value and ID stored in the row are replaced with the larger element resulting from the comparison. In just *log*$_2$*(number of elements)* levels of the tree, the result is computed and stored in place of the last element. Additional information regarding the SIMPLY-based implementation is available in [11, 31].

4.1.5 Batch normalization operation/bias term addition

Batch normalization layers are often used in neural networks to speed up the training of the network parameters and consist in the scaling of the inputs of a layer using learned parameters. In BNNs, instead, batch normalization is introduced between the MAC operation and the *sign()* activation function, as discussed by Courbariaux et al. [9]. Thus, during training, the average ($\overline{m}$) and the standard deviation (σ) of the results of the MAC operations of a layer are computed, while a scaling (γ) and a bias terms (β) are learned and optimized through each iteration. In forward passes of the neural network these parameters are used to regularize the results of the MAC operations (a), so that the input to the *sign()* function is $\hat{a} = (a - \overline{m}) \bigodot \gamma \bigodot \frac{1}{\sigma} + \beta$.

As discussed in Section 4.1.3, in the SIMPLY-based implementation the activation function is implemented as the comparison with a threshold (th), that is equivalent to half the number of inputs of a layer if no bias term is added to the result of MAC operations. Thus, the effect of the batch normalization parameters can be introduced in the SIMPLY-based neural network by adjusting the value that is mapped to the threshold of the activation function for each neuron ($\hat{th} = \frac{th-\beta}{\gamma}\sigma + m$).

In the output layer of a neural network, the hard max activation function is used to infer the output of the network. Thus, batch normalization or equivalently the addition of a bias term, needs to be performed also on the SIMPLY-based architectures. These operations consist in the addition of a fixed value to the results of the MAC operation, and can be implemented on the SIMPLY-based architecture by adding the bias term, that is stored on the memory array, using the ripple-carry adder described in Section 3.2.1 and shown in **Figure 11c**.

4.1.6 Array level implementation

As discussed in the previous sections, the SIMPLY-based implementation of BNN inference accelerators requires mapping to the resistive state of RRAM devices of memory arrays, the inputs, the neurons weights, the thresholds of the activation functions, the IDs of the output layer, and devices for storing partial results of computations. Ideally, infinitely large arrays would simplify this task by storing all the parameters of a single layer of the neural network onto a single array and the parameters of a single neuron all on a single row of that array. However, real memory arrays have a limited size due to nonideal effects, such as line parasitic resistance [44]. Thus, in practical applications, the parameters of a neural network are split onto multiple arrays, and the parameters of a single neuron onto multiple adjacent rows of the same sub-array, as shown in **Figure 13a**. When the parameters related to a single neuron are split onto multiple rows of the array, as in **Figure 13b**, each row of the array must still contain an equal number of devices for storing the inputs, the weights, their complement, and the outputs, for computing the partial results of the accumulation operations. Also, sufficient space on the array must be left to store support devices and other network parameters (e.g., the thresholds of the activation functions). After partial results are

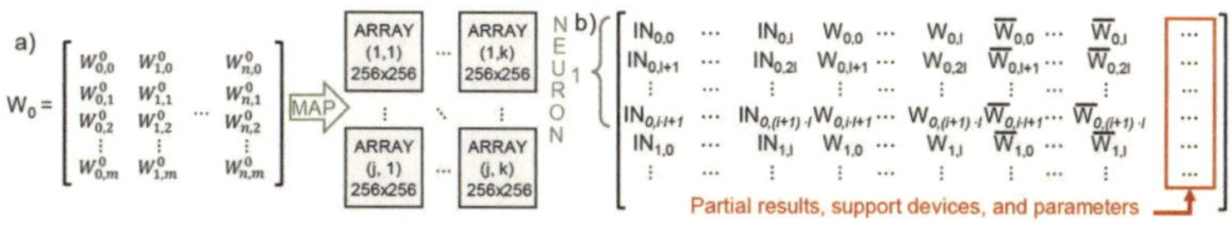

Figure 13.
(a) Mapping of the neural network parameters on 256x256 memory array. The number of arrays (i.e., j k) depends on the number of inputs (m) and outputs (n) of each layer of the neural network. (b) Description of the mapping strategy used in a single array. Each row of the array stores portions of the input activations (IN), the weights, and their complement, for a single neuron. Additional space to store partial results of the computations and additional parameters needs to be left and appropriately sized. The parameters and inputs of a single neuron may be split onto multiple rows of the array based on the number of available columns in the array.

computed on each row, these must be copied to adjacent rows to compute the final result. Despite the increased complexity in the mapping of the neural network weights and the increased chip area occupied by the peripheral circuitry, the use of multiple arrays enables increasing the computation parallelism, reducing the computation latency. Results relative to different neurons stored on different sub-arrays can be computed in parallel. Also, the intrinsic parallelism of 2D array implementations of the SIMPLY architecture enables the parallel computation of partial results relative to a single neuron when its parameters are split onto multiple rows on the same array.

However, the number of devices on the same column of an array that can be programmed in parallel is constrained by the maximum current that can flow on a single array line, potentially decreasing the maximum parallelism that can be achieved. Indeed, this effect is connected to the current compliance used to program RRAM devices in the array, and the use of RRAM technologies that can reliably store binary values when programmed at low current compliance values would alleviate this constraint and enable the parallel programming of devices located onto multiple rows of the array.

4.2 BNN analog vector matrix multiplication accelerators

4.2.1 State of the art on RRAM-based analog BNN vector matrix multiplication accelerators

Thanks to their intrinsic reconfigurability, LIM computing architectures such as SIMPLY, are an effective solution for resource-constrained devices deployed at the edge of the network. However, the high level of reconfigurability comes at the cost of suboptimal performance when specific tasks need to be accelerated. For instance, considering BNNs inference tasks, most of the computations performed are required to compute the result of dot products [45]. Thus, designing optimized accelerators for the VMM operation would provide a considerable improvement of the overall accelerator performance. Thus, several analog/mixed-signal RRAM-based accelerators of the VMM used in BNNs have been proposed in the literature [6, 8, 42, 46, 47]. To encode the −1 and +1 weights of BNNs, a common solution adopted in analog/mixed-signal RRAM-based accelerators is to use a pair of devices (i.e., either 1T1R or 1R devices) programmed in complementary resistive states for each weight of the BNN. Driving these pairs of devices with appropriate voltage schemes enables to compute the result of XNOR operations using only positive RRAM conductances. However, different RRAM implementations vary depending on the array implementation, the sensing scheme (i.e., current sensing or voltage sensing), and peripheral circuitry. The main approaches that can be found in the literature are summarized in **Figure 14**. Specifically, **Figure 14a** shows a solution exploiting 1T1R arrays and a current sensing scheme to compute the bitwise XNOR and accumulation operations. In this approach, the weights of the BNN are encoded in pairs of devices located in adjacent rows of the same column of the array. The select lines of the selector transistors encode the input activations using complementary signals, see **Figure 14a**. In the array periphery, a transimpedance amplifier, implemented with an operational amplifier (OPA), ensures a virtual ground at the end of the column of the array, so that the current flowing in each column (i.e., I_{sum} in **Figure 14a**) is linearly proportional to the number of +1 results from the bitwise XNOR operations in that column. Also, the transimpedance amplifier converts the current to a voltage that is then digitized using an analog to digital converter (ADC). However, this solution is inefficient in terms of energy consumption and chip area, meaning that the transimpedance amplifier and the ADC must be shared among adjacent columns by means of analog multiplexers. A similar approach that is more efficient in terms of area occupation and energy efficiency was proposed by Yin et al. in [8], and the core elements of their solution are sketched in **Figure 14b**. The working principle is similar to the previous approach, with the main difference being that no OPA is used in the array periphery, thus saving considerable area and energy. However, without the OPA, no virtual ground is provided at the end

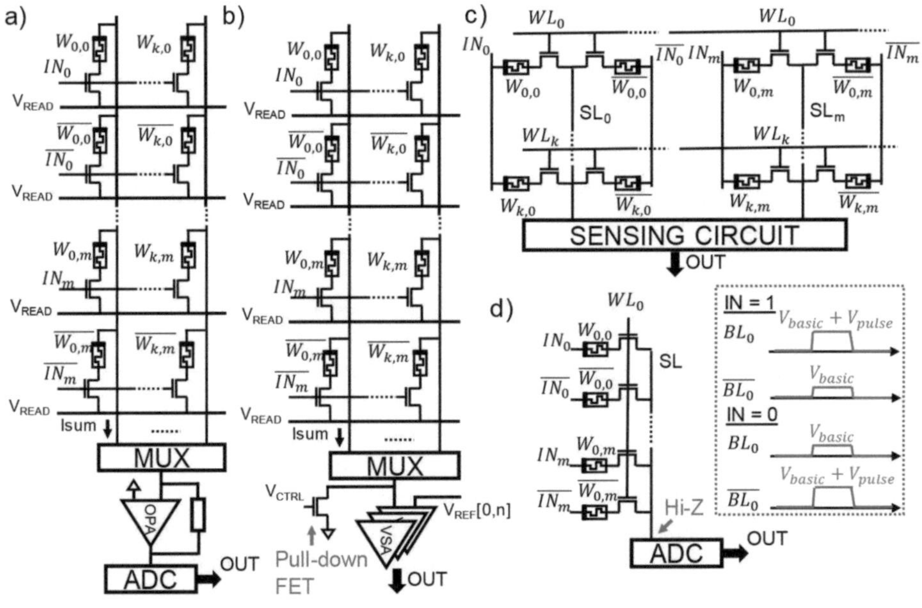

Figure 14.
Different implementations of the VMM operation with RRAM arrays. (a) 1T1R implementation in which each weight of the network is encoded as complementary resistive state of RRAM devices located in adjacent rows. Weights associated with the same neuron are stored in the same column of the array. An OPA provides a virtual ground node at the end of each column so that the current flowing in each column is equivalent to the sum of all the positive results of the bitwise XNOR operation. (b) Similar 1T1R implementation from [8], which uses a pull-down FET device and a flash ADC implemented with multiple VSAs instead of introducing a virtual ground node. (c) Sketch of the 2T2R implementation from [46]. Differently from a) and (b), the weights associated with the same neuron are located on the same row. A single vector-vector product is computed at a time by activating the desired word line (WL). Appropriate capacitive sensing circuits convert the results of the bitwise XNOR operations encoded in the select lines (SL) voltages to the output result. (d) 1T1R array implementation from [47], which exploits the select line capacitance to compute the result of the vector-vector multiplication. The SL is kept in hi-Z so that during a read-out step the voltage at the input of the ADC is proportional to the number of positive results of the bitwise XNOR operation.

of the columns of the array. The authors instead, introduced a pull-up or pull-down FET device, which creates a voltage divider with the equivalent resistance of the column which depends on the result of the bitwise XNOR operations. The voltage drop on the pull-down (pull-up) device increases (decreases) with the number of +1 from the bitwise XNOR operations in that column. Such voltage can be compared with a threshold using a VSA to compute the output activation. As discussed by Yin et al. in [8], multiple VSA and threshold voltages can be used to implement a flash ADC for improving the accuracy of the computation. The compact area and faster response of the VSAs compared to OPAs, not only reduce the area occupation but also increase the speed of the computation. In fact, the VSAs must be shared between fewer columns of the array compared with OPA-based solutions, thus increasing the throughput of the computation. A drawback of this approach is the nonlinear increment of the voltage at the input of the VSAs for an increasing number of positive results, which may require a fine tuning of the voltage thresholds to retain high accuracy in BNN inference tasks. Alternatively, approaches based on voltage mode sensing schemes have been recently proposed in the literature [46, 47], and shown to achieve high accuracy and energy efficiency. Specifically, Ezzadeen et al. proposed a solution based on 2T2R arrays which is sketched in **Figure 14c**. In this approach, the weights of each neuron are stored on the same row of the array, and the results for each neuron are computed one at a time. To do so, a single WL (see **Figure 14c**) is turned on at a time, and the input activations and their complements are applied to the corresponding pairs of devices which encode the neuron's weights. As a result, the voltage at each SL line (see **Figure 14c**) encodes the result of a single XNOR operation. In their paper, the authors proposed a capacitive sensing scheme to compute the result of the accumulation. Such a scheme requires a smaller chip area compared to other solutions, however at the cost of reduced computing speed since the result for a single neuron is computed at each step. Another voltage mode sensing scheme, sketched in **Figure 14d**, was proposed by Zhao et al. [47]. In this approach, the weights associated with the same neuron are encoded in adjacent RRAM devices programmed in complementary resistive states that are connected to the same SL (see **Figure 14d**). Input activations are encoded as a fixed bias (V_{bias}) term applied to both devices encoding a single neuron's weight, and an additional voltage contribution with amplitude *Vpulse* to the first or second line depending on the polarity of the input activation, as shown in **Figure 14d**. During a VMM, the WLs are activated, and the SLs are kept in Hi-Z. By exploiting the intrinsic capacitance of the array lines, the voltage on the SL lines is linearly proportional to the number of positive bitwise XNOR operations. Such voltage is then digitized by means of an ADC. Compared with the more common current mode sensing scheme, this approach is more energy efficient since smaller currents flow in the circuit.

Overall, each different scheme provides different tradeoffs, in terms of chip area (i.e., cost), computing speed, and energy. Thus, it is up to circuit designers to choose the design that better suits a specific application.

4.2.2 Circuit design tradeoffs: analysis of a case of study

In this section, we analyze in detail the tradeoffs existing in the BNN analog VMM accelerator shown in **Figure 14b** proposed by Yin et al. [8], since it can provide high-throughput combined with high energy efficiency, and a compact peripheral circuit design. An example of the core elements of a BNN VMM accelerator based on this approach is sketched in **Figure 15a**, where additional FET devices and the required control logic are also represented. As discussed in the previous section, the voltage at the input of the VSA increases nonlinearly with an increasing number of positive results of bitwise XNOR operations due to the existence of a voltage divider between the pull-down device and the equivalent resistance of the devices connected to the same column of the array. The reliability of this accelerator is linked to the linearity of the transfer characteristic and the read margin available at the input of the VSA. Indeed, a higher read margin over the whole span of possible inputs (i.e., the number of positive bitwise XNOR operations) would reduce the error rate of the VSA. Different design parameters can influence the linearity of the transfer characteristic and the available read margin at the input of the VSA. Specifically, the pull-down resistance (i.e., R_{PD}) changes the linearity of the transfer characteristic. As shown in **Figure 15b**, low R_{PD} values increase the linearity of the transfer, but too low values reduce the dynamic range, and thus the read margin, at the input of the VSA. Also, very low R_{PD} values may require excessively large FET devices which would reduce the area efficiency. Conversely, larger R_{PD} values increase the nonlinearity of the transfer characteristic, and too large values can quickly saturate the input of the VSA when just a few positive results are accumulated. Also, C2C variability can further reduce the read margin available at the input of the VSA, introducing possible overlaps between the distributions of the voltage at the input of the VSA for consecutive numbers of accumulated XNOR results, as shown **Figure 15c**. To increase the read margin higher power consumption can be traded by increasing the read voltage, provided that devices in HRS are not corrupted during a read step. All these circuit parameters must be optimized to ensure a reliable circuit operation. Still, a limited number of devices can be reliably read in parallel, introducing the need for the input split method described in [7]. In this approach, the computation of the complete sum of products is split into multiple steps, in which a partial number of the sum of products is computed, and the results of each step are accumulated using digital circuits. As suggested in [8], to improve the accuracy of the accelerator multiple VSA amplifiers and appropriately designed thresholds can be employed for each column, implementing a flash ADC converter. Specifically, in their work [8], the authors achieved high inference accuracy by using a 3-bit flash ADCs implemented with seven VSA, for computing operations with 64 inputs. Even when increasing the number of VSAs in the array

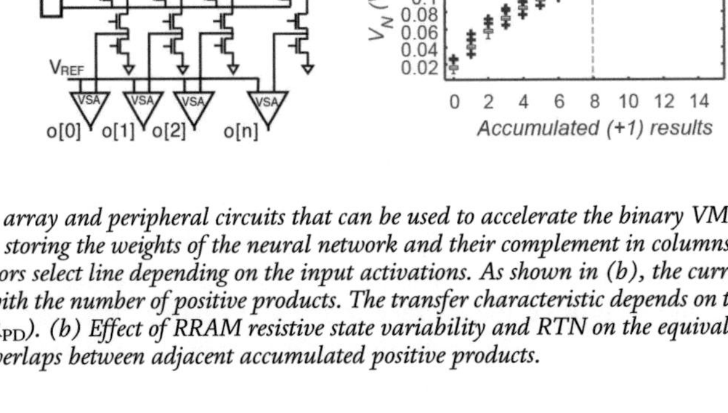

Figure 15.
(a) 1T1R memory array and peripheral circuits that can be used to accelerate the binary VMM operation in the analog domain, by storing the weights of the neural network and their complement in columns of the array, and controlling transistors select line depending on the input activations. As shown in (b), the current flowing in each column increases with the number of positive products. The transfer characteristic depends on the value of the pull-down resistance (R_{PD})*. (b) Effect of RRAM resistive state variability and RTN on the equivalent output voltage, showing possible overlaps between adjacent accumulated positive products.*

periphery, this approach can provide considerable efficiency improvement with respect to LIM-based accelerators since only read operations are performed without the need to use more energy-hungry programming steps.

4.3 Reconfigurable hybrid BNN SIMPLY/analog accelerator

To combine the reconfigurability of the SIMPLY architecture described in Section 4.1 and the efficiency of the BNN analog VMM accelerator discussed in Section 4.2.2, a hybrid architecture merging the two approaches on the same array was proposed in [28]. Here we discuss a variation of the architecture proposed in [28] which is less sensitive to sneak paths and requires fewer VSAs in the array periphery. Such architecture is shown in **Figure 16** and is similar to the SIMPLY-based architecture shown in **Figure 9c**. IMPLY and FALSE operations can be performed in the array as described in Section 4.1. Conversely, the architecture is slightly different to the one implementing the analog BNN VMM accelerator, shown in **Figure 15a**. Still the working principle is the same, with the only differences being that the currents encoding the results of multiplications flow in the rows of the array instead of the columns, that the columns of the array are biased with the read voltage, and that the pair of devices in complementary state encoding a single weight are encoded in adjacent columns instead of rows. To enable both computing paradigms the control logic and peripheral circuits need to be adapted, so that the reference voltage of the VSAs, and the bias voltage of the FET implementing either R_G or R_{PD} can be changed when performing computations in the SIMPLY or the analog BNN VMM frameworks, respectively. Also, if multiple VSA amplifiers are used to implement a flash ADC to improve the accuracy of the analog VMM operation, multiple rows of the array may share the same group of VSAs, thus reducing the maximum parallelism achievable, compared to the SIMPLY architecture in **Figure 9c**.

Despite the increased complexity of the peripheral circuits and the possible reduced parallelism, the proposed hybrid accelerator provides a more efficient exploitation of the hardware resources that are often scarce in low-power embedded devices.

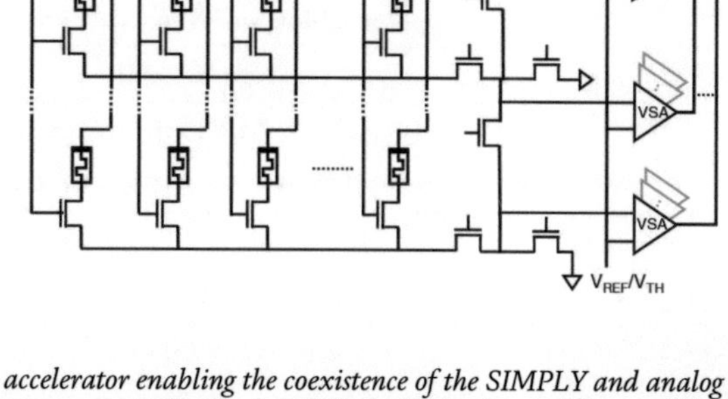

Figure 16.
*Sketch of the hybrid IMC accelerator enabling the coexistence of the SIMPLY and analog BNN VMM frameworks shown in **Figure 9c** and **15a**, respectively. Shaded VSAs highlight the possible use of flash ADCs in the architecture.*

4.4 Performance benchmark

In this section, the performance of the architectures discussed in sections 4.1–4.3, are estimated on a BNN inference task, and benchmarked against an implementation running on conventional embedded system hardware [48]. The performance was estimated on a classification task on the MNIST handwritten digits dataset. The 20x20 pixels images of the dataset were converted to black and white images before training. The adopted neural network consists of a single fully connected hidden layer with 1000 neurons and 10 output neurons and was trained on 9500 sample images using the DoReFa-Net algorithm [49]. Other 2500, and 2000 images were used for validation, and testing, respectively, and the trained neural network achieves an accuracy of 91.4% [29], which is comparable with other results in the literature and the BNN implementation considered as a reference [48]. The performance of each accelerator was estimated considering that the computations were executed on an appropriate number of 256x256 1T1R memory arrays. In the case of the SIMPLY-based implementations, the network parameters were mapped to memory arrays as discussed in Section 4.1.6, while in the hybrid architectures, the network weights and their complements were mapped to memory arrays, filling them completely, and parameters and devices required for computing logic operations in the SIMPLY framework were mapped onto other memory arrays. The two approaches require a similar number of arrays since most of the memory is used to store the neural network parameters. Energy and latency estimates were computed by mapping all the IMPLY, FALSE, and VMM operations required to implement the BNN inference task to the memory arrays. For the SIMPLY accelerator, two different implementations were studied to evaluate the performance improvement provided by the binary tree adder accumulator with respect to the HA chain implementation. All the test set was randomly shuffled and classified by the implemented neural network, preserving the memory states between classification tasks. In the case of SIMPLY-based implementations, the worst-case energy consumption for n-IMPLY and FALSE operations was estimated by considering the data of the maximum energy consumption for each input combination available in **Table 2** that were estimated by means of circuit simulations performed considering a 500 MHz clock frequency, meaning that IMPLY and FALSE operations are computed in 4 ns, and including the comparator overhead. The energy for BNN analog VMM operations was estimated considering RRAM devices approximated as resistors and computing the equivalent resistance for each active row of the array. To estimate the latency, when possible, operations are parallelized among different arrays, while inside each array two different degrees of parallelism were considered. Specifically, in the "serial" (see **Table 3**) case no

Operation	Input combination/Worst-case Energy
FALSE	(0/12 fJ), (1/190 fJ)
2-IMPLY	(00/509 fJ), (01/6.19 fJ), (11/6.5 fJ)
3-IMPLY	(000/409 fJ), (001/11.6 fJ), (011/13.1 fJ), (111/13.3 fJ)
4-IMPLY	(0000/409 fJ), (0001/11.6 fJ), (0011/13.1 fJ), (0111/13.3 fJ), (1111/13.7 fJ)

The effects of variability, RTN, and the comparator overhead, are included.

Table 2.
Worst-case energy estimates from circuits simulations of FALSE and n-IMPLY operations executed on the SIMPLY architecture.

Implementation	# 256x256 Arrays	# Steps	Latency	Energy	EDP Improvement
Embedded System [48]	NA	NA	17.35 ms	5.37 mJ	1
n-SIMPLY Serial HA chain	29	10,153,416	40.6 ms	3.27 μJ	$7{\cdot}10^{2}$
n-SIMPLY Parallel HA chain	29	724,306	2.9 ms	3.27 μJ	$9.8{\cdot}10^{3}$
n-SIMPLY Serial Binary Tree Adder	29	1,647,124	6.6 ms	1.69 μJ	$8.4{\cdot}10^{3}$
n-SIMPLY Parallel Binary Tree Adder	29	406,280	1.6 ms	1.69 μJ	$3.4{\cdot}10^{4}$
Hybrid Serial	28	214,795	859 μs	907 nJ	$1.2{\cdot}10^{5}$
Hybrid Parallel	28	112,518	450 μs	907 nJ	$2.3{\cdot}10^{5}$

(NA = not applicable).

Table 3.
Benchmark of different SIMPLY-based BNN accelerators against a state-of-the-art embedded system implementation from the literature.

additional parallelism was introduced, and operations were all executed sequentially. In the "parallel" (see **Table 3**) case the read step is executed in parallel on the rows of the array but, due to the high I_c of the considered technology (i.e., I_C = 100 μA), the conditional programming step is performed sequentially on the rows of the array. For the computation of BNN VMM operation in the hybrid accelerators, the same design choice performed in [8] was considered. Up to 128 lines were activated in parallel, 3-bit flash ADCs were used to digitize the partial results of the products, and each ADC was shared among 8 rows of the array. The digitized partial results were written onto other arrays, grouping on the same rows of the array the partial results related to the same neuron. Logic operations required to accumulate the results and to compute activation functions were executed in the SIMPLY framework.

The results of the simulations are reported in **Table 3** and show that all the proposed accelerators provide a > $7{\cdot}10^{2}$ EDP improvement with respect to the reference embedded system implementation [48]. All the accelerators considerably reduce the energy consumption, and except for the "n-SIMPLY Serial HA chain" implementation, they can also considerably reduce the computing time. In the SIMPLY-based accelerators, the adoption of the binary tree adder accumulator reduces considerably the number of computing steps leading to considerable latency and energy savings. As expected, additional improvements can be achieved by performing VMM in the analog domain using the hybrid accelerators, which can achieve an EDP improvement of $>10^{5}$ and > 3.5 with respect to the embedded system implementation and the SIMPLY-based implementations, respectively.

The latency of SIMPLY-based implementations could be further improved by adopting RRAM technologies with lower current compliance provided that a sufficient read margin and high retention are preserved. Adopting lower current compliances would potentially increase the maximum number of devices that can be written in parallel. Also, lower current compliance values would reduce the energy consumption in both computing approaches further highlighting the advantages of both approaches.

In addition, to compare the performance of the n-SIMPLY and hybrid BNN accelerators, the performance and characteristics of RRAM-based BNN inference accelerators from the literature [6, 8, 42, 46, 47, 50, 51] are summarized in **Table 4**. In general, these works employ the different schemes discussed in Section 4.2.1 to

Author(s)	VMM Implementation	RRAM Array Topology (Size)	Sim./Exp.[&]	TOPS/W (For MAC ops.)	MNIST Inference Energy	MNIST Inference Latency	MNIST Inference EDP (J•s)
This Work n-SIMPLY Binary Tree Adder (Serial/Parallel)	LIM	1T1R (256x256)	Sim.	/	1.69 μJ	6.6 ms/ 1.6 ms	$\approx 1 \cdot 10^{-8}$/ $2.6 \cdot 10^{-9}$
This Work Hybrid (Serial/Parallel)	Analog + LIM (3-bits ADC + 8:1 MUX for the columns)	1T1R (256x256)	Sim.	/	907 nJ	859 μs/ 450 μs	$\approx 7.8 \cdot 10^{-10}$/ $4 \cdot 10^{-10}$
Minguet et al. [50]	Analog	1S1R	Sim. (28 nm CMOS)	/	42.5 μJ (Reported)	103 μs[$$] (Estimated)	$4.4 \cdot 10^{-9}$ (Estimated)
Yin et al. [8]	Analog (3-bits ADC + 8:1 MUX for the columns)	1T1R (128x64)	Exp. (90 nm CMOS)	24.1[$] (Reported)	/	/	/
Yu et al. [42]	Analog (Online training)	1T1R (32 x 512x1024)	Exp. (130 nm CMOS)	/	2.4 mJ[$$$] (Reported)	2.7 s[$$$] (Reported)	$6.5 \cdot 10^{-3}$ (Estimated)
Zhao et al. [47]	Analog (Low-power voltage mode sensing scheme)	1T1R (32x32)	Sim. (180 nm CMOS)	19.7 (Reported)	/	/	/
Zhao et al. [47]	Analog (Low-power voltage mode sensing scheme)	1T1R (32x32)	Sim. (40 nm CMOS)	63 (Reported)	/	/	/
Pal Chowdhury [6]	Analog (Voltage sensing)	2R	Sim. (45 nm CMOS)	/	/	/	/
Ezzadeen et al. [46]	Analog (Capacitive neuron)	2T2R (32x32)	Exp. (130 nm CMOS)	/	/	/	/
Lopez et al. [51]	Analog (Subthreshold read scheme)	1S1R (32x32)	Sim. (28 nm CMOS)	/	≈ 62 nJ[$$$$] (estimated)		

[&]*Exp. is used to indicate that at least one array prototype was fabricated, and its performance experimentally measured.*[$]*Best-case for a 64x64 layer, so that no external logic is required and included in the energy-efficiency estimate.*[$$]*For arrays of arbitrary size, one neuron output computed per read operation (t_{READ} = 100 ns as reported). The ADC and external logic overhead and the limitation of the maximum current in a column is not considered.*[$$$]*Data from [52].*[$$$$]*Estimated considering the reported energy per single bit read operation (i.e., 76 fJ/bit).*

Table 4.
Performance comparison of different BNN inference accelerators from the literature.

accelerate in the analog domain the BNN VMM. However, each work uses different arrays sizes, topologies, devices, and technologies, thus complicating the comparison between different accelerators. Ideally, for a direct comparison between different implementations the same task should be executed on all the accelerators, however these data are rarely reported. Some works focus only on demonstrating the feasibility of their proposed implementation of the VMM [46], while other studies simulate a different inference task [6], or report different metrics that cannot be easily used to estimate the performance on a specific inference task. Specifically, in [8, 47] the TOPS/W metric is reported, however this metric indicates the maximum performance that can be achieved in specific conditions, which are not necessarily the one achieved by the circuit in a generic inference task. Still, some works [42, 50–52] directly report the performance, or sufficient data that can be used to estimate the performance of their accelerator on an MNIST handwritten digits classification task. Among these works, the results reported by Yu et al. in [42, 52] show the worst performance, however their solution was optimized for training directly on chip the entire network parameters, thus introducing additional overheads. Compared with the results reported by Minguet et al. in [50], the n-SIMPLY Binary Tree Adder implementation achieves similar EDP performance while the Hybrid implementation can further reduce the EDP. The lowest energy consumption for an MNIST inference task, was estimated from the work of Lopez et al. [51], where the authors proposed a subthreshold read scheme in which 1S1R devices are read using a read voltage that is lower than the threshold voltage of the selector device, achieving a read energy of 76 fJ/bit.

5. Conclusions

In this chapter, we discussed the implementation of different RRAM-based IMC accelerators describing an appropriate methodology for their analysis and design. The methodology is enabled by the use of physics-based compact models and consists in studying, by means of circuit simulations which also include devices nonideal effects, the performance and reliability of the core elements of IMC frameworks. The results of these simulations are then used to project the performance to larger architectures implemented on multiple arrays. An analysis of the reliability and performance of two LIM frameworks based on the material implication logic was presented. Due to its better performance and reliability, the SIMPLY architecture was used to implement a BNN hardware inference accelerator. The limitations of this approach were discussed and the benefits of using analog accelerators for the VMM operations were examined. Also, the implementation of an architecture which combines both approaches on the same hardware was discussed. The EDP estimates of these different RRAM-based BNN inference accelerators were compared against a state of the arts embedded system implementation, demonstrating considerable energy improvements. Overall, these hardware accelerators, by providing high reconfigurability and energy efficiency, represent a valuable solution for the implementation of future ultra-low power hardware.

Acknowledgements

Project funded under the National Recovery and Resilience Plan (NRRP), Mission 04 Component 2. Investment 1.5 – NextGenerationEU, Call for tender n. 3277 dated 30/12/2021. Award Number: 0001052 dated 23/06/2022.

Author details

Tommaso Zanotti*, Paolo Pavan and Francesco Maria Puglisi
Dipartimento di Ingegneria "Enzo Ferrari," Università di Modena e Reggio Emilia, Modena, Italy

*Address all correspondence to: tommaso.zanotti@unimore.it

References

[1] Xiao TP, Bennett CH, Feinberg B, Agarwal S, Marinella MJ. Analog architectures for neural network acceleration based on non-volatile memory. Applied Physics Reviews. 2020; **7**(3):031301. DOI: 10.1063/1.5143815

[2] Kvatinsky S, Belousov D, Liman S, Satat G, Wald N, Friedman EG, et al. MAGIC—Memristor-aided logic. IEEE Transactions on Circuits and Systems II: Express Briefs. 2014;**61**(11):895-899. DOI: 10.1109/TCSII.2014.2357292

[3] Kvatinsky S, Satat G, Wald N, Friedman EG, Kolodny A, Weiser UC. Memristor-based material implication (IMPLY) logic: Design principles and methodologies. IEEE Transactions on Very Large Scale Integration (VLSI) Systems. 2014;**22**(10):2054-2066. DOI: 10.1109/TVLSI.2013.2282132

[4] Borghetti J, Snider GS, Kuekes PJ, Yang JJ, Stewart DR, Williams RS. 'Memristive' switches enable 'stateful' logic operations via material implication. Nature. 2010;**8**(464):873. DOI: 10.1038/nature08940

[5] Zanotti T, Puglisi FM, Pavan P. Reliability-aware design strategies for Stateful logic-in-memory architectures. IEEE Transactions on Device and Materials Reliability. 2020;**20**(2):278-285. DOI: 10.1109/TDMR.2020.2981205

[6] Pal Chowdhury A, Kulkarni P, Nazm BM. MB-CNN: Memristive binary convolutional neural networks for embedded Mobile devices. Journal of Low Power Electronics and Applications. 2018;**8**(4):38. DOI: 10.3390/jlpea8040038

[7] Yin S, Kim Y, Han X, Barnaby H, Yu S, Luo Y, et al. Monolithically integrated RRAM- and CMOS-based In-memory computing optimizations for efficient deep learning. IEEE Micro. 2019;**39**(6):54-63. DOI: 10.1109/MM.2019.2943047

[8] Yin S, Sun X, Yu S, Seo JS. High-throughput In-memory computing for binary deep neural networks with monolithically integrated RRAM and 90-nm CMOS. IEEE Transactions on Electron Devices. 2020;**67**(10):4185-4192. DOI: 10.1109/TED.2020.3015178

[9] Courbariaux M, Hubara I, Soudry D, El-Yaniv R, Bengio Y. Binarized neural networks: Training deep neural networks with weights and activations constrained to+ 1 or-1. arXiv preprint. 2016. Available from: https://arxiv.org/abs/1602.02830

[10] Puglisi FM, Zanotti T, Pavan P. Unimore resistive random access memory (RRAM) Verilog-a model. nanoHUB. 2019. DOI: 10.21981/15GF-KX29. Available from: https://nanohub.org/publications/289/about/1#citethis

[11] Zanotti T, Pavan P, Puglisi FM. Comprehensive physics-based RRAM compact model including the effect of variability and multi-level random telegraph noise. Microelectronic Engineering. 2022;**27**:111886. DOI: 10.1016/j.mee.2022.111886

[12] Yu S, Wu Y, Chai Y, Provine J, Wong HSP. Characterization of switching parameters and multilevel capability in HfOx/AlOx bi-layer RRAM devices. In: Proceedings of 2011 International Symposium on VLSI Technology, Systems and Applications. Hsinchu, Taiwan: IEEE; 2011. pp. 1-2

[13] Zahoor F, Azni Zulkifli TZ, Khanday FA. Resistive random access memory (RRAM): An overview of materials, switching mechanism,

performance, multilevel cell (mlc) storage, modeling, and applications. Nanoscale Research Letters. 2020;**15**(1): 90. DOI: 10.1186/s11671-020-03299-9

[14] Wong HSP, Lee HY, Yu S, Chen YS, Wu Y, Chen PS, et al. Metal–Oxide RRAM. Proceedings of the IEEE. 2012; 100(6):1951–1970. DOI: 10.1109/JPROC.2012.2190369

[15] Kozicki MN, Barnaby HJ. Conductive bridging random access memory—Materials, devices and applications. Semiconductor Science and Technology. 2016;**31**(11):113001. DOI: 10.1088/0268-1242/31/11/113001

[16] Bersuker G, Gilmer DC, Veksler D, Yum J, Park H, Lian S, et al. Metal oxide RRAM switching mechanism based on conductive filament microscopic properties. In: 2010 International Electron Devices Meeting, San Francisco, CA, USA. New York City, USA: IEEE; 2010. pp. 19.6.1-19.6.4

[17] Celano U, Fantini A, Degraeve R, Jurczak M, Goux L, Vandervorst W. Scalability of valence change memory: From devices to tip-induced filaments. AIP Advances. 2016;**6**(8):085009. DOI: 10.1063/1.4961150

[18] Puglisi FM, Zagni N, Larcher L, Pavan P. Random telegraph noise in resistive random access memories: Compact modeling and advanced circuit design. IEEE Transactions on Electron Devices. 2018;**65**(7):2964-2972. DOI: 10.1109/TED.2018.2833208

[19] Panda D, Sahu PP, Tseng TY. A collective study on modeling and simulation of resistive random access memory. Nanoscale Research Letters. 2018;**13**(1):8. DOI: 10.1186/s11671-017-2419-8

[20] Zanotti T, Pavan P, Puglisi FM. Self-consistent automated parameter extraction of RRAM physics-based compact model. In: ESSDERC 2022 - IEEE 52nd European Solid-State Device Research Conference (ESSDERC). 2022. pp. 316-319

[21] Li H, Jiang Z, Huang P, Wu Y, Chen H, Gao B, et al. Variation-aware, reliability-emphasized design and optimization of RRAM using SPICE model. In: 2015 Design, Automation Test in Europe Conference Exhibition (DATE), Grenoble, France. New York City, USA: IEEE; 2015. pp. 1425-1430

[22] Bengel C, Siemon A, Cuppers F, Hoffmann-Eifert S, Hardtdegen A, von Witzleben M, et al. Variability-aware modeling of filamentary oxide-based bipolar resistive switching cells using SPICE level compact models. IEEE Trans Circuits Syst I. 2020;**67**(12):4618-4630. DOI: 10.1109/TCSI.2020.3018502

[23] Guan X, Yu S, Wong HP. A SPICE compact model of metal oxide resistive switching memory with variations. IEEE Electron Device Letters. 2012;**33**(10): 1405-1407. DOI: 10.1109/LED.2012.2210856

[24] Lehtonen E, Laiho M. Stateful implication logic with memristors. In: 2009 IEEE/ACM International Symposium on Nanoscale Architectures, San Francisco, CA, USA. New York City, USA: IEEE; 2009. pp. 33-36

[25] Xie L, Du Nguyen HA, Yu J, Kaichouhi A, Taouil M, AlFailakawi M, et al. Scouting logic: A novel Memristor-based logic Design for Resistive Computing. In: 2017 IEEE Computer Society Annual Symposium on VLSI (ISVLSI). 2017. pp. 176-181

[26] Zanotti T, Puglisi FM, Pavan P. A smart logic-in-memory architecture for low-power non-von Neumann

computing. IEEE Journal of the Electron Devices Society. 2020;**8**:1-1. DOI: 10.1109/JEDS.2020.2987402

[27] Lehtonen E, Poikonen JH, Laiho M. Two memristors suffice to compute all Boolean functions. Electronics Letters. 2010;**46**(3):230-231. DOI: 10.1049/el.2010.3407

[28] Zanotti T, Puglisi FM, Pavan P. Energy-efficient non-Von Neumann computing architecture supporting multiple computing paradigms for logic and Binarized neural networks. Journal of Low Power Electronics and Applications. 2021;**11**(3):29. DOI: 10.3390/jlpea11030029

[29] Zanotti T, Puglisi FM, Pavan P. Reliability and performance analysis of logic-in-memory based Binarized neural networks. IEEE Transactions on Device and Materials Reliability. 2021;**21**:1-1. DOI: 10.1109/TDMR.2021.3075200

[30] Lehtonen E, Poikonen J, Laiho M. Implication logic synthesis methods for memristors. In: 2012 IEEE International Symposium on Circuits and Systems, Seoul, Korea (South). New York City, USA: IEEE; 2012. pp. 2441-2444

[31] Zanotti T, Pavan P, Puglisi FM. Multi-input logic-in-memory for ultra-low power non-Von Neumann computing. Micromachines. 2021; **12**(10):1243. DOI: 10.3390/mi12101243

[32] Stine JE, Castellanos I, Wood M, Henson J, Love F, Davis WR, et al. FreePDK: An open-source variation-aware design kit. In: 2007 IEEE International Conference on Microelectronic Systems Education (MSE'07). 2007. pp. 173-174

[33] Kvatinsky S, Wald N, Satat G, Kolodny A, Weiser UC, Friedman EG. MRL — Memristor Ratioed logic. In: 2012 13th International Workshop on Cellular Nanoscale Networks and their Applications. 2012. pp. 1-6

[34] Ali KA, Rizk M, Baghdadi A, Diguet JP, Jomaah J. MRL crossbar-based full adder design. In: 2019 26th IEEE International Conference on Electronics, Circuits and Systems (ICECS). 2019. pp. 674-677

[35] Hoffer B, Rana V, Menzel S, Waser R, Kvatinsky S. Experimental demonstration of Memristor-aided logic (MAGIC) using valence change memory (VCM). IEEE Transactions on Electron Devices. 2020;**67**(8):3115-3122. DOI: 10.1109/TED.2020.3001247

[36] Escudero López M. Reliability-Aware Circuit Design to Mitigate Impact of Device Defects and Variability in Emerging Memristor-Based Applications [Thesis]. TDX (Tesis Doctorals en Xarxa). Catalonia, Spain: Universitat Politècnica de Catalunya; 2020

[37] Yu J, Du Nguyen HA, Abu Lebdeh M, Taouil M, Hamdioui S. Enhanced scouting logic: A robust Memristive logic design scheme. In: 2019 IEEE/ACM International Symposium on Nanoscale Architectures (NANOARCH), Qingdao, China. New York City, USA: IEEE; 2019. pp. 1-6

[38] Talati N, Gupta S, Mane P, Kvatinsky S. Logic design within Memristive memories using Memristor-aided loGIC (MAGIC). IEEE Transactions on Nanotechnology. 2016; **15**(4):635-650. DOI: 10.1109/TNANO.2016.2570248

[39] Siemon A, Drabinski R, Schultis MJ, Hu X, Linn E, Heittmann A, et al. Stateful three-input logic with Memristive switches. Scientific Reports. 2019;**9**(1):1-13. DOI: 10.1038/s41598-019-51039-6

[40] Cheng L, Zhang MY, Li Y, Zhou YX, Wang ZR, Hu SY, et al. Reprogrammable logic in memristive crossbar for in-memory computing. Journal of Physics D: Applied Physics. 2017;**50**(50):505102. DOI: 10.1088/1361-6463/aa9646

[41] Junsangsri P, Han J, Lombardi F. Logic-in-memory with a nonvolatile programmable metallization cell. IEEE Transactions on Very Large Scale Integration (VLSI) Systems. 2016;**24**(2): 521-529. DOI: 10.1109/TVLSI.2015.2411258

[42] Yu S, Li Z, Chen P, Wu H, Gao B, Wang D, et al. Binary neural network with 16 Mb RRAM macro chip for classification and online training. In: 2016 IEEE International Electron Devices Meeting (IEDM). 2016. pp. 16.2.1-16.2.4

[43] Zanotti T, Puglisi FM, Pavan P. Reconfigurable smart In-memory computing platform supporting logic and Binarized neural networks for low-power edge devices. IEEE Journal on Emerging and Selected Topics in Circuits and Systems. 2020;**10**(4):478-487. DOI: 10.1109/JETCAS.2020.3030542

[44] Yu S, Chen PY, Cao Y, Xia L, Wang Y, Wu H. Scaling-up resistive synaptic arrays for neuro-inspired architecture: Challenges and prospect. In: 2015 IEEE International Electron Devices Meeting (IEDM), Washington, DC, USA. New York City, USA: IEEE; 2015. pp. 17.3.1-17.3.4

[45] Welser J, Pitera JW, Goldberg C. Future computing hardware for AI. In: 2018 IEEE International Electron Devices Meeting (IEDM). 2018. pp. 1.3.1-1.3.6

[46] Ezzadeen M, Majumdar A, Bocquet M, Giraud B, Noël JP, Andrieu F, et al. Low-overhead implementation of Binarized neural networks employing robust 2T2R resistive RAM bridges. In: ESSCIRC 2021 - IEEE 47th European Solid State Circuits Conference (ESSCIRC). 2021. pp. 83-86

[47] Zhao Y, Yu J, Zhang D, Hu Q, Liu X, Jiang H, et al. A 0.02% accuracy loss voltage-mode parallel sensing scheme for RRAM-based XNOR-net application. IEEE Transactions on Circuits and Systems II: Express Briefs. New York City, USA: IEEE. 2022;**69**(6):2697-2701. DOI: 10.1109/TCSII.2022.3157767

[48] McDanel B, Teerapittayanon S, Kung HT. Embedded Binarized neural networks. In: Proceedings of the 2017 International Conference on Embedded Wireless Systems and Networks. Uppsala, Sweden: Junction Publishing; 2017. pp. 168-173

[49] Zhou S, Wu Y, Ni Z, Zhou X, Wen H, Zou Y. DoReFa-net: Training low Bitwidth convolutional neural networks with low Bitwidth gradients. arXiv preprint. 2016. Available from: https://arxiv.org/abs/1602.02830

[50] Minguet Lopez J, Hirtzlin T, Dampfhoffer M, Grenouillet L, Reganaz L, Navarro G, et al. OxRAM + OTS optimization for binarized neural network hardware implementation. Semiconductor Science and Technology. 2022;**37**(1): 014001. DOI: 10.1088/1361-6641/ac31e2

[51] Lopez JM, Rummens F, Reganaz L, Heraud A, Hirtzlin T, Grenouillet L, et al. 1S1R sub-threshold operation in crossbar arrays for low power BNN inference computing. In: 2022 IEEE International Memory Workshop (IMW), Dresden, Germany. New York City, USA: IEEE; 2022. pp. 1-4

[52] Yu S. Binary Neural Network and Its Implementation with 16 Mb RRAM Macro Chip [Internet]. Available from: https://www.src.org/calendar/e006125/yu-presentation2.pdf

Chapter 4

Enabling Neuromorphic Computing for Artificial Intelligence with Hardware-Software Co-Design

Bojing Li, Duo Zhong, Xiang Chen and Chenchen Liu

Abstract

In the last decade, neuromorphic computing was rebirthed with the emergence of novel nano-devices and hardware-software co-design approaches. With the fast advancement in algorithms for today's artificial intelligence (AI) applications, deep neural networks (DNNs) have become the mainstream technology. It has been a new research trend to enable neuromorphic designs for DNNs computing with high computing efficiency in speed and energy. In this chapter, we will summarize the recent advances in neuromorphic computing hardware and system designs with non-volatile resistive access memory (ReRAM) devices. More specifically, we will discuss the ReRAM-based neuromorphic computing hardware and system implementations, hardware-software co-design approaches for quantized and sparse DNNs, and architecture designs.

Keywords: neuromorphic computing, ReRAM, deep neural network, processing-in-memory, hardware-software co-design

1. Introduction

The explosive growth of data and the increasing scale and complexity of deep learning models have made traditional von Neumann architectures inefficient in terms of speed and energy for AI processing. As a result, novel computer architecture and hardware are gaining increasing attention due to their potential to break the "memory wall" constraints in traditional von Neumann architectures. Neuromorphic computing, which is inspired by the functioning of the human brain and now refers to the hardware implementation of artificial neural networks, has been extensively explored and has demonstrated great potential to revolutionize AI processing. By leveraging new nano-devices such as resistive memory (ReRAM), significant progress has been made in this area. Artificial synapses [1–12] and neurons [13–19], and corresponding neuromorphic circuits and hardware systems have been successfully investigated [11, 20–29].

The two-terminal ReRAM device forms a high-density crossbar structure, enabling highly efficient vector-matrix computation naturally [30]. It is worth noting that challenges still exist in applying the ReRAM crossbar to neural network implementation due to factors such as inherent sneak-path leakage, signal noise, and limited conductance states, which can degrade computing accuracy. To address these challenges, various hardware-software co-design methodologies have been proposed [31–41]. For example, researchers explored specialized circuits and algorithms to tolerate the sneak-path current and guarantee the computing accuracy [11, 25–29]. Hardware-adaptive neural network pruning was also been investigated to promote computational efficiency with reduced hardware design and computing energy costs. In addition, several works have focused on improving the computing reliability and security in ReRAM-based neuromorphic computing systems.

The ReRAM, along with its neuromorphic designs, has inspired the development of ReRAM-based in-memory processing accelerators. These accelerators, including PRIME [42], ISAAC [43], and PipeLayer [44], were designed to accelerate the training and inference of convolution neural networks (CNNs). Based on these fundamental explorations, researchers have designed accelerators for various neural networks and applications, such as ReRAM-based accelerators for GANs [45], RNNs [46, 47]. These accelerators demonstrated improved speed and energy efficiency compared to traditional computing platforms such as CPUs and GPUs. With the continuous advancement of ReRAM technology, the ReRAM-based neuromorphic engines are being applied in broader domains [48–61].

In this chapter, we will summarize recent research in ReRAM-based neuromorphic computing. The aforementioned research areas, including hardware implementation, hardware-software co-design, and accelerator architecture, will be covered.

2. ReRAM-based neural network implementation

The emerging ReRAM device has shown superior performance in neuromorphic computing. The ReRAM device is promising to enable highly parallel, ultra-low-power computing in memory for AI applications owing to its structural simplicity, low power consumption, and ease of integration. Hardware implementations of artificial neurons and synapses play an important role in neuromorphic computing, attracting considerable attention over the past few decades. The fundamental function of artificial neurons is to emulate potential accumulation processes and spike generation functions. Meanwhile, artificial synapses are designed to implement various synaptic plasticities and learning weight signals. These plasticities are crucial for the learning function in neuromorphic computing. Based on the artificial neurons and synapses, neuromorphic hardware systems are built for the whole neural network functions. In this section, we will provide an overview of the hardware implementation of ReRAM-based neuromorphic synapses, neurons, and hardware systems.

2.1 ReRAM-based artificial synapses

The performance of synaptic devices directly impacts the learning accuracy and efficiency of a neuromorphic computing system. ReRAM devices are widely utilized to implement synaptic plasticities, such as signal weighting, short-term potentiation/depression (STP/STD), long-term potentiation/depression (LTP/LTD), and spiking-time dependent plasticity (STDP).

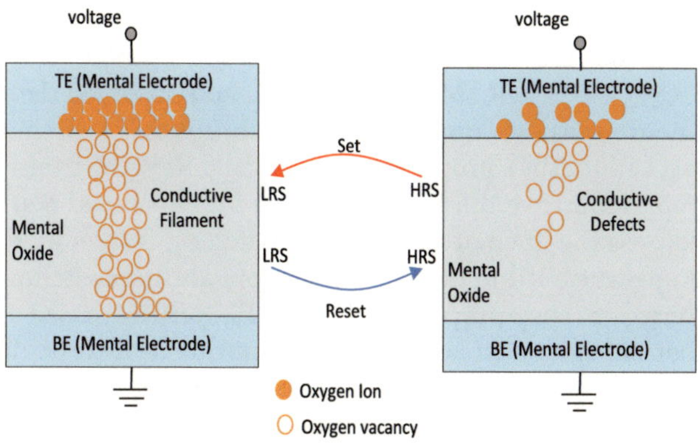

Figure 1.
The typical structure of ReRAMs devices and their switching states [62].

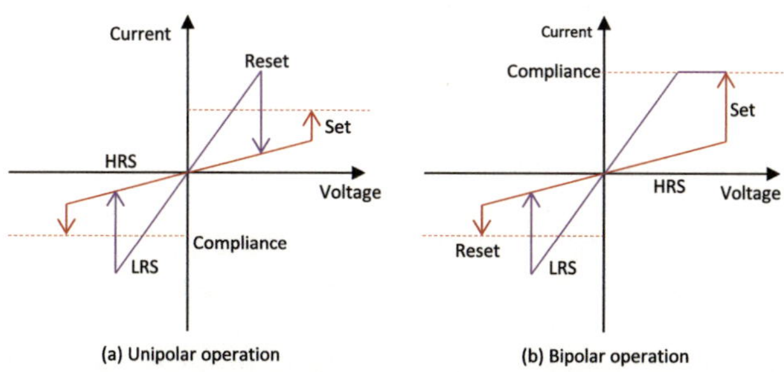

Figure 2.
I-V curves for ReRAM devices on unipolar and bipolar operations [62].

The typical ReRAM device is in a metal/insulator/metal (MIM) structure [62] as is depicted in **Figure 1**, where the device consists of the bottom electrode (BE), top electrode (TE), and oxide layers. As illustrated in **Figure 2**, ReRAM devices can be switched between a high resistance state (HRS) and a low resistance state (LRS) with the unipolar or bipolar operation. This nonlinear feature resembles biological synapses, whose weight changes in response to stimuli passing from pre- to post-synaptic neurons.

In 2010, Lu proposed the first resistive synapse and demonstrated its function of STDP [1]. To achieve the synaptic function of STDP, well-designed shapes of neuronal pulses were also required. Therefore, Yu et al. proposed a metal oxide ReRAM-based synapse and an energy-efficient signal scheme for synapse programming, which involved tuning the pulse amplitude in each time slot [2]. As a further study, Ohno et al. proposed an inorganic synapse using Ag_2S [3], showing that a single proposed synapse device exhibits both time-dependent STP and LTP features of a biological synapse by adjusting the repetition rate of input stimuli without the special design needs of neuronal pulses. Later, Li et al. demonstrated a chalcogenide resistive electronic synapse with an Ag/AgInSbTe/Ag structure to implement STDP function [5], in which the synaptic weights were modified with the cooperation of pre- and post-synaptic spikes and the growth of the weights was more stable by utilizing synaptic saturation mechanism. To eliminate the resistant fluctuations issue [6], Gao et al. proposed an oxide-based 3D vertical structure synapse and simulated a single synapse by using the mean value of a group of resistive switching devices.

Challenges in applying the ReRAM devices in neuromorphic computing were further explored and novel ReRAM-based synapses were proposed. Woo et al. analyzed the TiN/HfO_2/Ti/TiN ReRAM-based synapse and observed deteriorated accuracy drop of the neuromorphic computing under abrupt SET switching operation [8]. Based on this observation, they proposed an optimized ReRAM-based synapse with an Al electrode on top to improve the accuracy of the RerAM-based neuromorphic computing. To address the issue of abrupt SET switching, Wu et al. found that increasing the temperature led to the transition from abrupt switching to analog switching [9]. Therefore, they proposed a HfO_x/TEL ReRAM-based synapse, which incorporated a thermal enhanced layer (TEL) to confine heat for realizing analog switching. Kim et al. proposed a Ni/SiN_x/AlO_y/TiN ReRAM-based synapse, where the AlO_y layer reduces current overshoot, resulting in a smooth reset switching transition to enhance analog switching performance [10]. Sun et al. proposed an XNOR-ReRAM synapse that enables equivalent XNOR and bit-counting operations to be carried out in parallel in binary neural networks (BNN) [11]. Roy et al. improved the reliability of HfO_2 ReRAM devices through processes of Al-doping, ozone treatment, and post-deposition annealing [12].

2.2 ReRAM-based artificial neurons

Researchers have started to investigate ReRAM-based neuron designs with low design and computation costs for a few years. By leveraging the threshold transition characteristic of ReRAM devices, the dynamic process of pulse generation and resetting of neurons can be accurately represented. In addition, the ReRAM devices hold unique nonlinear features and flexible structures to achieve the rich kinetic properties of neurons. Furthermore, the nanoscale properties of ReRAM devices enable neurons with high power efficiency with the capability of achieving neuron functions with one or a few devices.

The ReRAM-based analog integrate-and-fire (I&F) neurons have been extensively explored. In 2016, Mehonic et al. implemented the Hodgkin-Huxley (H-H) neural model and leaky integrate-and-fire (LIF) neural model using silicon oxide ReRAM devices [13]. As is depicted in **Figure 3**, the charging and discharging process of neurons in the LIF model was successfully implemented. Compared to traditional hybrid analog/digital CMOS silicon neurons, the proposed ReRAM neurons

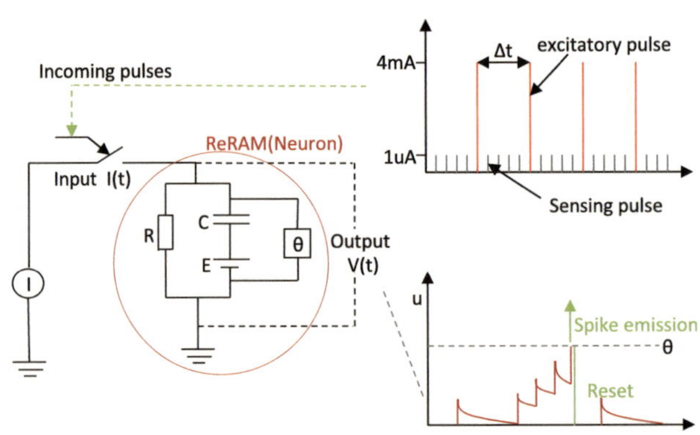

Figure 3.
The hardware implementation and equivalent circuits of LIF neural model with silicon oxide ReRAMs [13].

significantly simplified the hardware design. Similarly, Kwon et al. proposed a ReRAM-based analog integrate-and-fire (I&F) neuron circuit without capacitors, which greatly reduces power consumption, delay, and neuron size [15]. Zhang et al. implemented a new I&F neuron based on an $Ag/SiO_2/Au$ threshold switching ReRAM device [16]. The proposed neuron can achieve four critical neuron functions, including all-or-nothing spiking, threshold-driven spiking, a refractory period, and a strength-modulated frequency response. As the first attempt, Lashkare et al. proposed a $Pr_{0.7}Ca_{0.3}MnO_3$ (PCMO) ReRAM device to implement an integrate-and-fire (I&F) neuron with all the neuronal functionalities—integration, reset, threshold comparison, spike, etc., by only utilizing a single device without any external control circuits [17].

In addition, the ReRAM-based neuron implementations have been extended to a large scale. For example, spiking neural networks on a large scale were achieved based on PCMO ReRAM-based neurons [17]. Lin et al. implemented neurons with a one-transistor-one-ReRAM (1T1R) structure device and integrated the neurons within the synaptic crossbar to build more dense and high-throughput process element [18].

2.3 Neuromorphic system implementation

The CMOS-based neuromorphic chips have several limitations that hinder their further development and applications [20–23]. Firstly, on-chip memory based on SRAM is area inefficient and can hardly store the weights of a large-scale neural network. In addition, the power consumption of off-chip storage using DRAM is more than 100 times higher than that of on-chip memory. The ReRAM device, on the other hand, offers advantages such as low programming voltage, fast switching speed, high integration density, and excellent performance scalability, which make it a promising candidate for neuromorphic computing system implementation.

In the last 10 years, ReRAM-based neuromorphic computing systems for artificial neural network implementation have been extensively studied. In 2012, Hu et al. proposed a neuromorphic hardware system using ReRAM crossbar arrays to achieve recall functions of Brain-State-in-a-Box (BSB) network [24]. In the hardware implementation, the resistive crossbar naturally performs the intensive vector-matrix multiplication, which is the basic computation in the neural network model in parallel. The signal strength was represented by the voltage amplitude and analog neurons with circuitry (e.g., amplifier, analog-to-digital converter) were utilized. Two years later, Hu et al. further implemented the training functions of the BSB model with ReRAM crossbar arrays and alleviated the noise issues found in the previous works, and impacted the computing accuracy [25]. They introduced an iterative scheme for training that uses the sign of inputs and magnitude differences between outputs and inputs, which significantly reduced the circuit design complexity. The analog designs of the neuromorphic systems are vulnerable to signal variations and introduce extremely heavy hardware design and area costs due to the large analog circuit components. With the target of solving the above challenges, in 2015, Liu et al. proposed spiking neuromorphic systems, where input signals were represented by spikes and the need for large-scale analog circuits was significantly reduced [26]. In their works, spiking neuromorphic systems with 1T1R resistive crossbar arrays were implemented for feedforward and Hopfield networks on digital image recognition with IBM 130 nm technology. The 1T1R structure is utilized to control the impact of the sneak path and thus guarantee computing accuracy and a novel Integrate-and-Fire Circuit (IFC) with high-speed and low-power consumption was developed.

With the continuous growing of artificial neural networks in scale and complexity, there is an urgent need to improve computing efficiency in speed and energy. Neuromorphic systems for large-scale and complex neural networks were developed accordingly. For example, Yakopcic et al. implemented the first deep convolutional neural network (CNN) on ReRAM-based crossbars with all data represented in 16 values without classification accuracy degradation [27]. The proposed neuromorphic hardware also achieved parallel computing of the convolution operations. Wen et al. implemented a long short-term memory (MLSTM) network using ReRAM crossbars [29]. The neural network training process is in extremely high demand of computing resources, while its speed and energy efficiency are highly constrained. Therefore, researchers also explored neuromorphic computing systems that can support neural network training besides inference. For example, Yao et al. implemented parallel online training for gray-scale face classification using an integrated 1024-cell array [28].

3. Neuromorphic hardware-software Co-design

3.1 Algorithm-driven neuromorphic hardware optimizations

The accuracy of data representation in neural network models has a significant impact on computing accuracy. It is worth noting that ReRAM devices face challenges in providing high-precision computing due to manufacturing and hardware design costs although analog conductance states can be ideally achieved. Therefore, ReRAM devices with the limited physical precision pose a severe challenge to neuromorphic computing accuracy.

Researchers have attempted to alleviate the accuracy loss caused by the limited precision through hardware-software co-optimization. For example, Wang et al. proposed a new quantization regularization according to the computing characteristics of ReRAM devices and leverage different levels of regularization for different network layers [31]. They also minimized the impact of quantization by dynamical bias tuning under the fixed weights. Their quantization method achieved minimal accuracy loss under the limited resolution of synaptic weights. Yang et al. investigated a novel approach that processed quantization and training concurrently by optimizing the calculation of continuous weights and quantized weights in stochastic gradient descent [32]. Researchers found that quantizing partial sums was an effective approach to performing high-precision calculations in the ReRAM-based neuromorphic computing system. A team from the University of Illinois at Urbana-Champaign developed a comprehensive quantization approach that considered inputs, weights, and partial sums [33]. They developed a deep reinforcement learning-based search method that can automatically discover the best-mixed configuration to identify the optimal precision configurations of these three types of data. **Table 1** presents a comparison of the accuracy achieved by these quantization methods.

DNN models are becoming increasingly complex and involve tremendous parameters. It was noted that sparsity exists in the neural networks, which indicates that a significant number of parameters are redundant and can be pruned without causing accuracy loss [34, 35]. In the ReRAM-based neuromorphic designs, pruned neural networks can ideally reduce hardware costs and improve computation speed and energy efficiency. Therefore, identifying and eliminating redundancies is crucial to enhance computational efficiency while maintaining accuracy. SNrram [63]

	Dataset	Model	Base-quantization	Ideal Acc	Base Acc	Proposed quantization
Wang et al. [31]	MNIST	MLP	1-level	98.39%	95.97%	98.00%
	MNIST	LeNET	1-level	99.15%	90.77%	98.96%
	CIFAR-10	CNN	1-level	82.12%	17.80%	76.59%
Yang et al. [32]	MNIST	MLP	2-level	97.51%	85.98%	94.53%
	MNIST	LeNET1	2-level	99.00%	69.24%	98.75%
	CIFAR-10	LeNET2	2-level	76.99%	26.20%	73.04%
Huang et al. [33]	—	LeNET	mix-level	97.27%	—	96.09%

Table 1.
The accuracy comparison of quantization methods.

	Software design	Hardware design	Performance
Snrram [63]	Normalizing sparsity in weight and activation into the same format and eliminate redundancy	IRU* units store sparse indexes and corresponding peripheral circuit design	**Pruned connection(%):** 62.7, 82.7, 96.2 in VGG, MLP, Lenet **Saved resource(%):** 53.2, 81.3, 75 in VGG, MLP, Lenet **Speedup:** 1.53x, 3.41x in VGG, Lenet
Recom [64]	• SWOF† positively finds and eliminates sparse redundancy in weights and activations • IPMC‡ to improve pipeline computing efficiency	Column fetching unit for SWOF† and corresponding peripheral circuit design	**Energy saving:** 3.7x, 3.07x, 1.59x in Lenet, Alexnet, Caffenet **Speedup:** 4.81x, 4.40x, 2.25x in Lenet, Alexnet, Caffenet
Pim-prune [65]	Dividing into blocks and use structural pruning to avoid dislocation problem when focusing on row sparsity and column sparsity simultaneously	• Modules to store sparsity table for both row and column • Explore optimal energy-efficiency hardware design	**Pruned accuracy(%)(lost):** 93.84 (−0.30), 93.23(−0.47), 68.72(−1.04) in Resnet, VGG, Resnet **Compression rate:** 24.85x, 26.85x, 3.56x in Resnet, VGG, Resnet
Sme [66]	• Quantization and encoding scheme to increase bit-level sparsity • Bit-slicing scheme to accumulate the 0-bits to the same crossbars and decouple crossbar structure • Squeeze-out pruning method to eliminate redundancy	• Buffer connection to support squeeze-out algorithm and peripheral circuits design to support activation results splicing • Low overhead and orthogonal hardware design that can combine with other research	**Pruned accuracy(%)(lost):** 93.6 (+0.1), 94.19(+0.05), 76.03(−0.1), 71.57(−0.31) in VGG, Resnet, Resnet, Mobilenet **Register reduction(%):** 77.80, 96.80 compared to Pim-Prune, SRE **Energy efficiency:** 2.3x in Resnet **Area efficiency:** 6.1x in Resnet

IRU represents indexing register unit.† SWOF represents structurally compressed weight oriented fetching.‡ IPMC represents in-layer pipeline for memory and computation.

Table 2.
The comparison of pruning methods.

introduced a sparsity transfer algorithm to standardize sparsity in weights and activation and an indexing register unit was designed to store sparsity indexes and parse the data. Later, researchers proposed ReCom, another hardware and software co-design accelerator for high-sparsity networks based on ReRAM [64]. ReCom utilized a group lasso algorithm to standardize the shape of filters and accomplish pruning by compressing sparse weights through the regularization of each layer. Other methods besides regularization for ReRAM-based network pruning have been proven to be effective [65, 66]. **Table 2** illustrates a comparison of these pruning methods.

3.2 Hardware-driven hardware-software co-optimization

In ReRAM-based neuromorphic systems, process variation [67, 68], circuit noise [69, 70], retention issues, and endurance issues [71–73] greatly impact its practical applications in real world [74].

Bit failure in resistive devices is a common fault in high-density crossbar arrays. The ReRAM device is unable to switch its conductivity in response to the writing voltage when the failure occurs, which causes fixed resistive devices that are expressed as constant weights in neural network computation. These fixed resistive devices may destroy the overall neural network accuracy. To solve this challenge, Liu et al. first proposed a novel hardware and software co-design approach to improve computational accuracy in neuromorphic designs with high defects [75]. The proposed approach had two strategies: network retraining (software) and redundant resistive devices (hardware). The network retraining consisted of a standard weight-tuning process and a retraining method that prevented the weights at defective resistive devices from being updated. The redundant resistive device strategy deployed additional columns for highly significant weights with defects. Similarly, researchers at Northeastern University proposed a hierarchical progressive pruning method to improve the fault tolerance of ReRAM computing under stuck-off defects and a corresponding differential mapping scheme to support their method for both stuck-on and stuck-off defects [76].

A series of research works have utilized network retraining techniques to minimize the accuracy loss introduced by the imperfect hardware [77–79]. Nevertheless, as the location and quantity of defective resistive devices can differ for each ReRAM-based chip, it is necessary to retrain the network for every instance, leading to an enormous computational burden. To address this challenge, researchers at Purdue University proposed CxDNN, a solution that combines hardware-software compensation techniques for DNNs [80]. CxDNN consists of three optimization steps: a quantization and conversion algorithm, a re-training method, and hardware compensation. The quantization and conversion algorithm extracts a fixed-point neural network from open-access weights based on the accuracy of inputs, weights, and ADC/DAC components. The re-training process mitigates accuracy loss resulting from the nonlinear representation of components such as ADC/DAC, and leverages the available weights to accelerate calculation. Finally, the hardware compensation mechanism adjusts the compensation factor of each column in crossbar arrays based on relative and absolute errors to reduce accuracy loss caused by hardware limitations.

In addition to the aforementioned permanent defects, ReRAM-based systems can also experience instabilities during processing, including noise [81], drifting [82], and programming errors [83]. To mitigate these issues during computation, researchers proposed FTNNA, a ReRAM-customized advanced error-correcting output code (ECCO) scheme [84]. FTNNA applied collaborative logistic classifiers to replace the classic softmax function and adjusts the weights of these classifiers through transfer learning. Furthermore, they designed a variable-length decode-free coding scheme to reduce neural competition [84]. This approach resulted in significant accuracy improvements without the need for any hardware-specific calibration. Researchers at Tsinghua University proposed a re-configurable redundancy scheme to rescue accuracy degradation caused by stocked resistive devices [85]. Many other studies are dedicated to addressing various aspects of ReRAM technology, such as stuck failures [86, 87], temperature [88], IR-drop [89, 90], and other factors, to enhance the error tolerance of ReRAM-based neuromorphic computing.

4. ReRAM-based in-memory computing architectures

In today's DNN-based AI, the concept of neuromorphic computing has been expanded to non-von Neumann architecture computing paradigms that integrate processing and memory on a single chip. The successful hardware implementation of synapses and neurons with ReRAM devices has enabled the prosperous development of ReRAM-based in-memory accelerators for various DNN applications, including convolutional neural networks (CNNs), graph neural networks (GNNs), etc. These accelerators have demonstrated significant computing speedup and energy efficiency in AI applications such as image processing and language processing. In this chapter, we discuss early exploration of ReRAM-based accelerators and computing architectures for various DNNs and applications.

4.1 ReRAM-based in-memory computing accelerators

With the successful exploration of ReRAM-based neuromorphic hardware, researchers also started to explore architecture-level innovations by developing novel ReRAM-based in-memory processing accelerators.

In 2016, researchers proposed a ReRAM-based neural network accelerator for DNN applications—PRIME [42] with novel in-memory processing architecture. As is shown in **Figure 4**, unlike previous in-memory processing architectures that integrate additional processing units to memory, PRIME has full-function (FF) subarrays that

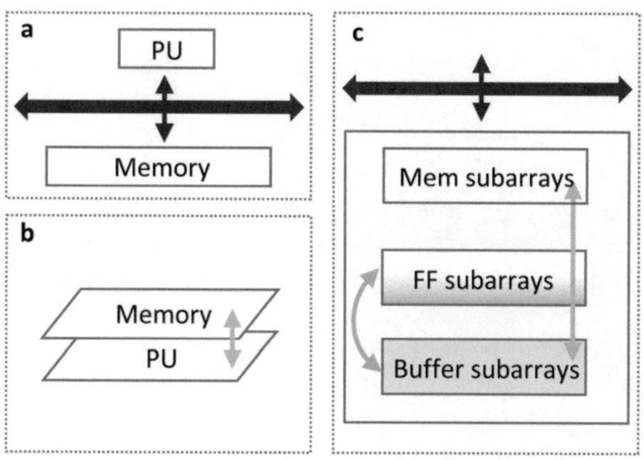

Figure 4.
(a) Traditional shared memory-based processor-coprocessor architecture, (b) processing in-memory approach using 3D integration technologies, (c) PRIME design [42].

can act as a memory unit for data storage or perform matrix multiplication in DNN computation. The FF subarrays mainly consist of the components: (1) decoders and drivers that provide analog inputs to ReRAM crossbar arrays by controlling voltage through amplifiers, latches, etc.; (2) multiplexers that are used to support subtraction and the sigmoid activation function; (3) sense amplifiers to sense computing results from the ReRAM devices and support ReLU functions with the additional hardware units. PRIME maximum reuses the periphery circuits to enable the switch of ReRAM between storage and computing with reduced design costs. Moreover, PRIME proposed high-precision NN acceleration based on its architecture and provided software/hardware interfaces for developers.

In the same year, researchers at the University of Utah introduced ISAAC, a comprehensive ReRAM-based accelerator, aimed at offering high-speed and energy-efficient computations of convolutional neural networks (CNNs) [43]. The ISAAC is a multi-stage architecture, which includes tiles, IMAs, and ReRAM crossbar arrays, as illustrated in **Figure 5**. The fundamental component used for deploying a neural network layer is IMAs, where several IMAs share eDRAM activation components and other peripheral circuits. ISAAC presented an inter-layer pipeline coupled with a buffer optimization strategy. ISAAC processed all layers in a CNN concurrently and redeployed layers to maintain the pipeline's continuity, and hence improved overall throughput and computational efficiency were achieved. The subsequent layer started processing as soon as the output feature maps of the current layer meet the size requirement of the convolution kernels of the next layer instead of waiting for the final results. This strategy significantly reduced the capacity requirement and on-chip space occupation of eDRAM. In addition, ISAAC developed an encoding scheme to reduce the ADC resolution needs to 1-bit, which can significantly reduce hardware costs and improve computing efficiency. The ISAAC achieved significant improvement in computing throughput, energy, and computational density compared to previous designs.

Although previous architectures demonstrated high computing efficiency, limitations still exist. For example, they were designed only for inference and did not

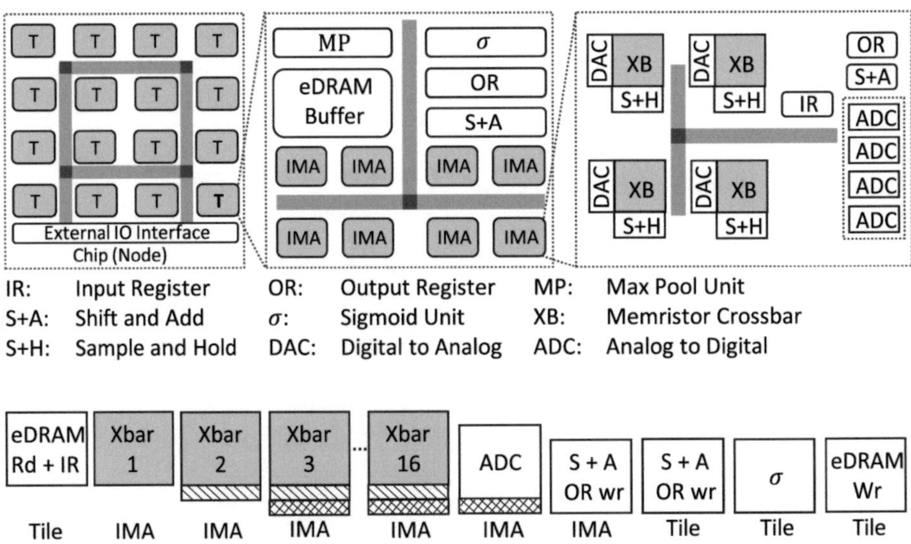

Figure 5.
Architecture and pipeline of ISAAC [43].

	CE* (GOPS/s/mm²)	PE† (GPOS/s/W)	Large data mapping	Data precision	Support Train/Inf
PRIME [42]	1230	2100	Crossbar (Analog)	6-bit (In.) & 8-bit (Wt.)	Inf
ISAAC [43]	479.0	380.7	Crossbar (Analog)	16-bit Fixed	Inf
PipeLayer [44]	1485	142.9	Bitline-wise Keral/X-bar	16-bit	Train & Inf

**CE represents computation efficiency.†PE represents power efficiency.*

Table 3.
The comparison of ReRAM-based in-memory computing accelerators.

support the training process. The computing pipelines were deep and may introduce pipeline bubbles. Therefore, in 2017, a novel ReRAM-based in-memory computing architecture Pipelayer was proposed aiming to solve this challenge [44]. The proposed architecture is based on PRIME and ISAAC and can support the training process. In addition, PipeLayer has optimized inter-layer and intra-layer pipelines to achieve high throughput in both training and inference. The same as PRIME, the ReRAM subarrays were utilized for both storage and computing with the capability of functionality shifting. During the training process, the error is obtained after comparing the labels and loss function in the last forward cycle. In the error backpropagation process, weights and errors, which were also calculated in the ReRAM morphable subarrays and saved in storage subarrays, were updated by the layer errors and partial derivatives and passed to the previous layer. It is noted that the depth of logical cycles was independent of the number of NN layers, preventing potential issues arising from deep pipelines, such as stall, bubble, and latency. In addition, PipeLayer introduced a parameter to repeatedly deploy operators, building a trade-space-for-time intra-layer parallel pipeline. This intra-layer pipeline, together with the inter-layer pipeline, supports the high throughput of PipeLayer. The comparison of these ReRAM-based in-memory computing accelerators is shown in **Table 3**.

4.2 ReRAM-based accelerators for various DNNs and applications

After the great success of the ReRAM-based neuromorphic accelerators, customized ReRAM accelerators tailored for specific types of neural networks and applications are also emerging. For example, generative adversarial networks (GANs) [91] require frequent data transmission between the generative and adversarial networks, which makes them highly demanding in terms of memory, computational resources, and data transfer. ReGAN, a ReRAM-based accelerator for GANS, has been designed to leverage the advantages of processing in-memory and ReRAM crossbar arrays, as well as optimized parallel computation and data dependency, to achieve significantly high performance and energy efficiency [45]. To expedite the execution of recurrent neural networks (RNNs), ReRAM accelerators tailored to various RNN configurations were proposed [46].

In graph processing tasks, high data transfer costs have led researchers to focus on the optimization of memory access [92]. ReRAM architecture's inherent *in situ* computing can minimize these overheads. GraphR was the first to introduce a ReRAM-based architecture for graph computation [51]. The feasibility of ReRAM for graph processing was analyzed, and sparse matrix-vector multiplications were applied to data blocks of compressed representation. This approach designed ReRAM-based graph engine to accelerate the network with parallel computation, while a drawback of additional useless multiplications with zero due to sparsity still exists. Subsequently, Spara presented a novel vertex mapping strategy to address this challenge [52]. There are also ReRAM-based architectures for graph processing focus on sparsity [53, 54], three-dimensional architecture [55, 56], regularization, redundant computation [57], etc.

Transformers, one of the most advanced models in current natural language processing (NLP), present several challenges to the ReRAM-based neuromorphic computing [93]. For example, one layer's input includes the results of the previous layer due to the self-attention mechanism, generating data dependencies that result in bubbles in the traditional interlayer pipeline. The calculation of scaled dot-product attention significantly differs from traditional multiply-accumulate operations in CNNs. ReTransformer [58] and ReBERT [59] are dedicated to addressing these challenges. ReTransformer mathematically decomposed matrix multiplication into two steps, reducing the pipeline bubbles by performing the first initialization step and the second calculation step sequentially in two separate cycles. ReBERT concentrated on the attention mechanism and proposed a window self-attention mechanism with a corresponding window size search algorithm to support the ReRAM crossbar arrays and achieved significant speedup and energy performance. There are also other ReRAM-based architectures for NLP, including sparsity attention [60], and BERT deployment [61], etc.

5. Conclusions

In this chapter, we gave an overview of the implementation of ReRAM-based neuromorphic computing engines in the last several years. The overview covers hardware designs of synapses and neurons, neural network implementation, hardware and software co-design, and novel architectural designs. The ReRAM-based synapses and neurons give a rebirth of neuromorphic computing through its ultra-small scale, inherent synaptic plasticity property, and capability to enable high-parallelism neural network operations. Accordingly, neuromorphic computing systems for different neural networks were implemented with significantly high computing efficiency in speed and energy compared to the traditional computing platform. Hardware-software co-design is a widely utilized approach to overcome the ReRAM-based hardware limitations such as limited resistive states, stuck-at-fault, signal fluctuation. Novel in-memory computing architectures are also extensively explored based on the ReRAM-based neuromorphic engine and various techniques are explored to overcome the challenges in neural network computing and optimize computing efficiency. With the prosperous advancement of ReRAM-based neuromorphic computing, specialized accelerators for various neural networks and applications are also under extensive study.

Acknowledgements

This work is supported by NSF CNS-1939380. Any opinions, findings, and conclusions or recommendations expressed in this material are those of the authors and do not necessarily reflect the views of grant agencies or their contractors.

Author details

Bojing Li[1†], Duo Zhong[1†], Xiang Chen[2] and Chenchen Liu[1*]

1 University of Maryland Baltimore County, Baltimore, USA

2 George Mason University, Fairfax, USA

*Address all correspondence to: ccliu@umbc.edu

† These authors contributed equally.

References

[1] Jo SH, Chang T, Ebong I, Bhadviya BB, Mazumder P, Wei L. Nanoscale memristor device as synapse in neuromorphic systems. Nano Letters. 2010;**10**(4):1297-1301

[2] Shimeng Y, Yi W, Jeyasingh R, Kuzum D, Philip H-S, Wong. An electronic synapse device based on metal oxide resistive switching memory for neuromorphic computation. IEEE Transactions on Electron Devices. 2011; **58**(8):2729-2737

[3] Ohno T, Hasegawa T, Tsuruoka T, Terabe K, Gimzewski JK, Aono M. Short-term plasticity and long-term potentiation mimicked in single inorganic synapses. Nature Materials. 2011;**10**(8):591-595

[4] Shimeng Y, Gao B, Fang Z, Hongyu Y, Kang J, Philip H-S, et al. A low energy oxide-based electronic synaptic device for neuromorphic visual systems with tolerance to device variation. Advanced Materials. 2013; **25**(12):1774-1779

[5] Li Y, Zhong Y, Zhang J, Lei X, Wang Q, Sun H, et al. Activity-dependent synaptic plasticity of a chalcogenide electronic synapse for neuromorphic systems. Scientific Reports. 2014;**4**(1):4906

[6] Gao B, Bi Y, Chen H-Y, Liu R, Huang P, Chen B, et al. Ultra-low-energy three-dimensional oxide-based electronic synapses for implementation of robust high-accuracy neuromorphic computation systems. ACS Nano. 2014; **8**(7):6998-7004

[7] Kim S, Chao D, Sheridan P, Ma W, Choi SH, Lu WD. Experimental demonstration of a second-order memristor and its ability to biorealistically implement synaptic plasticity. Nano Letters. 2015;**15**(3): 2203-2211

[8] Woo J, Moon K, Song J, Lee S, Kwak M, Park J, et al. Improved synaptic behavior under identical pulses using AlOx/HfO 2 bilayer rram array for neuromorphic systems. IEEE Electron Device Letters. 2016;**37**(8):994-997

[9] Wei W, Huaqiang W, Gao B, Deng N, Shimeng Y, Qian H. Improving analog switching in HfO x-based resistive memory with a thermal enhanced layer. IEEE Electron Device Letters. 2017; **38**(8):1019-1022

[10] Kim S, Kim H, Hwang S, Kim M-H, Chang Y-F, Park B-G. Analog synaptic behavior of a silicon nitride memristor. ACS Applied Materials & Interfaces. 2017;**9**(46):40420-40427

[11] Sun X, Yin S, Peng X, Liu R, Seo J-s, Yu S. Xnor-rram: A scalable and parallel resistive synaptic architecture for binary neural networks. In: 2018 Design, Automation & Test in Europe Conference & Exhibition (DATE). Dresden, Germany: IEEE; 2018. pp. 1423-1428

[12] Roy S, Niu G, Wang Q, Wang Y, Zhang Y, Heping W, et al. Toward a reliable synaptic simulation using Al-doped HFO2 RRAM. ACS Applied Materials & Interfaces. 2020;**12**(9): 10648-10656

[13] Mehonic A, Kenyon AJ. Emulating the electrical activity of the neuron using a silicon oxide rram cell. Frontiers in Neuroscience. 2016;**10**:57

[14] Babacan Y, Kaçar F, Gürkan K. A spiking and bursting neuron circuit based on memristor. Neurocomputing. 2016;**203**:86-91

[15] Kwon M-W, Kim S, Kim M-H, Park J, Kim H, Hwang S, et al. Integrate-and-fire (I&F) neuron circuit using resistive-switching random access memory (rram). Journal of Nanoscience and Nanotechnology. 2017;**17**(5): 3038-3041

[16] Zhang X, Wang W, Liu Q, Zhao X, Wei J, Cao R, et al. An artificial neuron based on a threshold switching memristor. IEEE Electron Device Letters. 2017;**39**(2):308-311

[17] Sandip Lashkare S, Chouhan TC, Bhat A, Kumbhare P, Ganguly UJIEDL. Pcmo rram for integrate-and-fire neuron in spiking neural networks. IEEE Electron Device Letters. 2018;**39**(4): 484-487

[18] Lin J, Yuan J-S. A scalable and reconfigurable in-memory architecture for ternary deep spiking neural network with ReRAM based neurons. Neurocomputing. 2020;**375**:102-112

[19] Suhas Kumar R, Williams S, Wang Z. Third-order nanocircuit elements for neuromorphic engineering. Nature. 2020;**585**(7826):518-523

[20] Markram H. The blue brain project. Nature Reviews Neuroscience. 2006; **7**(2):153-160

[21] Merolla PA, Arthur JV, Alvarez-Icaza R, Cassidy AS, Sawada J, Akopyan F, et al. A million spiking-neuron integrated circuit with a scalable communication network and interface. Science. 2014;**345**(6197):668-673

[22] Benjamin BV, Gao P, McQuinn E, Choudhary S, Chandrasekaran AR, Bussat J-M, et al. Neurogrid: A mixed-analog-digital multichip system for large-scale neural simulations. Proceedings of the IEEE. 2014;**102**(5): 699-716

[23] Davies M, Srinivasa N, Lin T-H, Chinya G, Cao Y, Choday SH, et al. Loihi: A neuromorphic many core processor with on-chip learning. IEEE Micro. 2018;**38**(1):82-99

[24] Hu M, Li H, Wu Q, Rose GS. Hardware realization of BSB recall function using Memristor crossbar arrays. In: Proceedings of the 49th Annual Design Automation Conference. New York, USA: ACM; 2012. pp. 498-503

[25] Miao H, Li H, Chen Y, Qing W, Rose GS, Linderman RW. Memristor crossbar-based neuromorphic computing system: A case study. IEEE Transactions on Neural Networks and Learning Systems. 2014;**25**(10):1864-1878

[26] Liu C, Yan B, Yang C, Song L, Zheng L, Liu B, et al. A spiking neuromorphic design with resistive crossbar. In: Proceedings of the 52nd Annual Design Automation Conference. New York, USA: ACM; 2015. pp. 1-6

[27] Yakopcic C, Alom MZ, Taha TM. Memristor crossbar deep network implementation based on a convolutional neural network. In: 2016 International Joint Conference on Neural Networks (IJCNN). Vancouver, BC, Canada: IEEE; 2016. pp. 963-970

[28] Yao P, Huaqiang W, Gao B, Eryilmaz SB, Huang X, Zhang W, et al. Face classification using electronic synapses. Nature Communications. 2017;**8**(1):15199

[29] Wen S, Wei H, Yang Y, Guo Z, Zeng Z, Huang T, et al. Memristive LSTM network for sentiment analysis. IEEE Transactions on Systems, Man, and Cybernetics: Systems. 2019;**51**(3): 1794-1804

[30] Joshua Yang J, Strukov DB, Stewart DR. Memristive devices for

computing. Nature Nanotechnology. 2013;**8**(1):13-24

[31] Wang Y, Wen W, Song L, Li HH. Classification accuracy improvement for neuromorphic computing systems with one-level precision synapses. In: 2017 22nd Asia and South Pacific Design Automation Conference (ASP-DAC). Chiba, Japan: IEEE; 2017. pp. 776-781

[32] Yang Q, Li H, Qing W. A quantized training method to enhance accuracy of ReRAM-based neuromorphic systems. In: 2018 IEEE International Symposium on Circuits and Systems (ISCAS). Florence, Italy: IEEE; 2018. pp. 1-5

[33] Huang S, Ankit A, Silveira P, Antunes R, Chalamalasetti SR, El Hajj I, et al. Mixed precision quantization for ReRAM-based dnn inference accelerators. In: 2021 26th Asia and South Pacific Design Automation Conference (ASP-DAC). New York, USA: ACM; 2021. pp. 372-377

[34] Changpinyo S, Sandler M, Zhmoginov A. The power of sparsity in convolutional neural networks. arXiv preprint arXiv:1702.06257. 2017

[35] Han S, Mao H, Dally WJ. Deep compression: Compressing deep neural networks with pruning, trained quantization and huffman coding. arXiv preprint arXiv:1510.00149. 2015

[36] Wang Y, Jin S, Li T. A low cost weight obfuscation scheme for security enhancement of ReRAM based neural network accelerators. In: 2021 26th Asia and South Pacific Design Automation Conference (ASP-DAC). Tokyo, Japan: IEEE; 2021. pp. 499-504

[37] Zhang J, Wang C, Cai Y, Zhu Z, Kline D, Yang H, et al. Wesco: Weight-encoded reliability and security co-design for in-memory computing systems. In: 2022 IEEE Computer Society Annual Symposium on VLSI (ISVLSI). Nicosia, Cyprus: IEEE; 2022. pp. 296-301

[38] Cai Y, Chen X, Lu T, Yu W, Yang H. Enabling secure in-memory neural network computing by sparse fast gradient encryption. In: 2019 IEEE/ACM International Conference on Computer-Aided Design (ICCAD). Westminster, CO, USA: IEEE; 2019. pp. 1-8

[39] Zou M, Zhou J, Cui X, Wang W, Kvatinsky S. Enhancing security of memristor computing system through secure weight mapping. In: 2022 IEEE Computer Society Annual Symposium on VLSI (ISVLSI). Nicosia, Cyprus: IEEE; 2022. pp. 182-187

[40] Zou M, Du N, Kvatinsky S. Review of security techniques for memristor computing systems. arXiv preprint arXiv:2212.09347. 2022

[41] Yang C, Liu B, Li H, Chen Y, Barnell M, Qing W, et al. Thwarting replication attack against memristor-based neuromorphic computing system. IEEE Transactions on Computer-Aided Design of Integrated Circuits and Systems. 2020;**39**(10):2192-2205

[42] Chi P, Li S, Xu C, Zhang T, Zhao J, Liu Y, et al. Prime: A novel processing-in-memory architecture for neural network computation in ReRAM-based main memory. In: 2016 ACM/IEEE 43rd Annual International Symposium on Computer Architecture (ISCA). New York, NY, USA: ACM; 2016. pp. 27-39

[43] Shafiee A, Nag A, Muralimanohar N, Balasubramonian R, Strachan JP, Hu M, et al. Isaac: A convolutional neural network accelerator with in-situ analog arithmetic in crossbars. In: 2016 ACM/IEEE 43rd Annual International Symposium on Computer Architecture

(ISCA). New York, NY, USA: ACM; 2016. pp. 14-26

[44] Song L, Qian X, Li H, Chen Y. Pipelayer: A pipelined ReRAM-based accelerator for deep learning. In: 2017 IEEE International Symposium on High Performance Computer Architecture (HPCA). 2017. pp. 541-552

[45] Chen F, Song L, Chen Y. Regan: A pipelined ReRAM-based accelerator for generative adversarial networks. In: 2018 23rd Asia and South Pacific Design Automation Conference (ASP-DAC). Jeju, Korea (South): IEEE; 2018. pp. 178-183

[46] Long Y, Na T, Mukhopadhyay S. ReRAM-based processing-in-memory architecture for recurrent neural network acceleration. IEEE Transactions on Very Large Scale Integration (VLSI) Systems. 2018; **26**(12):2781-2794

[47] Hochreiter S, Schmidhuber J. Long short-term memory. Neural Computation. 1997;**9**(8):1735-1780

[48] Liu X, Zhou M, Ausavarungnirun R, Eilert S, Akel A, Rosing T, et al. FPRA: A fine-grained parallel rram architecture. In: 2021 IEEE/ACM International Symposium on Low Power Electronics and Design (ISLPED). Boston, MA, USA: IEEE; 2021. pp. 1-6

[49] Yuan G, Behnam P, Li Z, Shafiee A, Lin S, Ma X, et al. Forms: Fine-grained polarized ReRAM-based in-situ computation for mixed-signal DNN accelerator. In: 2021 ACM/IEEE 48th Annual International Symposium on Computer Architecture (ISCA). 2021. pp. 265-278

[50] Jin H, Liu C, Liu H, Luo R, Jiahong X, Mao F, et al. REHY: A ReRAM-based digital/analog hybrid PIM architecture for accelerating CNN training. IEEE Transactions on Parallel and Distributed Systems. 2022;**33**(11): 2872-2884

[51] Song L, Zhuo Y, Qian X, Li H, Chen Y. Graphr: Accelerating graph processing using ReRAM. In: 2018 IEEE International Symposium on High Performance Computer Architecture (HPCA). 2018. pp. 531-543

[52] Zheng L, Zhao J, Yu H, Wang Q, Zeng Z, Xue J, et al. Spara: An energy-efficient ReRAM-based accelerator for sparse graph analytics applications. In: 2020 IEEE International Parallel and Distributed Processing Symposium (IPDPS). 2020. pp. 696-707

[53] Yu H, Zheng L, Liao X, Jin H, Yao P, Gui C. Ragra: Leveraging monolithic 3d ReRAM for massively-parallel graph processing. In: 2019 Design, Automation Test in Europe Conference Exhibition (DATE). Florence, Italy: IEEE; 2019. pp. 1273-1276

[54] Yang T, Li D, Ma F, Song Z, Zhao Y, Zhang J, et al. PASGCN: An ReRAM-based PIM design for GCN with adaptively sparsified graphs. IEEE Transactions on Computer-Aided Design of Integrated Circuits and Systems. 2022; **42**(1):150-163

[55] Arka AI, Doppa JR, Pande PP, Joardar BK, Chakrabarty K. Regraphx: NOC-enabled 3d heterogeneous ReRAM architecture for training graph neural networks. In: 2021 Design, Automation Test in Europe Conference Exhibition (DATE). Grenoble, France: IEEE; 2021. pp. 1667-1672

[56] Choudhury D, Barik R, Rajam AS, Kalyanaraman A, Pande PP. Software/hardware co-design of 3d NOC-based GPU architectures for accelerated graph

computations. ACM Transactions on Design Automation of Electronic Systems (TODAES). New York, USA: ACM; 2022;**27**(6):1-22

[57] Chen C, Li K, Li Y, Zou X. REGNN: A redundancy-eliminated graph neural networks accelerator. In: 2022 IEEE International Symposium on High-Performance Computer Architecture (HPCA). 2022. pp. 429-443

[58] Yang X, Yan B, Li H, Chen Y. Retransformer: ReRAM-based processing-in-memory architecture for transformer acceleration. In: 2020 IEEE/ACM International Conference on Computer Aided Design (ICCAD). 2020. pp. 1-9

[59] Kang M, Shin H, Kim L-S. A framework for accelerating transformer-based language model on ReRAM-based architecture. IEEE Transactions on Computer-Aided Design of Integrated Circuits and Systems. 2022;**41**(9): 3026-3039

[60] Li H, Jin H, Zheng L, Yu H, Liao X, Chen D, et al. CPSAA: Accelerating sparse attention using crossbar-based processing-in-memory architecture. arXiv preprint arXiv:2210.06696. 2022

[61] Kang M, Shin H, Shin J, Kim L-S. A framework for area-efficient multi-task Bert execution on ReRAM-based accelerators. In: 2021 IEEE/ACM International Conference on Computer Aided Design (ICCAD). Munich, Germany: IEEE; 2021. pp. 1-9

[62] Akinaga H, Shima H. Resistive random access memory (ReRAM) based on metal oxides. Proceedings of the IEEE. 2010;**98**(12):2237-2251

[63] Wang P, Yu J, Hong C, Lyu Y, Wang D, Xie Y. SNRRAM: An efficient sparse neural network computation architecture based on resistive random-access memory. In: 2018 55th ACM/ESDA/IEEE Design Automation Conference (DAC). 2018. pp. 1-6

[64] Ji H, Song L, Jiang L, Li H, Chen Y. Recom: An efficient resistive accelerator for compressed deep neural networks. In: 2018 Design, Automation Test in Europe Conference Exhibition (DATE). 2018. pp. 237-240

[65] Chu C, Wang Y, Zhao Y, Ma X, Ye S, Hong Y, et al. Pim-prune: Fine-grain dcnn pruning for crossbar-based process-in-memory architecture. In: 2020 57th ACM/IEEE Design Automation Conference (DAC). San Francisco, CA, USA: IEEE; 2020. pp. 1-6

[66] Liu F, Zhao W, He Z, Wang Z, Zhao Y, Yang T, et al. SME: ReRAM- based sparse-multiplication-engine to squeeze-out bit sparsity of neural network. In: 2021 IEEE 39th International Conference on Computer Design (ICCD). Storrs, CT, USA: IEEE; 2021. pp. 417-424

[67] Chen A, Lin M-R. Variability of resistive switching memories and its impact on crossbar array performance. In: 2011 International Reliability Physics Symposium. 2011. pp. MY.7.1-MY.7.4

[68] Dongale TD, Patil KP, Mullani SB, More KV, Delekar SD, Patil PS, et al. Investigation of process parameter variation in the memristor based resistive random access memory (RRAM): Effect of device size variations. Materials Science in Semiconductor Processing. 2015;**35**:174-180

[69] Ambrogio S, Simone Balatti A, Cubeta AC, Ramaswamy N, Ielmini D. Understanding switching variability and random telegraph noise in resistive ram. In: 2013 IEEE International Electron Devices Meeting. Washington, DC, USA: IEEE; 2013. pp. 31-35

[70] Choi S, Yang Y, Wei L. Random telegraph noise and resistance switching analysis of oxide based resistive memory. Nanoscale. 2014;**6**(1):400-404

[71] Beckmann K, Holt J, Manem H, Van Nostrand J, Cady NC. Nanoscale hafnium oxide RRAM devices exhibit pulse dependent behavior and multi-level resistance capability. MRS Advances. 2016;**1**(49):3355-3360

[72] Chen YY, Goux L, Clima S, Govoreanu B, Degraeve R, Kar GS, et al. Endurance/retention trade-off on HfO_2/Metal cap 1t1r bipolar rram. IEEE Transactions on Electron Devices. 2013; **60**(3):1114-1121

[73] Wong H-SP, Lee H-Y, Yu S, Chen Y-S, Wu Y, Chen P-S, et al. Metal–oxide rram. Proceedings of the IEEE. 2012; **100**(6):1951-1970

[74] Chen Y, Xie Y, Song L, Chen F, Tang T. A survey of accelerator architectures for deep neural networks. Engineering. 2020;**6**(3):264-274

[75] Liu C, Hu M, Strachan JP, Li H. Rescuing memristor-based neuromorphic design with high defects. In: 2017 54th ACM/EDAC/IEEE Design Automation Conference (DAC). New York, USA: ACM; 2017. pp. 1-6

[76] Yuan G, Liao Z, Ma X, Cai Y, Kong Z, Shen X, et al. Improving DNN fault tolerance using weight pruning and differential crossbar mapping for ReRAM-based edge AI. In: 2021 22nd International Symposium on Quality Electronic Design (ISQED). 2021. pp. 135-141

[77] Chakraborty I, Roy D, Roy K. Technology aware training in memristive neuromorphic systems for nonideal synaptic crossbars. IEEE Transactions on Emerging Topics in Computational Intelligence. 2018;**2**(5): 335-344

[78] Chen L, Li J, Chen Y, Deng Q, Shen J, Liang X, et al. Accelerator- friendly neural-network training: Learning variations and defects in RRAM crossbar. In: Design, Automation & Test in Europe Conference & Exhibition (DATE), 2017. Lausanne, Switzerland: IEEE; 2017. pp. 19-24

[79] Chen P-Y, Lin B, Wang I-T, Hou T-H, Ye J, Vrudhula S, et al. Mitigating effects of non-ideal synaptic device characteristics for on-chip learning. In: 2015 IEEE/ACM International Conference on Computer-Aided Design (ICCAD). IEEE; 2015. pp. 194-199

[80] Jain S, Raghunathan A. CXDNN: Hardware-software compensation methods for deep neural networks on resistive crossbar systems. ACM Transactions on Embedded Computing Systems (TECS). New York, USA: ACM; 2019;**18**(6):1-23

[81] Yide D, Jing L, Fang H, Chen H, Cai Y, Wang R, et al. Exploring the impact of random telegraph noise-induced accuracy loss on resistive ram-based deep neural network. IEEE Transactions on Electron Devices. 2020; **67**(8):3335-3340

[82] Chang T, Jo S-H, Wei L. Short-term memory to long-term memory transition in a nanoscale memristor. ACS Nano. 2011;**5**(9):7669-7676

[83] Liu T, Wen W, Jiang L, Wang Y, Yang C, Quan G. A fault-tolerant neural network architecture. In: 2019 56th ACM/IEEE Design Automation Conference (DAC). 2019. pp. 1-6

[84] Liu T, Liu Z, Lin F, Jin Y, Quan G, Wen W. Mt-spike: A multilayer time-

based spiking neuromorphic architecture with temporal error backpropagation. In: 2017 IEEE/ACM International Conference on Computer-Aided Design (ICCAD). IEEE; 2017. pp. 450-457

[85] Xia L, Huangfu W, Tang T, Yin X, Chakrabarty K, Yuan Xie Y, et al. Stuck-at fault tolerance in rram computing systems. IEEE Journal on Emerging and Selected Topics in Circuits and Systems. 2018;**8**(1):102-115

[86] Zhang B, Uysal N, Fan D, Ewetz R. Handling stuck-at-fault defects using matrix transformation for robust inference of DNNS. IEEE Transactions on Computer-Aided Design of Integrated Circuits and Systems. 2020; **39**(10):2448-2460

[87] Yeo I, Chu M, Gi S-G, Hwang H, Lee B-G. Stuck-at-fault tolerant schemes for memristor crossbar array-based neural networks. IEEE Transactions on Electron Devices. 2019;**66**(7): 2937-2945

[88] Beigi M, V, Memik G. Thermal-aware optimizations of ReRAM-based neuromorphic computing systems. In: 2018 55th ACM/ESDA/IEEE Design Automation Conference (DAC). 2018. pp. 1-6

[89] Liang J, Philip Wong H-S. Cross-point memory array without cell selectors—Device characteristics and data storage pattern dependencies. IEEE Transactions on Electron Devices. 2010; **57**(10):2531-2538

[90] Huang C, Nuo X, Qiu K, Zhu Y, Ma D, Fang L. Efficient and optimized methods for alleviating the impacts of ir-drop and fault in RRAM based neural computing systems. IEEE Journal of the Electron Devices Society. 2021;**9**:645-652

[91] Goodfellow I, Pouget-Abadie J, Mirza M, Bing X, Warde-Farley D, Ozair S, et al. Generative adversarial networks. Communications of the ACM. 2020;**63**(11):139-144

[92] Ham T, J, Lisa W, Sundaram N, Satish N, Martonosi M. Graphicionado: A high-performance and energy-efficient accelerator for graph analytics. In: 2016 49th Annual IEEE/ACM International Symposium on Microarchitecture (MICRO). Taiwan, China: IEEE; 2016. pp. 1-13

[93] Vaswani A, Shazeer N, Parmar N, Uszkoreit J, Jones L, Gomez AN, et al. Attention is all you need. In: 2017 31st Conference on Neural Information Processing Systems (NIPS). Long Beach, CA, USA: Neural Information Processing Systems Foundation; 2017. pp. 5998-6008

Section 3

Neuromorphic Circuits

Chapter 5

Neuromorphic Computing between Reality and Future Needs

Khaled S. Ahmed and Fayroz F. Shereif

Abstract

Neuromorphic computing is a one of computer engineering methods that to model their elements as the human brain and nervous system. Many sciences as biology, mathematics, electronic engineering, computer science and physics have been integrated to construct artificial neural systems. In this chapter, the basics of Neuromorphic computing together with existing systems having the materials, devices, and circuits. The last part includes algorithms and applications in some fields.

Keywords: neuromorphic computing expectation, neuromorphic systems, neuromorphic chips, applications, algorithms

1. Introduction

Neuromorphic computing, technology that mimic neuro-biological architectures present in the nervous system using electronic circuits, has attracted attention as next-generation computing due to its characteristics as they can process complex data with high efficiency, high speed and low power consumption. Neuromorphic computing importance increased in the industry because it efficiently executes artificial intelligence algorithms by imitating the brain nerve structure of humans. The conventional von Neumann computing using separated processors and memory systems is not efficient for machine learning due to processor-memory bottlenecks. Because the machine learning has a special workload that iterates simple computation with alot of data, there should be huge data traffic between processors and memory. For neuromorphic computing system, it consists of multiple neurons and synapses to compute and store data, and a neural network to communicate them. Therefore, this computing system can compute simple iterations efficiently for the training of machine learning. This chapter will discuss neuromorphic computing goals and challenges, showing how materials, devices, algorithms development researches lead to high expectations in neuromorphic computing field. We will discuss neuromorphic systems that are already existing today and expectations for what neuromorphic computing can achieve in the future.

2. What is neuromorphic computing?

2.1 Neuromorphic computing history

The development of the perceptron in 1958 served as the forerunner of the artificial neurons employed in modern neural networks. With the inadequate understanding of the brain's inner workings at the time, the perceptron was an amateurish attempt to imitate some aspects of organic neural networks. The U.S. Navy planned to use the perceptron as a piece of mechanical hardware that was specifically designed for picture identification. Before it was understood that the technology could not perform the required job, it was subject to a great deal of hype.

Carver Mead, a professor at Caltech, first proposed neuromorphic computing in the 1980s. Mead's description of the first analogue silicon retina presaged a brand-new class of physical calculations that were motivated by the neural paradigm. Mead is also stated as believing that, given a thorough grasp of how the nervous system functions, nothing the human nervous system does cannot be replicated by computers in an article about neural computation in analogue VLSI.

However, the ubiquitous and expanding usage of AI, machine learning, and neural networks in consumer and enterprise technologies might be partly blamed for the recent investment and enthusiasm surrounding neuromorphic research. It can also be mainly related to the perception among many IT specialists that Moore's law is coming to an end. According to Moore's Law, a chip can accommodate twice as many micro components each year while maintaining the same price.

Major chip manufacturers like IBM and Intel are paying close attention to neuromorphic computing because it has the potential to get around conventional architectures and achieve radically higher levels of efficiency. In fact, Intel launched Moore's Law which was about to come to an end in 2017.

Moore's Law by condensing Gordon Moore's ideas has been launched in 2013, was also quoted as saying, "Going to multicore chips helped, but now we are up to eight cores and it does not look like we can go much further. People have to crash into the wall before they pay attention." This belief supports the idea that there are ups and downs in the hype and popular discourse surrounding artificial intelligence, with periods of low interest sometimes referred to as AI winters and times of high interest frequently brought on by an urgent issue that needs to be resolved, in this case the end of Moore's Law. the integration of neuromorphic chips and smart cockpits and explore related fields with BMW has been announced by SynSense The world's leading neuromorphic intelligence and solutions provider.

2.2 Neuromorphic computing architecture

Neuromorphic computing is a method of computer engineering in which elements of a computer are modelled after systems in the human brain and nervous system. The phrase describes the creation of both software and hardware components in computers. To construct artificial neural systems that are inspired by biological architecture, neuromorphic engineers depend on a variety of fields, including computer science, biology, mathematics, electronic engineering, and physics.

Researchers have investigated neuromorphic computing throughout the years to create various types of machines that think and learn like humans. Studies have been done to imitate how the human brain learns and computes by using computer hardware in the form of an artificial neural network to simulate human learning abilities.

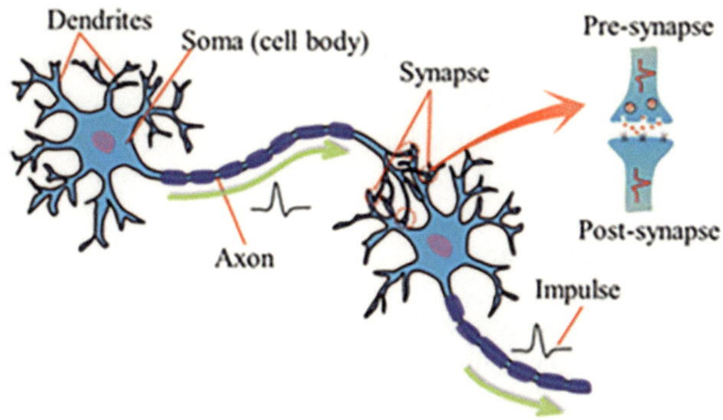

Figure 1.
Three interconnected neurons. A presynaptic neuron transmits the signal toward a synapse, whereas a postsynaptic neuron transmits the signal away from the synapse [1].

The anatomy of the human brain, which has over one billion neurons and trillions of synapses, is extremely intricate. As shown in **Figure 1**, neurons are made up of a cell body, an axon that generates neural signal impulses, and a dendrite that receives signals from other neurons. A synapse is a physical component that enables one neuron to transmit an electrical signal to an additional neuron.

As seen in **Figure 2b**, the neuromorphic hardware typically comprises of neurons and synapses to imitate the human brain's neural network. Each neuron functions as a data processing centre in the neuromorphic hardware, and neurons are linked together in parallel by synapses to convey data [2–4]. The single signal bus present in the neuromorphic hardware does not result in a von Neumann bottleneck. Instead of using regular CMOS devices, artificial synaptic devices that reflect the properties of bio-synapses must be developed in order to execute this in practical design. The block diagrams for the traditional von Neumann architecture and the emerging neuromorphic design are shown in **Figure 2**.

Significant improvements have recently been made in the field of neuromorphic computing. Three major steps are used to categorise the most recent improvement [6–8]. A GPU-Centric system, which is primarily used for learning and supports artificial intelligence by using a graphics processing unit (GPU), is the initial phase. An ASIC-Centric system, which is now the next stage, is a hot topic of research. An effective and low-power application-specific integrated circuit (ASIC) for machine learning is anticipated to be created by this trend. As a result, numerous semiconductor firms are creating ASIC chips [9–12]. However, it is anticipated that neuromorphic computing will be developed 1 day into neuromorphic hardware, enabling ultra-low power and ultra-high speed computation to support for general-purpose artificial intelligence.

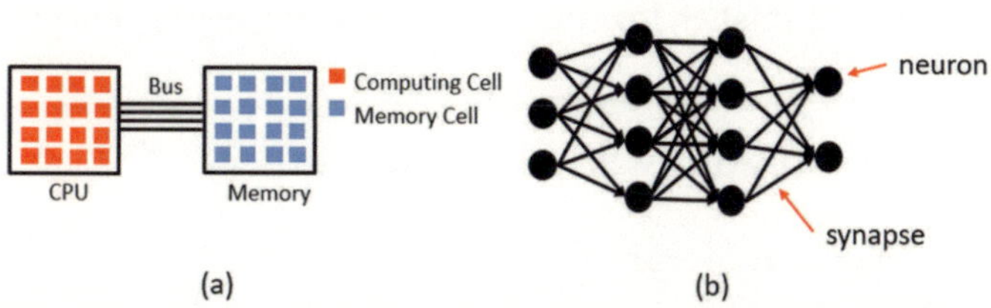

Figure 2.
Block diagram of computing systems: (a) von Neumann architecture; (b) neuromorphic architecture [5].

The neuromorphic-centric hardware must be able to handle massive amounts of data in parallel while using incredibly little power. Furthermore, compared to current technology that uses ordinary CMOS components, the neuromorphic semiconductor chip demands a quicker rate of computing. This suggests that creating an emerging synaptic device is essential to the development of neuromorphic-centric hardware.

2.3 Neuromorphic computing goals

Everything appears to be becoming "smarter" these days. Artificial intelligence (AI) is being used by an increasing number of goods and services, from industrial machinery to residential appliances, to comprehend user requests, analyse data, and spot patterns. It's simple to understand why AI-powered goods are so well-liked. Instead of using buttons or touchscreens, smart interfaces enable voice and gesture control of devices. It's a lot more instinctive to use a device in this way. Additionally, AI has the potential to make products more autonomous, freeing us from laborious or repetitive tasks. Smart items can also enable ongoing optimization and data analysis, which can be used to monitor our health and send us alerts, as well as to anticipate when a piece of equipment needs to be serviced or replaced. The popularity of smart devices today is increasing demand for increasingly complex AI-powered experiences. And we are beginning to push the capabilities of the hardware that is now available. Today's smart devices actually use a lot of processing that is done remotely in the cloud or a data center, where there is sufficient computer power to conduct the required algorithms. This means that a network connection is necessary, and when data is transmitted back and forth, this can also cause latency to grow. When sending particular types of data to the cloud, there are additional real and perceived data privacy factors to take into account.

These factors suggest benefits of increasing the amount of smart computing housed within the device itself. Since the processing is carried out in edge network devices rather than a centralised cloud, this is referred to as "edge AI." However, because many edge processors are mobile, they frequently rely on batteries for power. How can we run low-power edge devices with power-hungry AI algorithms? To do that, we are going to need to reevaluate how AI hardware is created.

This is the goal of neuromorphic computing. The information processing methods used by a biological brain are the foundation of neuromorphic computing design. Consider the 80–100 billion neurons found in the average human brain, each of which functions incredibly effectively and asynchronously to offer huge parallel processing. We are able to be so intelligent without constantly ingesting enormous quantities of energy because to this power and efficiency combination. Spiking neural networks are one of the most effective ways to simulate how a real neuron fires or "spikes" to relay a signal before going back to being silent. The end result is a system that uses much less power than artificial neural networks, which are currently employed in the majority of AI systems. Additionally, such efficiency makes it possible for considerably more AI processing to be done on smaller, low-power devices at the network edge.

3. Neuromorphic computing working principles

3.1 Von Neumann vs. neuromorphic

Non-von Neumann computers that are inspired by the structure and operation of brains and are made up of neurons and synapses are referred to as neuromorphic computers. Von Neumann computers consist of distinct CPUs and memory units, with the latter housing data and instructions. On the other hand, with a neuromorphic computer, the neurons and synapses control both processing and memory. In contrast to von Neumann computers, which use explicit instructions to define programmes, neuromorphic computers instead use the characteristics and structure of the neural network. In addition, while von Neumann computers encode information as numerical values represented by binary values, neuromorphic computers receive spikes as input, where the associated time at which they occur, their magnitude and their shape can be used to encode numerical information. Binary values can be turned into spikes and vice versa, but the precise way to perform this conversion is still an area of study in neuromorphic computing.

These are some fundamental operational differences between Neuromorphic computers and Von Neumann:

- *Highly parallel operation:* All of the neurons and synapses in neuromorphic computers are potentially capable of operating simultaneously due to their inherent parallelism; nevertheless, compared to parallelized von Neumann systems, the computations carried out by neurons and synapses are comparatively straightforward.

- *Collocated processing and memory:* In neuromorphic hardware, the idea of a distinction between processing and memory is absent. In many implementations, neurons and synapses both do processing and store information, despite the fact that neurons are sometimes thought of as processing units and synapses as memory. The von Neumann bottleneck relating to the processor/memory separation, which slows down the maximum throughput that may be attained, is lessened by collocating processing and memory. Additionally, this collocation aids in avoiding main memory data accesses, which are common in conventional computing systems and use a lot more energy than compute energy.

- *Inherent scalability:* Since adding more neuromorphic chips implies increasing the potential number of neurons and synapses, neuromorphic computers are designed to be intrinsically scalable. In order to run larger and larger networks, it is possible to treat a number of physical neuromorphic chips as a single huge neuromorphic implementation. Several large-scale neuromorphic hardware devices, like SpiNNaker and Loihi, have been used to successfully achieve this.

- *Event-driven computation:* neuromorphic computers leverage event-driven computation and temporally sparse activity to allow for extremely efficient computation. Neurons and synapses only perform work when there are spikes to process, and typically, spikes are relatively sparse within the operation of the network.

- *Stochasticity:* To account for noise, neuromorphic computers can incorporate a notion of randomness, such as in the firing of neurons.

The literature makes extensive note of the characteristics of neuromorphic computers and provides reasons for their use and implementation. The extraordinarily low power consumption of neuromorphic computers for computation is one of its most appealing qualities; they frequently consume orders of magnitude less power than conventional computing systems. Because they operate in enormous parallel and are event-driven, only a small percentage of the system is normally active at any given moment, with the remainder being idle. Energy efficiency alone is a compelling incentive to examine the use of neuromorphic computers, given the rising energy cost of computing and the rising number of applications (such as edge computing applications) that have energy limits. Furthermore, neuromorphic computers provide a suitable platform for many of the artificial intelligence and machine learning applications of today since they naturally incorporate neural network-style processing. Additionally, neuromorphic computers hold potential for utilising their inherent computational capabilities to carry out a wide range of different forms of computing.

It is unclear whether these are the only parts of biological brains that are significant for computation, despite the fact that each of these qualities of neuromorphic computers is modelled after traits of the brain and has received priority in recent years. For instance, glial cells are one of many additional types of neural components that may be beneficial for computation, despite the fact that neurons and synapses have been chosen as the main computational units of neuromorphic computers. Furthermore, neurons and synapses have proved a useful level of abstraction for neuromorphic computers, but it is still unclear whether they are the best level of abstraction.

Contrary to some of the upcoming computing technologies, the research community already has access to various physical realisations of neuromorphic hardware that are now being developed. Numerous large-scale neuromorphic computers have been created with various methodologies and objectives. SpiNNaker and BrainScaleS were created with funding from the Human Brain Project of the European Union in order to facilitate large-scale neuroscience simulations. It has also been suggested to use slightly more complicated neuron models using an improved digital neuromorphic processor known as the online-learning digital spiking neuromorphic (ODIN). The Tianjic chip, a platform that supports both neuromorphic spiking neural networks and conventional artificial neural networks for various issue types, is one of the neuromorphic platforms aiming for more broad computations for wider classes of applications [13]. Both business and academia are interested in neuromorphic systems; examples from business include IBM's TrueNorth [14] and Intel's Loihi [15], while other academic initiatives exist as well, including DYNAPs [16], Neurogrid [17], IFAT [18], and BrainScales-2 [19]. In order to optimise learning-to-learn scenarios—situations where an optimization method is used to specify how learning occurs—for spiking neural networks, running at considerably faster timescales than biological timescales, neuromorphic hardware such as BrainScales-2 has been shown to be useful [20].

However, there is a lot of research being done in the neuromorphic community to create new types of materials for neuromorphic implementations, such as phase-change, ferroelectric, non-filamentary, topological insulators, or channel-doped biomembranes. All of the large-scale neuromorphic computers mentioned above are silicon-based and implemented using conventional complementary metal oxide semiconductor technology [21–23]. Memristors are frequently utilised in the literature as the fundamental device to have resistive memory to collocate processing and memory [24, 25], although other types of devices, such as optoelectronic devices, have also been employed to create neuromorphic computers [26]. Neuromorphic computers can be implemented using a variety of devices and materials, each of which has its own operating properties, including speed of operation, energy consumption, and biological likeness. Because neuromorphic hardware may now be implemented using a variety of tools and materials, it is possible to tailor its properties to a particular application.

The majority of current research in the field of neuromorphic computing focuses on the aforementioned hardware systems, devices, and materials; however, in order to best utilise neuromorphic computers in the future, take advantage of all of their special computational properties, and influence their hardware design From this vantage point, we give an overview of the state-of-the-art in neuromorphic algorithms and applications and offer a look ahead at the possibilities for the development of neuromorphic computing in computer science and computational science. It is important to note that a variety of various sorts of technology have all been referred to as neuromorphic computing.

3.2 Neuromorphic computing and artificial general intelligence (AGI)

Artificial general intelligence (AGI) is the term used to describe AI that demonstrates intelligence comparable to that of humans. One could say it's the holy grail of all AI. That degree of intelligence has not yet been attained by machines and might never be. However, neuromorphic computing presents brand-new opportunities for advancing in that direction.

For example, the Human Brain Project which features the neuromorphic supercomputer SpiNNaker aims to produce a functioning simulation of the human brain and is one of many active research projects interested in AGI.

The criteria for determining whether a machine has achieved AGI are debated, but these points need to be discussed if the machine can reason and make judgements under uncertainty. If it can plan, learn, communicate using natural language, represent knowledge including common-sense knowledge, and if it can integrate these skills in the pursuit of a common goal.

The ability to imagine, subjective experience, and self-awareness are occasionally included. The well-known Turing Test and the Robot College Student Test, in which a machine enrols in classes and earns a degree just like a human would, are two more suggested techniques for verifying AGI.

There are disagreements over how it should be handled ethically and legally if a machine ever attained human intelligence. Some contend that it ought to be regarded under the law as a nonhuman animal. These debates have been going on for years, in part because we still do not fully understand consciousness as a whole.

4. Challenges of neuromorphic computing

Conventional CMOS-based neuromorphic systems have shown promise in delivering brain-like features including pattern recognition, adaptive learning, and sophisticated sensing. However, because of insurmountable problems inherent in the material qualities of conventional semiconductors, their future potential is constrained. For creating a system that imitates a biological brain, quantum materials and technologies are anticipated to function as a ground-breaking, next-generation computational platform. This field is at a point where further fundamental investigation into every facet of the issue would be extremely beneficial. This issue is devoted to evaluating the state of the art in the field and exhibiting fresh, possibly ground-breaking concepts.

Ab initio theoretical calculations combined with state-of-the-art synthesis and nanoscale structural, electrical, magnetic, and optical characterisation are some of the techniques that will be used to better understand the properties of quantum materials, the impact of defects, and their ultimate effect on devices and systems. It is crucial to conduct a thorough, quantitative, interdisciplinary analysis of the relevant materials under intense (electrical, thermal, magnetic, and physical) stresses. Modern technologies for synthesis and characterisation are now at a point where they can give precise control and information about a material's structure and how it affects its physical properties.

With a full understanding of the properties of the materials, new design ideas that go beyond typical semiconductor devices based on the charge or even the spin of the electron are being developed. For instance, Mott physics offers a way to simulate technology that is inspired by the brain. Synaptic devices encode a memory state by shifting and modifying the concentration of defects, whereas "neuronal" devices accumulate "metal phase" and modulate the conduction by metallic filamentary development (e.g., oxygen vacancies). In general, the combination of memristive phenomena with light-sensitive oxides is leading to exciting design ideas for memory, networks, and neuro-sensors. Polymer-gated transistors with numerous functionalities can be used to achieve synaptic behaviour, which calls for plasticity in a material's response function, as well as in photonic and magnetoelectric systems. Spin-based devices, on the other hand, benefit from pronounced nonlinearities in the magnetic responses of quantum materials. Optomagnetic neural networks can be adjusted to construct low-dissipation networks thanks to the low-energy plasticity and non-volatility of magnetic properties, while magnetic anisotropies are proposed to train networks. Nanoscale magnetic phenomena can also be used to create nanowires. Time-dependent responses that resemble synaptic and dendritic trees as well as neural spikes can be produced by superconducting Josephson devices. Many physical phenomena, such reservoir computing and in-memory computing, are being studied as avenues to duplicate the magnificent processes carried out by the brain. It is crucial to comprehend the ultimate physical limits of these phenomena, including the smallest sizes, shortest times, or closest physical proximity permitted by the physics of the materials and devices; all of these are directly related to scaling issues. Only then can these phenomena be incorporated into functional devices.

Emergent behaviour is at the core of biological neurons' and synapses' complexity and remarkable effectiveness. Numerous self-organising principles may be seen in the emergent properties of quantum materials and their nonlinearities, which offer a variety of static and dynamic states helpful for simulating sophisticated brain devices and networks. [27] A diversity of different memory states, fine-tuning of critical behaviour, and the formation of unique collective phenomena are possible thanks to the metastability provided by the complex energy landscape of the quantum system [26, 28]. It is crucial to take into account the network in which the devices are located in order to approach the problem of building a brain-like machine. The network is the system, to adapt Herbert Kroemer's aphorism to bio-inspired neuromorphic computing.

5. Existing neuromorphic systems

5.1 The Tianjic Chip

This chip is the world's first hybrid-paradigm chip for brain-inspired computing to facilitate the development of artificial general intelligence (AGI) (**Figure 3**).

Two key approaches to the creation of AGI are computer science- and neuroscience-based. Both have benefits and drawbacks of their own. The combination of these two methods is currently thought to be one of the most promising approaches to

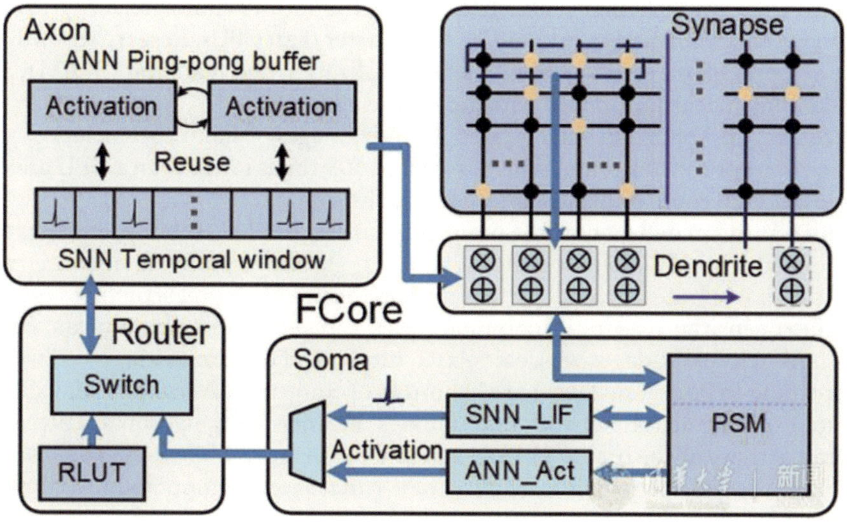

Figure 3.
Tianjic chip architecture [29].

artificial general intelligence (AGI), and a strong hardware foundation that supports hybrid-paradigm computing is one of its key pillars.

The CBICR team suggests a novel brain-inspired computing architecture, a hybrid-paradigm Tianjic chip, which can simultaneously support computer-science-oriented and neuroscience-oriented neural networks and capitalise on their strengths, such as artificial neural networks and spiking neural networks. This architecture is based on the computing principles of brain science. The second-generation Tianjic chip was introduced in 2017 and was built upon the first-generation chip created in 2015. The modern Tianjic achieves versatile functionality, fast speed, and low power after continual design refinement. Tianjic has 20% higher density, 10 times the speed, 100 times the internal bandwidth, better flexibility/scalability, and more complete functionality than IBM's TrueNorth chip. The group has also created a first-generation software tool chain that facilitates compilation and model mapping automatically.

As a flexible and scalable AGI development platform, an unmanned bicycle was built. Using just one chip, it was able to perform a variety of complex tasks, including speech recognition, object tracking and detection, balance control, obstacle avoidance, and decision making. This device outperformed a comparable GPU 160 times faster and 120,000 times more effectively thanks to its 40,000 neurons and 10 million synapses. This book offers a fresh perspective and a new venue for academic study of AGI, which will facilitate its advancement and effects on business.

5.2 Intel's Loihi Chips and Intel's Pohoiki Beach computers

Intel improved its Pohoiki Beach neuromorphic system, paving the way for quicker processing to be employed in AI for the Internet of Things and autonomous vehicles and equipment. More than 60 partners will use the system to solve challenging challenges that need extensive number crunching. It is built to imitate 8 million neurons and runs 64 Loihi research chips.

Loihi, which was originally launched in 2017, has 10,000 times greater efficiency and can process information up to 1000 times faster than CPUs. Researchers will be able to scale up sparse coding, simultaneous localization and mapping (SLAM), and path planning to learn from new data inputs and adapt.

According to Applied Brain Research, the advantages of Loihi also include decreased power consumption, which is around 109 times lower than a GPU and five times lower than comparable IoT inference hardware.

Loihi was motivated by how the brain approaches problem-solving, which is the fundamental origin of all neuromorphic research. According to Intel, neuromorphic computing is computer technology that mimics the brain's neural structure and can use context-sensitive reasoning, common sense, and deal with ambiguity and contradiction. AI in its early stages was less robust, literal, and deterministic.

According to Prof. Konstantinos Michmizos of Rutgers University, "Loihi allows us to create a spiking neural network that imitates the underlying neuronal representations and activity of the brain." His lab was able to operate mobile robots precisely and with 100 times less energy while using a Loihi-run network as opposed to a commonly utilised CPU.

Neuromorphic computing-based AI has been touted by Intel as having applications in medical imaging and autonomous vehicles (AV). With AV, solutions could emerge to deal with uncertainties, like a ball rolling onto the roadway or an aggressive driver in another car on the road. AI can be used to highlight areas of a medical image where an ailment is more likely to be present.

Without providing a detailed roadmap, Intel stated that their research will eventually result in the commercialization of neuromorphic technology. Moore's Law and process-node computing cannot continue to expand, hence there is a critical need to sustain the gains in computing power and performance. According to Intel, specialised architectures like the Pohoiki Beach neuromorphic approach are required for particular developing applications like AV, smart homes, and cybersecurity (**Figure 4**).

5.3 IBM's TrueNorth chip

In 2014, IBM released the neuromorphic CMOS integrated circuit known as TrueNorth. With 4096 cores and 256 programmable simulated neurons per core for a

Figure 4.
Intel uses 64 Loihi research chips in its neuromorphic system called Pohoiki Beach [30].

total of just over a million neurons, it is a manycore processor network on a chip. The 256 programmable "synapses" that connect each neuron serve to transmit impulses between them. Thus, there are little over 268 million programmable synapses overall. It has 5.4 billion fundamental transistors.

TrueNorth avoids the von Neumann-architecture bottleneck and is extremely energy efficient, with IBM claiming a power consumption of 70 milliwatts and a power density that is 1/10,000th of conventional microprocessors. Memory, computation, and communication are handled in each of the 4096 neurosynaptic cores [31]. The SyNAPSE chip operates at lower temperatures and power consumption because it only uses the power required for computing. Skyrmions have been proposed as models of the synapse on a chip [32, 33].

The neurons are emulated using a Linear-Leak Integrate-and-Fire (LLIF) model, a simplification of the leaky integrate-and-fire model [34]. IBM claims that it lacks a clock [35] and only uses unary numbers, counting up to a maximum of 19 bits to perform computations [34, 36]. The said cores are event-driven by using both a synchronous logic and are interconnected through an asynchronous packet-switched mesh network on chip (NOC).

In order to programme and use TrueNorth, IBM has created an entirely new environment. It featured libraries, a simulator, a fresh programming language, and even an integrated programming environment. Due to the substantial concerns of vendor lock-in and other negative effects, this absence of backward compatibility with any prior technology (such as C++ compilers) may preclude it from becoming commercialization in the future (**Figure 5**) [35].

5.4 Human brain project

The European Union funded Human Brain Project (HBP), a 10-year endeavour that started in 2013, aims to better understand the brain through research in six areas, including neuromorphic computing. SpiNNaker and BrainScaleS are two significant neuromorphic university initiatives that were inspired by the HBP. The largest neuromorphic supercomputer in existence at the time, the million-core SpiNNaker system, was released in 2018; The University of Manchester plans to expand it up in the future to model one million neurons.

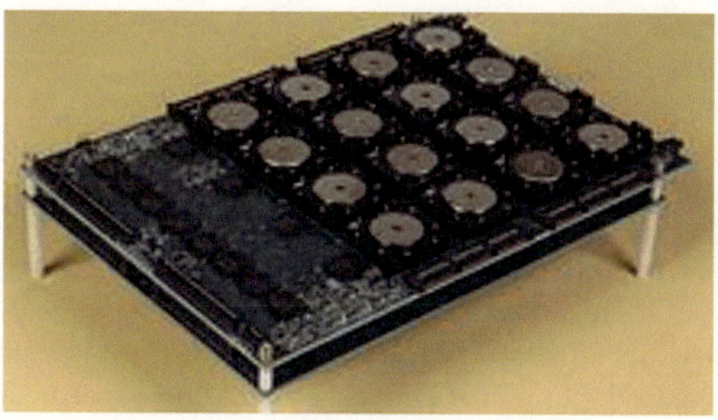

Figure 5.
Board with 16 TrueNorth chips.

5.5 BrainScaleS From Heidelberg University

BrainScaleS (Brain-inspired multiscale computation in neuromorphic hybrid systems) was an EU FET-Proactive FP7 funded research project. The undertaking began on January 1, 2011, and it was completed on March 31, 2015. It involved the cooperation of 19 research teams from 10 different European nations. The Human Brain Project's (HBP) Neuromorphic Computing Platform is where the hardware for neuromorphic computing systems is currently being developed. The goal of the BrainScaleS project is to comprehend and simulate how various spatial and temporal scales interact to process information in the brain.

6. Materials and devices

The production and characterisation of materials for neuromorphic systems has been one of the major areas of advancement in neuromorphic computing in recent years. We want to highlight the variety of innovative nanoscale devices and materials that the materials science community is developing and characterising for neuromorphic systems.

Two typical nano-scale devices that have been constructed with various materials that can result in various behaviours are atomic switches and CBRAM. A survey of many atomic switch types for neuromorphic devices is given in [37], but common atomic switch materials include Ag_2S [38–40], Cu_2S [41], Ta_2O_5 [42], and WO_3-x [43]. Under various conditions, many atomic switch materials can display various switching behaviours. As a result, the atomic switch's behaviour can be controlled by the material choice, which is likely to vary on the application. CBRAM has been implemented using GeS_2/Ag [44–50], HfO_2/GeS_2 [51], Cu/Ti/Al_2O_3 [52], Ag/$Ge_{0.3}Se_{0.7}$ [53–55], Ag_2S [56–58] and Cu/SiO_2 [54]. Similar to atomic switches, the material chosen affects the stability and dependability of the device as well as the switching behaviour of CBRAM devices,

Memristors can be used in a wide number of ways. Transition metal-oxide-based memristor systems may be the most widely used (TMOs). Many different types of materials are utilised to make metal-oxide memristors, including HfOx [59–67], TiOx [68–73], WOx [74–78], SiOx [79, 80], TaOx/TiOx [81, 82], NiOx [83–85], TaOx [86–88], FeOx [89], AlOx [90, 91], TaOx/TiOx [81, 82], HfOx/ZnOx [92], and PCMO [93–98]. The quantity and sorts of resistance states that can be created by various metal oxide memristor types determine the range of weight values that can be stored on the memristor. They also differ in terms of their capacities for endurance, stability, and dependability.

There have also been other more suggested materials for memristors. For instance, spin-based magnetic tunnel junction memristors based on MgO have been suggested for implementations of both synapses and neurons [99], though it has been highlighted that they have a restricted range of resistance levels, making them less suitable to store synaptic weights [88]. Synapses have also been implemented using chalcogenide memristors [100–102]. One of the justifications provided for doing so is the chalcogenide-based memristor's ultra-fast switching speeds, which allow for processes like STDP to take place at nanosecond scale [97]. Polymer-based memristors have been used because of their inexpensive cost and adjustable performance [68, 103–111]. There have also been suggestions for organic memristors, which include organic polymers [69, 112–123].

Ferroelectric materials have been taken into consideration for the construction of analogue memory for synaptic weights [124–129], and synaptic devices [130–133], including those based on ferroelectric memristors [134–136]. Their main area of investigation has been three-terminal synaptic devices as opposed other implementations that may be two-terminal. Three-terminal synaptic devices do not need additional circuitry to implement learning processes like STDP because they can realise them in the device itself [130, 135].

Neuromorphic systems have more recently incorporated graphene to produce more compact circuits. It has been used in neuromorphic implementations as well as full synapse implementations [137, 138] for transistors [139–141] and resistors [142].

The carbon nanotube is another material under consideration for various neuromorphic applications. A variety of neuromorphic components, including dendrites on neurons [143–147], synapses [148–165], and spiking neurons [166–169], have been proposed to use carbon nanotubes. Carbon nanotubes have been used because they can provide the scale (number of neurons and synapses) and density (in terms of synapses) of neuromorphic systems that may be necessary for replicating or imitating biological neural systems. They have also been used to interface with living tissue proving that carbon-nanotube based devices may be helpful in prosthetic applications of neuromorphic systems [155].

Synaptic transistors with different architectures, such as silicon-based ones [170, 171], and oxide-based ones [172–184], have also been produced for neuromorphic applications. Synapses and other neuromorphic components have been built using organic electrochemical transistors [185–190] and organic nanoparticle transistors [191–194]. Organic transistors are being ordered for similar reasons as organic memristors: low processing costs and versatility. Additionally, they are ideally suited for the development of brain-machine interfaces as well as chemical or biological sensors [185]. It's interesting to note that various teams are working to create transistors within polymer-based membranes that can be applied to neuromorphic applications such biosensors [195–199].

Recently, there has been an increased interest in nonvolatile memory (NVM) technologies for neuromorphic computing, beyond their potential as DRAM replacement or as hybrid memory in shared memory systems. For neuromorphic architectures, nonvolatile devices offer a wide range of great qualities, including as memory retention, analogue behaviour, high integration density, faster read and program speeds, high energy efficiency, and programming voltages that are compatible with CMOS electronics. So we focus here on emerging materials for NVM devices:

6.1 Polymer

Polymers are long-chain molecules with repetitive units. They fall within the category of one-dimensional materials with features including biocompatibility, chemical sensitivity, and mechanical flexibility. You can use organic or inorganic materials to create polymers. A flexible wearable memristor is designed with ammonium polyphosphate (APP) in a stack of Au/APP/ITO [201]. The I-V characteristics suggested that the bidirectional voltage sweeps' memristive behaviour was caused by ion movement in the APP. Even under harsh humid, temperature, or radiation settings, the proposed structure has demonstrated stable function. However, a lot of research has been done on organic semiconductors (OSC) for use in neuromorphic devices. The two-terminal OSCs can utilise filament formation, charge trapping, and ion migration to facilitate the integration into ReRAM, PCM, or FeRAM. Fuller et al.

present a polymer-based redox transistor combined with a CBRAM synaptic device [200] whose conductance change is initiated by reversible electrochemical reactions. The authors also show an array of 1024 × 1024 organic polymer memristors arranged for performance characteristics simulation. The main difficulties with OSCs are speed and density. Due to low mobilities of carriers and defects, OSC speed is impacted [201]. Due to OSCs' incompatibility issues with various solvents, the patterning of these devices through photolithography is limited, restricting the fabrication of dense networks.

6.2 2D materials

Fundamental 2D material research has received a great deal of attention during the past 10 years. The 2D materials maintain a stable mono-layer structure with special chemical and physical properties suitable for simulating synapses. They are recognised for their inter-layer weak van der Waals forces. Haigh et al. demonstrate that the high mobility of 2D synapses allows them to exhibit fast switching speed at low operating voltages [202]. The alteration in electrical, photonic, and electrochemical properties is another distinguishing quality of 2D material synaptic devices [203]. Shi et al. use a CBRAM-based h-BN memristor to exhibit STP and LTP properties [204]. The creation and deformation of the conductive filament caused by ion migration between Cu or Ag electrodes controls weight update. The h-BN exhibits boron vacancies that stimulate resistance changes.

Wang et al. demonstrate how externally introduced oxygen atoms occupy intrinsic defects that can be adjusted as MoS_2 sulphur vacancies, causing resistance changes [16]. In [205], authors use bilayer MoS_2 vertical memristors to demonstrate STDP features at 0.1–0.2 V voltage. Apart from CBRAM, 2D materials have been integrated as PCM 2-terminal synaptic elements, which have the advantage of better reliability. $MoTe_2$ exhibits an electric field-induced amorphous to crystalline phase transition in TMD materials [206]. Multilevel programming resistance could be facilitated by further device engineering to the device stack. On the other hand, A 3-terminal device uses the gate and the channel as presynaptic and postsynaptic inputs, exhibits superior stability and effective channel conductance control. Chen et al. demonstrate how a graphene and ferroelectric insulator (polyvinylidene fluoride, i.e., PVDF) synapse can imitate the synaptic behaviour as a FeFET device [207]. The ferroelectric material's changing polarisation state affects the carrier concentration in graphene. The polarisation shift is induced when the gate voltage is raised above the threshold voltage. Future research might also look toward Li + ion gated synaptic transistors [208] and various hetero synaptic plasticity implementations [201, 209, 210].

6.3 Quantum dot

Quantum dots (QD) are zero-dimensional memristors. Semiconducting quantum dots which are small particles with clearly defined energy levels, exhibit electrical and optical features based on quantum mechanics. Josephson junctions are the foundation for how a QD functions as a memristor. The phase difference between quasi-particles is employed ss a state variable [211]. Here, a hybrid construction made of QDs and a memristor is used. In [212], Lv et al. demonstrate that RRAM devices with QD-film as their insulator can be switched in response to an external signal. The memristive characteristic of QD-RRAMs is catalysed by ion migration, charge trapping, or redox reaction. Qi et al. show how to fabricate

RRAM with Carbon QD for usage as a LED [213]. In [214], Roychowdhury et al. use QD arrays to show quantum neuromorphic computing. There are a lot of potential opportunities that are yet to be explored.

6.4 Carbon nanotube

A carbon nanotube (CNT) is a cylindrical rolled up, often single-walled carbon in a tube shape of nanometre diameter. They have a metallic or semiconducting character because of their achiral connections. These fall within the category of one-dimensional materials that resemble axons structurally. Due to their great charge mobility, semiconductor CNTs can be employed as conducting channels in FETs. CNTFET is a CNT replacing semiconductor channel between the source and drain. With voltage applied, a Schottky barrier that has formed at the metal-CNT contact is alleviated. The ON/OFF state of memory cells is determined by the interaction between CNTs. The gate and source of the single-walled CNT matrix network are respectively coupled to presynaptic and integrate-and-fire (IF) postsynaptic neurons by Feldmann et al. [215]. The channel conduct anceto store synaptic weights is controlled by varying voltage pulses at the pre- and post-neuron. All of the post synaptic neuron spikes are gathered to fire back the CNT if the output hits a threshold level. In order to choose the sign and magnitude of the weight update for an STDP implementation, the channel conductance change when the gate and source voltages are correlated. Kim et al. describe p-type CNTFET-based models of excitatory and inhibitory neurons [158], where the neurons exhibit STP accumulative current. The very lateral geometry of CNTFETs is not viable for larger integration, though. CNT TFTs have therefore become a popular substitute for the same.

The materials science community is doing a lot of intriguing work to create devices for neuromorphic systems out of novel materials in order to create smaller, faster, and more effective neuromorphic devices. Diverse materials can have radically different properties, even when used in the same device. These variations will have an impact on the rest of the community, up through the device, high-level hardware, supporting software, models, and algorithms levels of neuromorphic systems. Therefore, it is crucial that we as a community comprehend the potential effects that various materials may have on functionality. Going forward, tight partnerships with the materials science community will undoubtedly be necessary.

7. Neuromorphic circuits

The creation of brain-inspired computing systems necessitates the careful design of circuit blocks that exhibit specific elements of neuromorphic functions, such as plasticity and learning, in addition to device engineering. The individual synapses, for instance, must act as electrical connections between two neurons and change their weight in accordance with particular learning criteria that are inspired by the brain. This is one of the biggest challenges with neuromorphic computing is trying to replicate the human "connectome," or the physical wiring of the nervous system. Reverse-engineering the brain on solid-state devices (SSD) is still a long way off, despite advances in our understanding of how the wiring of the brain carries out higher-level processes (**Figure 6**).

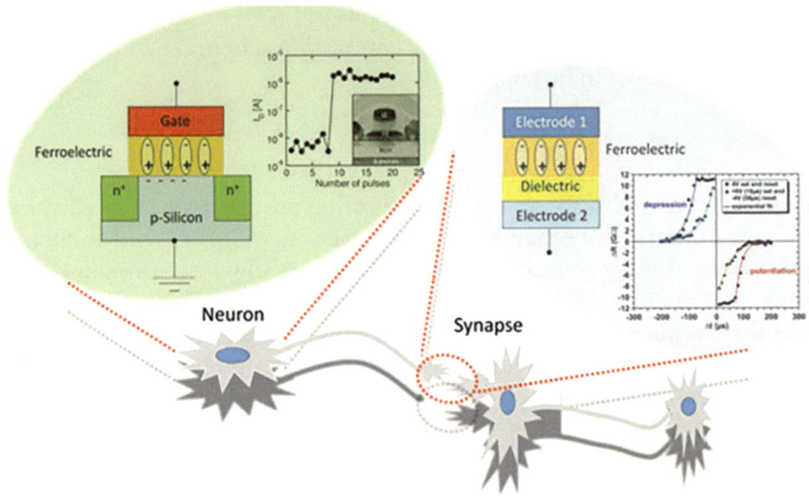

Figure 6.
Key elements of a neural network [216].

By digitally transferring biological neuronal network models onto electronic devices, neuromorphic computing explicitly aims to appropriate the biological connectome of the brain. Neuro-electronic interfaces are brain-computer interfaces that transmit data from the brain to outside equipment. Because there are currently no suitable neuro-electronic interfaces, digital brain mimicry is difficult.

On the other hand, it has been demonstrated that time has a significant impact on synaptic plasticity in the brain. For instance, the human brain's spike-timing dependent plasticity (STDP) is a weight update mechanism where the amount of time between pre- and postsynaptic spikes determines the weight change's magnitude and sign, i.e., potentiation for the pre-synaptic spike coming before the postsynaptic spike and depression for the postsynaptic spike coming before the pre-synaptic spike [217]. Time calculation between a pair or triplet of spikes is a key component of other, more complex synaptic weight updating algorithms [218–220]. These plasticity rules typically call for the development of intricate circuits utilising CMOS devices [221–223] or nanoscale devices, such as PCM [224–226] or RRAM [24, 64, 224, 227–229] technology. Plastic synapses based on PCM and RRAM often include one or more transistors in a hybrid design to enable a controllable computation of time within the circuit block, although examples of time-sensitive synapses that have been described [230, 231].

7.1 Hybrid STDP synapses

Figure 7a shows a hybrid RRAM-CMOS synapse example with a 2T1R design that consists of one RRAM element and two transistors [228]. The top electrode of the RRAM device and the gate of one transistor known as the communication transistor are both driven by the pre-synaptic neuron. It displays the voltage VTE applied to the top electrode and VCG applied to the communication gate's gate, both of which are applied during a spike event by the pre-synaptic neuron. A synaptic current that spikes as a result of the applied voltage spikes in the figure, which is proportional to the conductance of the RRAM, acts as a storage component for synaptic weight. According to the schematic circuit of **Figure 7c**, synaptic current travels via the

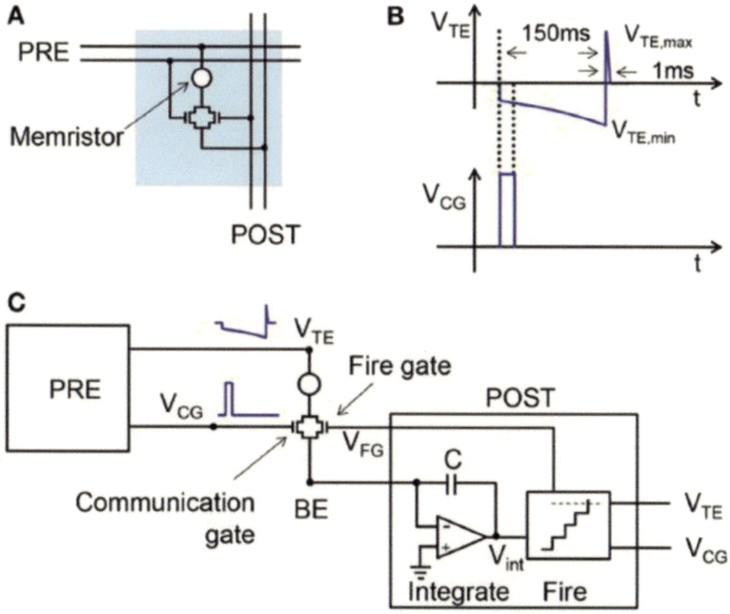

Figure 7.
Hybrid RRAM-CMOS synapse with 2T1R configuration (a), voltage waveforms for VCG and VTE applied by the pre-synaptic neuron in the spike event (b), and overall circuit sketch including the synapse and the pre- and post-synaptic neurons [232].

synaptic circuit and is fed into the input terminal of the post-synaptic neuron, where integration and firing occur. The input node of the post-synaptic neuron is a virtual ground, guaranteeing a zero voltage at the bottom electrode of the 2T1R synapse and acting as a summing input for a practically infinite number of synaptic channels. The post-synaptic neuron fires when the integrated current exceeds a predetermined threshold, sending spikes to the subsequent neurons in the network and applying a feedback spike to the fire gate.

7.2 Learning with RRAM STDP synapses

A conceptual demonstration of learning cannot be provided by definitively demonstrating STDP in individual synapses; instead, simulations and experiments at the higher level of synaptic/neural networks are needed. A straightforward example of a feed-forward neural network, known as a perceptron [233–235], is shown in **Figure 8**. The network is made up of two layers: a presynaptic layer where synaptic channels receive neural spikes, and a second layer with just a single post-synaptic neuron to integrate and fire current spikes. Each pre-synaptic neuron is coupled to a post-synaptic neuron via a hybrid CMOS-RRAM synapse, indicating that the network is fully connected. Depending on the time Δt between pre- and post-synaptic spikes, the post-synaptic neuron transmits a feedback spike to each synapse at each fire event to enable LTP/LTD. As a result, submitted patterns, such as images, sounds, or speech, tend to be learned by the network because the synapses corresponding to the pattern channels are potentiated while all other synapses, commonly known as the background synapses, tend to become depressed, thus enabling on-line learning of submitted patterns [226, 228, 229].

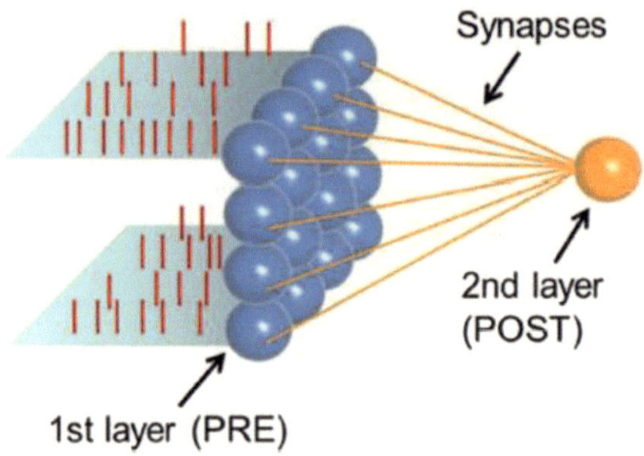

Figure 8.
Schematic illustration of a perceptron-like neural network with a 4 × 4 first layer, and a single post-synaptic neuron in the second layer. Each neuron in the first layer is connected to the post-synaptic neuron by synapses [232].

With the recent advancements in ferroelectric devices, nanowire networks, organic materials, and new memory hardware, duplicating the human neural network is now a more realistic possibility. The great ability of memory devices promise synaptic device that can handle massive amounts of data effectively while consuming extremely little power, which is necessary to develop artificial intelligence technologies. The detailed analysis of these memory devices, together with the present problems and future prospects, is covered in the following subsections.

7.2.1 PRAM: phase-change synaptic devices

Phase-change memory (PRAM), also called PCM, is a type of non-volatile random-access memory. A thesis on the viability of a phase-change memory device using chalcogenide film and a diode was published by Charles Sie in 1969 [236]. The phase-change memory process in chalcogenide glass involves electric-field-induced crystalline filament development, according to the following study from 1970 [237, 238].

PRAM uses the resistance difference between the crystalline phase's low resistivity and the high resistivity of the amorphous phase in phase change materials. A phase change material known as T-cell is typically used between two electrodes in PRAM structures. The phase of the material is changed after a high voltage/current is applied to the electrodes. A current pulse is used to anneal and quench the phase-change material to cause it to crystallise in order to set into the crystalline phase. The high resistance amorphous component can be steadily crystallised using a series of Set pulses to achieve the progressive Set function of PRAM devices. The programming region is first melted and then quickly quenched by using a strong current pulse for a brief period of time in order to reset into the amorphous phase.

The device has been investigated as a good candidate for artificial synapses in implementing the machine learning algorithm because of desirable properties as high speed, multi-level capability, and low energy consumption [239–241]. But issues with material quality and power consumption prohibited PRAM technology from being widely used. Resistance drift, which refers to how the resistance changes over time is a particular problem [242–244]. In phase-change materials, resistance drift is a common occurrence that severely hinders the development of PRAM and destroys stability. A structural relaxation, which is a locally rapid local rearrangement thermally induced at intervals the amorphous region, has been used to explain the drift phenomenon in amorphous chalcogenide materials [243, 245–247]. This implies that the resistance can be modified with time and that a lost memory resembles almost human memory. In addition to the significant reliability problem with amorphous chalcogenide materials, a low power neuromorphic device must overcome the high power need for melting the phase-change material. These reliability and power problems must be solved for PRAM technology to function well as a synaptic device.

7.2.2 Reram: filament type synaptic devices

One of the leading candidates to replace current memory technology as the next generation is resistive random access memory (ReRAM). The non-volatile nature of ReRAM is the main justification, among many others. Researchers and manufacturers are currently paying close attention to every non-volatile memory. This results in a race between several technologies that practically all offer the same advantages while aiming for the same standard level of performance goals. Each non-volatile memory is competing fiercely to prove that it is a worthy rival in the memory field. Other non-volatile memories of current interest are PCRAM (Phase Change), FeRAM (Ferroelectric), and MRAM (Magnetoresistive). ReRAM is distinguished and enabled to compete with these technologies because of its quick read and write speeds. Additionally, ReRAM's manufacturing is not as difficult as other technologies.

Low dependability and performance variations, on the other hand, are the obstacles preventing its market penetration. Therefore, it still requires significant development in terms of material fabrication, construction, circuit difficulties, and handling unwanted disturbances from the outside environment. ReRAM technology appears to have more benefits than drawbacks, and this good promise for the future of the technology.

ReRAM has a metal-insulator-metal structure, with the insulator positioned between two layers of metal. Both metals form a low resistive state when touched, and a high resistive state when not touched. A conducting filament serves as the structure's central functional part. The rupture of the filament depends on the applied voltage across the metal terminals (**Figure 9**).

ReRAM's ability to store digital data, or 0 s and 1 s, is a result of its structural design. ReRAM typically operates in a High Resistive State [HRS], or logic low referred to as 0; a voltage is applied across ReRAM to switch its state from the High Resistive State to a Low Resistive State [LRS], or logic high referred to as 1. There is

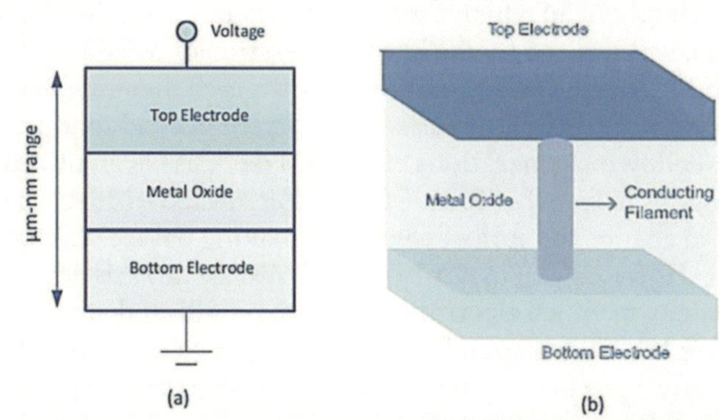

Figure 9.
(a) 2D structure of ReRam (b) 3D structure of ReRam [248].

no suitable path for the current to pass since the filament is disconnected during the HRS. The filament contacts the bottom terminal and produces a low resistance route when an appropriate voltage is placed across the top electrode (TE) and bottom electrode (BE). This is known as coming into LRS, which is also known as logic high, or 1. HRS is the OFF state, and LRS is the ON state. The "Set" voltage is the voltage used to switch the ReRAM state from HRS to LRS. In a similar vein, the voltage used to switch a device from LRS to HRS is known as the "Reset" voltage. After a voltage that is greater than the Reset or Set voltage is applied, the filament either forms or ruptures. This state change is the write operation during putting ReRAM in the desired state to hold a logic. A voltage that is less than the Reset or Set voltage is applied to read the operation, and the output is monitored to determine the state of the ReRAM. In accordance with the polarity of the applied voltage, there are two forms of resistive switching. It could be bipolar or unipolar. In unipolar switching, the voltage's polarity has no dependence on the switching process. In contrast, switching in bipolar switching is dependent on the voltage's polarity, going from HRS to LRS in one case and from LRS to HRS in the other [249].

7.2.3 MRAM: spintronic synaptic devices

Magneto-Resistive Random Access Memory (MRAM) is a kind of non-volatile memory (NVM) that can retain saved data even if the power is out or is unintentionally turned off. MRAM, commonly known as Magnetic RAM, is not brand-new. MRAM has been available on the market for more than 20 years, but recent advancements in the technology have allowed it to be employed successfully in both new and old applications.

Magnetic Tunnel Junction (MTJ), a part of MRAM technology, is a combination of two ferromagnetic layers and an insulating layer. MRAM stores a logic 1 or a logic 0 by altering the resistance of an MTJ. Two ferromagnetic layers' relative spin orientations determine the resistance of the MTJ, which has two resistance options: high resistance and low resistance.

According to **Figure 10**. The free layer of the MTJ can have its magnetic direction modified by applying polarised currents or magnetic fields, whereas the reference layer of the MTJ maintains a constant magnetic direction. The resistance of the MTJ is minimal when the directions of the reference layer and the free layer coincide (a logic 0 is stored). However, resistance is strong if they are going in opposite directions (a logic 1 is stored).

Toggle mode technology, in which a magnetic field modifies the electron spin to program/write bits, was utilised by the initial generation of MRAM devices. Although it was simpler to build toggle MRAM, scaling it down is challenging. As the write lines get smaller and other difficulties arise, the current needed to flip the bit does not decrease in value. Following generations of MRAM devices began utilising a different technique known as Spin-Transfer Torque (STT) MRAM. SST MRAM uses a spin-polarised current to change the spin of electrons, making device scaling down simpler and less expensive. However, the majority of STT MRAM devices now available use a perpendicular strategy in which electrons spin horizontally and change perpendicular to the plane (**Figure 11**). This perpendicular method provides the benefit of offering higher density memory products by permitting lower switching currents with fewer transistors and less power consumption.

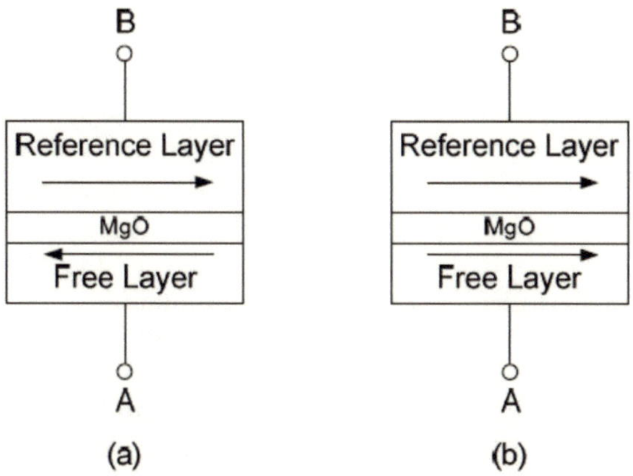

Figure 10.
Structure of a MTJ. (a) Anti-parallel (high resistance) and (b) parallel (low resistance) [250].

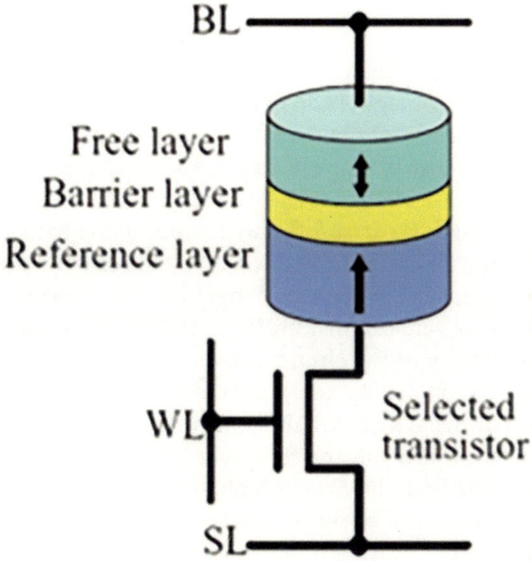

Figure 11.
Perpendicular MTJ diagram [251].

STT-MRAM products now on the market have the benefit of being faster and using less power than NVM flash memory devices. They could also compete with volatile memory units like SRAM and DRAM due to their potential speed as well as their ability to be scaled down even smaller than 10 nm. They are even more appealing to embedded memory applications because of this property. STT-MRAM has inherent limitations, much like many other technologies, and only time will tell if this technology will take the place of DRAM or SRAM. STT-MRAM technology offers intriguing applications in neuromorphic computing, automotive microcontrollers, and system-on-chips (SoCs).

7.2.4 ASN (atomic switch network): network-based synaptic device

Inorganic "atomic-switch" synapses embedded in a sophisticated network of silver nanowires make up the more than a billion interconnected synapses used in the self-assembling neuromorphic (brain-like) technology created by researchers in the U.S. and Japan. It has been demonstrated that the atomic switch, a freshly created nanoscale circuit component, displays synapse-like characteristics in an entirely inorganic device. Each square centimetre of the device's silver nanowire network contains a billion junctions that are intricately coupled to one another.

These atomic switch networks (ASN) produce emergent behaviours resembling memristors through distributed, collective interactions, just like biological neural networks. A key feature of biological brain networks and many other complex systems is these emergent behaviours. According to the researchers, experiments are currently being conducted that exploit these emergent behaviours for information processing with the goal of creating a new category of cognitive technology. The architecture of the ASN device is extremely linked and includes synaptic circuit components at each point of nanowire contact. According to the researchers, the collective interactions between these atomic switches provide distinctive emergent features that offer substantial potential for neuromorphic computing.

Comparing these emerging devices, MRAM features non-volatility, as the magnetic properties of ferromagnets do not disappear due to power failure. The magnetism of ferromagnets not only does not disappear due to power failure, but can be considered almost never to disappear, so MRAM can be rewritten infinitely as well as DRAM. The write time of MRAM is as low as 2.3 ns and the power consumption is extremely low, enabling instant on/off and extending the battery life of portable devices. MRAM cells can be easily embedded into logic circuit chips, requiring only one or two additional steps in the back-end metallization process, which requires a photolithographic mask version. In addition, MRAM cells can be completely fabricated in the metal layer of the chip, and even 2–3 layers of cells can be stacked, so it has the potential to construct large-scale memory arrays on logic circuits. However, the biggest drawback of MRAM is the interference between memory cells. When programming the target bit, the free layer in the non-target bit can be easily misprogrammed, especially in the case of high density, the overlap of the magnetic field between adjacent cells will be more serious. For PCM, it does not need to erase previous code or data before writing updated code, so its speed has an advantage over NAND, and the read and write time is more balanced. PCM read and write is non-destructive, so its write resistance far exceeds that of flash memory, and PCM is used to replace traditional mechanical hard drives with higher reliability. PCM has no mechanical rotation device, and no refresh current is required to save code or data, so PCM's power consumption is lower than HDD, NAND, and DRAM. Some PCMs use a non-transistor design, which can realise high-density storage. PCM storage technology is independent of the charged particle state of the material, so it has a strong resistance to space radiation and can meet the needs of national defence and space. While RRAM erase speed is determined by the pulse width of the trigger resistor transition, generally less than 100 ns. Some RRAM materials also have a variety of resistance states, making it possible to store multi-layer data when a memory cell, thus increasing the storage density. The memory matrix of RRAM can be divided into two types: passive matrix and active matrix. The memory cell of the passive matrix is connected by a resistive element and a nonlinear element (usually using a diode), the role of the latter is to make the resistive element get a suitable voltage division, so as to avoid the loss of the memory cell reading and writing information when the resistive element is in a low resistance state. The advantage of this method is that the design is relatively simple and the process is well miniaturised, but the use of a passive matrix will make the adjacent cells inevitably interfere with each other. Active cells are controlled by transistors to read and write and erase resistive components, although good isolation of adjacent cells interference, its design is more complex, and the device can be poorly miniaturised. In terms of capacity, these three new types of memory, MRAM up to 4 Gb, PRAM up to 8 Gb, and RRAM up to 32 Gb, are still very different from flash memory but all three are more than 1000 times faster than flash memory in terms of read and write speed.

8. Neuromorphic algorithms and applications

8.1 Neuromorphic algorithms

How to define an SNN for a certain application is frequently covered in algorithms for neuromorphic implementations. There are many different algorithmic strategies for neuromorphic computing systems, and they can be divided into two main groups: (1) algorithms for training or learning an SNN that will be used in a neuromorphic computer, and (2) non-machine learning algorithms where SNNs are manually built to complete a specific task. It's important to note that training and learning algorithms here refer to the mechanism of optimising.

It is difficult to learn in a spiking neural network. Traditional artificial neural networks have had great success with backpropagation-based gradient descent learning; however, training SNNs is challenging due to the nondifferentiable nature of spike events. Due to the interest in deep learning, a significant amount of research has been put into creating appropriate learning algorithms that can be used with multilayer SNNs. Unsupervised learning, supervised learning, conversion from learned ANNs, and evolutionary algorithms are the four main methods for training SNNs. The following subsections provide a quick overview of them.

8.1.1 Unsupervised learning

Learning without preexisting labels is referred to as unsupervised learning. The Hebbian rule, which unsupervised learning of SNNs is based on, entails adapting the synaptic connections of the network to the data received by the neurons [252]. Hebb's rule is applied through the spike-timing-dependent plasticity (STDP) algorithm. The STDP phenomenon, which has been observed in the brain, explains how the relative timing of presynaptic and postsynaptic spikes affects how effective a synapse is. The spike that reaches the neuron's synapse is referred to in this context as a presynaptic spike. The spike released by the neuron itself is known as a postsynaptic spike [253]. The idea behind STDP's mechanism is that the synapses that are most likely to have caused the neuron to fire should be reinforced. Similar to this, weakening should be done to synapses that did not participate or contributed negatively [254].

In unsupervised learning in SNNs, STDP is commonly employed as a learning method. A synaptic weight is said to be reinforced by STDP if a presynaptic neuron fires just before a postsynaptic neuron. Similar to that, the synaptic weight is diminished if the presynaptic spike follows the postsynaptic neuron for a brief period of time [255]. The most observed STDP rule is described by Eqs. (1) and (2):

$$\Delta w = \begin{cases} +A_{+} \exp\left(\frac{-\Delta t}{\tau}\right), if\, \Delta t > 0 \\ -A_{-} \exp\left(\frac{+\Delta t}{\tau}\right), if\, \Delta t \leq 0 \end{cases} \tag{1}$$

$$\Delta t = t_{post} - t_{pre} \tag{2}$$

where w is the synaptic weights, τ is the time constant, and $A+$ and $A-$ are constant parameters indicating the strength of potentiation and depression.

Significant study has been put on training SNNs with STDP in recent years. Two new hardware-friendly techniques, lateral inhibition and homeostasis, were created by Qu et al. [256]. These techniques lessen the number of inhibitory connections, which lowers the hardware overhead. On the MNIST data set, an STDP rule was utilised to modify the synapse weight between the input and learning layers to obtain 92% recognition accuracy. A hybrid learning system called deep CovDense SNN, put out by Xu et al. [255], combines the biological plausibility of SNNs and feature extraction from CNNs. They updated the parameters of their deep CovDenseSNN model, which is appropriate for implementation on neuromorphic hardware, using an unsupervised STDP learning method. Other STDP learning techniques include supervised learning and reinforcement learning [257].

A semisupervised method was put out by Lee et al. [258] to train a convolutional SNN with several hidden layers. The training approach consisted of two steps: initialise the network weights by unsupervised learning (namely, SSTDP), and the supervised gradient descent backpropagation (BP) algorithm to fine-tune the synaptic weight. Pretraining methods resulted in 99.28% accuracy on the MNIST database, quicker training times, and greater generalisation. An innovative technique for training multilayer spiking convolutional neural networks was created by Tavanaei et al. [259]. (SCNNs). Unsupervised (a unique STDP learning scheme for feature extraction) and supervised (a supervised learning scheme to train spiking CNNs (ConvNets)) components are both included in the training process.

8.1.2 Supervised learning

The SpikeProp algorithm by Bohte et al. [260] is one of the first to train SNNs utilising backpropagation errors. Using a three-layer design, this approach is successfully used to solve classification challenges. Spike train SpikeProp (ST-SpikeProp), a more recent and sophisticated variation of SpikeProp, trains single-layer SNNs using the output layer's weight update rule [261]. Wu et al. [262] proposed the spatiotemporal backpropagation (STBP) technique, which combines the timing-dependent temporal domain and the layer-by-layer spatial domain, to address the nondifferentiable problem of SNNs. The energy consumption of SNNs has been demonstrated to significantly decrease with supervised learning using temporal coding. Mostafa [263] created a direct training method using the temporal coding scheme and back propagation error. The preprocessing technique is not general, and his network lacks convolutional layers. By adding convolutional layers into the SNN, creating a new kernel operation, and suggesting a new method for preprocessing the input data, Zhou et al. [264] improved on Mostafa's work. They used fewer trainable parameters to attain good recognition accuracy with their SCNN. Using the stochastic gradient descent (SGD) algorithm and then converting it to an SNN, Stromatias et al. [265] developed a supervised technique for training a classifier. Zheng and Mazumder [266] suggested backpropagation-based learning for training SNNs in previous works. Their suggested learning approach can be used with neuromorphic hardware.

8.1.3 Conversion from trained ANN

The third method involves converting an offline-trained ANN to an SNN so that the modified network can benefit from an established, fully trained ANN model. This method is frequently referred to as "spike conversion" or "spike transcoding." There are many advantages of converting an ANN to SNN. First, simulating the precise spike dynamics in a big network can be expensive in terms of computing, especially if precise spike durations and high firing rates are needed. As a result, this method enables the use of SNNs to challenging benchmark tasks that demand massive networks, such ImageNet or CIFAR-10, with a minimal accuracy loss when compared to its formal ANNs [267, 268]. Second, we can use very effective training methods created for ANNs as well as many state-of-the-art deep networks for classification tasks for conversion to SNNs. The optimization approach can also be used to ANNs. This makes it possible to use of state-of-the-art optimization techniques and GPUs for training [269]. The conversion method's primary drawback is that it cannot support on-chip learning. Additionally, many SNN particularities that are not present in the equivalent ANNs cannot be taken into account during training. Because of this, the SNNs' inference performance is frequently inferior than that of the original ANNs [270].

An extensive amount of research has been done to successfully convert an ANN to an SNN with performance on the MNIST data set. Diehl et al. [269] proposed a method for transforming an ANN into an SNN that has the least amount of performance loss, and using the MNIST database, a recognition rate of 98.64% was attained. Rueckauer et al. [271] transformed continuous-valued deep CNN to its accurate spiking equivalent in another work. This network exhibits a 99.44% recognition rate on the MNIST data set and incorporates typical operations like softmax, max-pooling, batch normalisation, biases, and inception modules. An approach to conversion that is appropriate for mapping on neuromorphic hardware was put out by Xu et al. [272]. On the MNIST data set, they demonstrated a threshold rescaling strategy to lessen the loss and attained a maximum accuracy of 99.17%. To transform CNNs into spiking CNNs, Xu et al. [255] developed an effective and hardware-friendly conversion rule. On the MNIST data set, they suggested a "n-scaling" weight mapping method that delivers high accuracy and low latency classification. On the MNIST data set, Wang et al. [273] suggested a weights-thresholds balancing conversion method that uses less memory and delivers excellent recognition accuracy. They concentrated on the relationship between weights and thresholds of spiking neurons throughout the conversion process, as opposed to the existing conversion strategies, which concentrate on the approximation between the artificial neurons' activation values and the spiking neurons' firing rates.

8.1.4 Evolutionary spiking neural networks

Population-based metaheuristics are evolutionary algorithms (EAs). Historically, observations of natural evolution in biological populations served as the inspiration for their design. The network architecture, model hyperparameters, and synaptic weights and delays can all be directly optimised using these approaches [274, 275]. The synaptic weights of SNNs are currently learned using evolutionary algorithms like differential evolution (DE), grammatical evolution (GE), harmony search algorithm (HSA), and particle swarm optimization (PSO). The synaptic weights of a spiking neuron, such as integrate and fire, Izhikevich, and spike response model (SRM) models, can be trained by employing algorithms like DE, GE, and HSA to carry out classification tasks, as demonstrated by Vazquez [276], López-Vázquez et al. [277], and Yusuf et al. [278]. In both linear and nonlinear classification problems, Vazquez and Garro [279] used the PSO algorithm to train the synaptic weights of a spiking neuron. They found that firing rates are matched by input patterns of the same class. For the purpose of training supervised feedforward SNNs, Pavlidis et al. [280] presented the parallel differential evolution method. Their method is tested only with exclusive OR, which does not accurately reflect its advantages. Evolutionary algorithms can be an alternative to exhaustive search. They take a lot of time, especially because the fitness function requires expensive computation [281].

The deployment of appropriate learning and training algorithms, which have a significant impact on application accuracy and execution cost, is a challenge in the development of SNNs. The way that information is encoded by spikes presents another difficulty. Choosing the right training and learning algorithms to use can have a significant impact on the accuracy and cost of an application. Information encoding using spikes presents another difficulty. Although neural coding significantly improves the performance of SNNs, there are still problems about the ideal encoding strategy and how to create a learning algorithm that complements the encoding technique. It has become extremely difficult to create a learning algorithm that can train hidden neurons in a linked SNN. As neuromorphic computing is still in its infancy, significant work needs to be done to develop algorithms and hardware that can simulate human intellect.

8.2 Neuromorphic applications

8.2.1 Medicine

The ability of neuromorphic devices to receive and process information from their surroundings is particularly effective [282–286]. These devices can work with the human body when combined with organic components. Neuromorphic devices may enhance drug delivery techniques in the future. Due to their high responsiveness, they may release a medicine when they noticed a change in the body's conditions (i.e. varying insulin and glucose levels). The employment of neuromorphic computer technology in prostheses is also possible. Another advantage of this technology is its capability to efficiently accept and process an external signal. For those with prosthesis, using neuromorphic devices rather than conventional ones could result in a more natural, seamless experience [282].

8.2.2 Large-scale operations and product customization

Neuromorphic computing may also be useful for large-scale initiatives and product customization [282]. It could be used to handle massive amounts of data from environmental sensors more quickly. Depending on the requirements of the sector, these sensors could monitor water content, temperature, radiation, and other characteristics. By identifying patterns in these data, the neuromorphic computing framework might make it simpler to draw useful conclusions. Due to the characteristics of the materials used to construct them, neuromorphic devices may potentially facilitate product customisation. These substances can be turned into fluids that are simple to control. They can be processed through additive manufacturing in liquid form to produce devices that are especially suited to the requirements of the user [282].

8.2.3 Artificial intelligence

By definition, the goal of neuromorphic computing is to replicate how the human brain works. Neurons in the brain receive, process, and transmit impulses in a very quick and energy-efficient manner. As a result, it makes sense that technology experts, particularly those working in the area of artificial intelligence (AI), would be fascinated by this kind of computing. As the name implies, experts in the field of artificial intelligence (AI) concentrate on a certain aspect of brain intelligence. The capacity of the intellect to gather and use knowledge is known as intelligence. It would be advantageous for the two fields to work together moving forward because this concept is so closely related to neuromorphic computing [287]. The answer to creating computers with truly human-level intelligence may lie in concentrating on brain functionality at the level of an artificial neuron and smaller.

8.2.4 Imaging

Similar to how the human eye creates images, neuromorphic vision sensors do the same. They are event-based imaging devices [288]. This demonstrates that they create images in response to light intensity which is an external signal rather than an internal signal [289]. Furthermore, they move at a faster pace independent of conventional frame rates. In a neuromorphic sensor, each pixel functions independently of its neighbours. Additionally, the gadget almost instantly communicates changes to each pixel [288]. These mechanisms work together to make data utilisation significantly more effective. Like their traditional equivalents, these sensors do not exhibit motion blur or a delayed response to the environment. Incorporating neuromorphic vision sensors into virtual and augmented reality systems may be advantageous given these qualities.

8.2.5 Other applications

Neuromorphic computing may be suited for use on "the edge" due to its low energy consumption [290]. The boundary of a network where a device could connect to a cloud platform is referred to as "the edge." Since driverless cars must operate in this environment, neuromorphic computing may enable them to react to their surroundings more quickly. A system using neuromorphic computing could take control of these vehicles when they are not linked to a reliable internet source. This might increase the safety and environmental suitability of driverless cars. Neuromorphic computing's superior sensory capabilities may also enhance current "smart technologies" [283]. This modification might improve the effectiveness of the devices for a wider range of circumstances, similar to driverless cars. Expanding communication channels is another potential application that could be done using neuromorphic computing.

9. The gap between expectations and reality

Deep learning systems are used nowadays to process the majority of neuromorphic computing tasks on CPUs, GPUs, and FPGAs. However, none of these are optimised for neuromorphic processing. These functions were the exclusive focus of the creation of chips like Intel's Loihi. Because of this, Loihi was able to obtain the same outcomes with a much smaller energy profile, as demonstrated by ABR. The next generation of compact gadgets that require AI capabilities will depend heavily on this efficiency.

Many experts predict that during the next three to 5 years, commercial applications will start to appear seriously, but that will not be the end of it. For this reason, Samsung, for instance, declared in 2019 that it would double the size of its neuromorphic processing unit (NPU) business by 2030, going from 200 personnel to 2000. At the time, Samsung predicted that the market for neuromorphic chips will expand by 52% a year through 2023.

Developing common workloads and benchmarking approaches will be one of the upcoming difficulties in the neuromorphic space. Technology adopters have benefited greatly from the use of benchmarking tools like 3DMark and SPECint to match products to their needs. Although author Mike Davies of Intel Labs suggests a spiking neuromorphic system dubbed SpikeMark, there are sadly no such benchmarks in the neuromorphic space. Dmitri Nikonov and Ian Young, researchers at Intel, outline a number of guidelines and techniques for doing neuromorphic benchmarking in a technical article titled "Benchmarking Physical Performance of Neural Inference Circuits."

There is currently no practical testing tool on the market, though Intel Labs Day 2020 in early December made some significant advancements. When processing "Sudoku solver" issues, Intel, for instance, compared Loihi to its Core i7-9300K and demonstrated how fast Loihi's searching was.

Researchers solved Latin squares with a remarkable reduction in power consumption and experienced a similar 100× gain. What was perhaps the most significant finding was how various processor types performed versus Loihi for specific workloads.

Loihi competed against IBM's TrueNorth neuromorphic microprocessor in addition to traditional computers. On neuromorphic solutions like Loihi, deep learning feedforward neural networks (DNNs) clearly underperform. Data moves linearly from input to output with DNNs. Recurrent neural networks (RNNs) use feedback loops and behave more dynamically, making them more similar to how the brain functions. In RNN workloads, Loihi excels. As Intel stated: "The more bio-inspired properties we find in these networks, typically, the better the results are."

One could consider the aforementioned examples to be early benchmarks. They are an important first step in the direction of a tool that is widely used and runs typical workloads for the industry. New applications and use cases will ultimately appear, and testing holes will be filled. The deployment of these benchmarks and apps in response to urgent requirements will continue to be a focus of developers.

Research and development for neuromorphic computing is still ongoing. It's becoming more and more obvious which applications work best with neuromorphic computing. For these tasks, neuromorphic computers will be more faster and more energy-efficient than any current, conventional options. Neuromorphic computing will simply coexist with CPU and GPU computing to perform tasks more effectively than anything we have seen before. CPU and GPU computing will not go away.

10. Conclusion and outlook

In this chapter, we have summarised what neuromorphic computing reaches until today, discussing its materials, algorithms, circuits and applications. Although neuromorphic devices show promising characteristics, they still have some challenges to be solved to realise energy-efficient neuromorphic systems. Neuromorphic computing technology has created high expectations and its market size expands every day in different fields.

One of the markets for neuromorphic chips that is expanding the quickest is the auto business. All of the top vehicle producers are working hard to achieve Level 5 of vehicle autonomy, which is anticipated to result in enormous demand for AI-controlled neuromorphic processors. The self-governing driving industry needs constant advancements in AI calculations for maximum throughput and minimal force requirements. Neuromorphic chips are excellent for order-related tasks and could be applied in a few instances to self-driving cars. Compared to static deep learning arrangements, they are also proficient in loud climate like self-driving vehicles. According to Intel, four terabytes is the estimated amount of information that a standalone vehicle might produce of approximately 60 min and portion of driving, or the amount of time a person spends in their vehicle on average every day. The ability of self-ruling vehicles to effectively handle all of the data generated during these excursions will be put to the test.

The ADAS (Advanced Driver Assistance System) applications include picture learning and recognition work among other uses of neuromorphic chips in vehicles. It functions similarly to one of the standard ADAS features found in passenger vehicles, such as cruise control or intelligent speed assistance. By sensing the traffic data set aside on streets, such as crosswalks, school zones, street knocks, and so forth, it can control vehicle speed.

Some of the major market players, like Intel Corporation and IBM Corporation, are based in North America. Due to factors including governmental initiatives, speculative activities, and others, the market for neuromorphic chips is expanding in the area. For instance, the Department of Energy (DOE) reported funding of USD 2 million for five crucial exploration projects to advance neuromorphic registration in September 2020. The DOE's initiative supports the development of software and hardware for mind-enhanced neuromorphic registering.

The industry is also growing as a result of the scaling down of neuromorphic chips that help in numerous applications. For instance, MIT engineers expected to release a mind-on-chip in June 2020 that was smaller than a piece of confetti and was made using a sizable number of memristors. Such chips can be utilised in portable AI devices. The Canadian government is also focusing on artificial intelligence technology, which will support advancements in neuromorphic figuring throughout the coming years. For instance, the legislatures of Canada and Quebec joined together in June 2020 to foster the thoughtful development of AI. The focus will be on a variety of topics, including solid AI, commercialization, information management, and future work and development.

There is a tremendous amount of interest in organisations' innovative work exercises. For instance, Intel and Sandia National Laboratories collaborated in October 2020 to study the value of neuromorphic processing for escalating computational issues. Sandia National Laboratories is one of three National Nuclear Security Administration innovative work research centres in the United States.

The penetration of neural-based chipsets in well-known applications is another factor driving market expansion. For instance, one of the most important technology companies, Apple, released the M1 Chip in November 2020 with the specific purpose of being used with its Mac products. The M1 Chip speeds up AI assignments and transports Apple Neural Engine to the Mac. The 16-center architecture can complete 11 trillion tasks per second, enabling up to 15 times faster ML execution.

Author details

Khaled S. Ahmed[1*] and Fayroz F. Shereif[2]

1 Faculty of Engineering, Biomedical Department, Benha University, Egypt

2 Computers and Systems Department, Electronics Research Institute, Egypt

*Address all correspondence to: khaled.sayed@bhit.bu.edu.eg

References

[1] Sörnmo L, Laguna P. Bioelectrical Signal Processing in Cardiac and Neurological Applications. Vol. 8. Academic Press; 2005

[2] Zidan MA, Strachan JP, Lu WD. The future of electronics based on memristive systems. Nature Electronics. 2018;**1**:22-29

[3] Yu S. Neuro-Inspired Computing Using Resistive Synaptic Devices. Berlin/Heidelberg, Germany: Springer; 2017

[4] Goldberg DH, Cauwenberghs G, Andreou AG. Probabilistic synaptic weighting in a reconfigurable network of VLSI integrate-and-fire neurons. Neural Networks. 2001;**14**:781-793

[5] An H, Bai K, Yi Y. The Roadmap to Realize Memristive Three-Dimensional Neuromorphic Computing System. 2018. 10.5772/intechopen.78986

[6] Choi S, Ham S, Wang G. Memristor synapses for neuromorphic computing. In: Memristors-Circuits and Applications of Memristor Devices. London, UK: IntechOpen; 2019

[7] Camuñas-MesaLA,Linares-BarrancoB, Serrano-Gotarredona T. Neuromorphic spiking neural networks and their memristor-CMOS hardware implementations. Materials. 2019;**12**:2745

[8] Priestley A. Emerging technology analysis: Neuromorphic computing. Nanotechnology. 2018;**30**:032001

[9] Fowers J, Ovtcharov K, Papamichael M, Massengill T, Liu M, Lo D, et al. A configurable cloud-scale DNN processor for real-time AI. In: Proceedings of the 2018 ACM/IEEE Press 45th Annual International Symposium on Computer Architecture (ISCA); 1-6 June 2018; Los Angeles, CA, USA. pp. 1-14

[10] Ma D, Shen J, Gu Z, Zhang M, Zhu X, Xu X, et al. Darwin: A neuromorphic hardware co-processor based on spiking neural networks. Journal of Systems Architecture. 2017;**77**:43-51

[11] Jiao Y, Han L, Jin R, Su YJ, Ho C, Yin L, et al. 7.2 A 12nm Programmable Convolution-Efficient Neural-Processing-Unit Chip Achieving 825TOPS. In: Proceedings of the 2020 IEEE Press International Solid-State Circuits Conference-(ISSCC); 16-20 February 2020; San Francisco, CA, USA. pp. 136-140

[12] Corinto F, Civalleri PP, Chua LO. A theoretical approach to memristor devices. EEE Journal on Emerging and Selected Topics in Circuits and Systems. 2015;**5**:123-132

[13] Pei J et al. Towards artificial general intelligence with hybrid Tianjic chip architecture. Nature. 2019;**572**:106-111

[14] Merolla PA et al. A million spiking-neuron integrated circuit with a scalable communication network and interface. Science. 2014;**345**:668-673

[15] Davies M et al. Loihi: A neuromorphic manycore processor with on-chip learning. IEEE Micro. 2018;**38**:82-99

[16] Moradi S, Qiao N, Stefanini F, Indiveri G. A scalable multicore architecture with heterogeneous memory structures for dynamic neuromorphic asynchronous processors (DYNAPs). IEEE Transactions on Biomedical Circuits and Systems. 2017;**12**:106-122

[17] Benjamin BV et al. Neurogrid: A mixed-analog-digital multichip system for large-scale neural simulations. Proceedings of the IEEE. 2014;**102**:699-716

[18] Thakur CS et al. Large-scale neuromorphic spiking array processors: A quest to mimic the brain. Frontiers in Neuroscience. 2018;**12**:891

[19] Schemmel J, Billaudelle S, Dauer P, Weis J. Accelerated analog neuromorphic computing. 2020. Preprint at https://arxiv.org/abs/2003.11996

[20] Bohnstingl T, Scherr F, Pehle C, Meier K, Maass W. Neuromorphic hardware learns to learn. Frontiers in Neuroscience. 2019;**13**:483

[21] Islam R et al. Device and materials requirements for neuromorphic computing. Journal of Physics D. 2019;**52**:113001

[22] Nandakumar S, Kulkarni SR, Babu AV, Rajendran B. Building brain-inspired computing systems: Examining the role of nanoscale devices. IEEE Nanotechnology Magazine. 2018;**12**:19-35

[23] Najem JS et al. Memristive ion channel-doped biomembranes as synaptic mimics. ACS Nano. 2018;**12**:4702-4711

[24] Jo SH et al. Nanoscale memristor device as synapse in neuromorphic systems. Nano Letters. 2010;**10**:1297-1301

[25] Li Y, Wang Z, Midya R, Xia Q, Yang JJ. Review of memristor devices in neuromorphic computing: Materials sciences and device challenges. Journal of Physics D. 2018;**51**:503002

[26] Schuman CD et al. A survey of neuromorphic computing and neural networks in hardware. 2017. Preprint at https://arxiv.org/abs/1705.06963

[27] Markovic D, Grollier J. Quantum neuromorphic computing. Applied Physics Letters. 2020;**117**:150501

[28] Goteti US, Zaluzhnyy IA, Ramanathan S, et al. Low-temperature emergent neuromorphic networks with correlated oxide devices. Proceedings of the National Academy of Sciences. 2021;**118**:e2103934118

[29] Wang G, Ma S, Wu Y, Pei J, Zhao R, Shi LP. End-to-end implementation of various hybrid neural networks on a cross-paradigm neuromorphic Chip. Frontiers in Neuroscience. 2021;**15**:615279. DOI: 10.3389/fnins.2021.615279

[30] Available from: https://newsroom.intel.com/news/intels-pohoiki-beach-64-chip-neuromorphic-system-delivers-breakthrough-results-research-tests/#gs.jaqzr2

[31] How IBM Got Brainlike Efficiency from the TrueNorth Chip. Available from: https://spectrum.ieee.org/how-ibm-got-brainlike-efficiency-from-the-truenorth-chip

[32] Song KM, Jeong J-S, Pan B, Zhang X, Xia J, Cha S, et al. Skyrmion-based artificial synapses for neuromorphic computing. Nature Electronics. 2020;**3**(3):148-155. arXiv:1907.00957. DOI: 10.1038/s41928-020-0385-0. S2CID195767210

[33] Neuromorphic computing: The long path from roots to real life. 2020. Available from: https://venturebeat.com/ai/neuromorphic-computing-the-long-path-from-roots-to-real-life/

[34] The brain's Architecture, Efficiency... on a Chip. IBM Research Blog. 2016. Available from: https://www.ibm.com/blogs/research/2016/12/the-brains-architecture-efficiency-on-a-chip/

[35] IBM Research: Brain-inspired Chip. 2021. Available from: https://research.ibm.com/blog; www.research.ibm.com

[36] Andreou AG, Dykman AA, Fischl KD, Garreau G, Mendat DR,

Orchard G, et al. Real-time sensory information processing using the TrueNorth Neurosynaptic System. In: 2016 IEEE Press International Symposium on Circuits and Systems (ISCAS): 2911. 2016. DOI: 10.1109/ISCAS.2016.7539214. ISBN 978-1-4799-5341-7. S2CID 29335047

[37] Tsuruoka T, Hasegawa T, Aono M. Synaptic plasticity and memristive behavior operated by atomic switches. In: Cellular Nanoscale Networks and their Applications (CNNA), 2014 14th International Workshop on. IEEE; 2014. pp. 1-2

[38] Avizienis AV, Sillin HO, Martin-Olmos C, Shieh HH, Aono M, Stieg AZ, et al. Neuromorphic atomic switch networks. PLoS One. 2012;7(8):e42772

[39] Hasegawa T, Ohno T, Terabe K, Tsuruoka T, Nakayama T, Gimzewski JK, et al. Learning abilities achieved by a single solid-state atomic switch. Advanced Materials. 2010;**22**(16):1831-1834

[40] Stieg AZ, Avizienis AV, Sillin HO, Aguilera R, Shieh H-H, Martin-Olmos C, et al. Self-organization and emergence of dynamical structures in neuromorphic atomic switch networks. In: Memristor Networks. Springer; 2014. pp. 173-209

[41] Nayak A, Ohno T, Tsuruoka T, Terabe K, Hasegawa T, Gimzewski JK, et al. Controlling the synaptic plasticity of a cu2s gap-type atomic switch. Advanced Functional Materials. 2012;**22**(17):3606-3613

[42] Tsuruoka T, Hasegawa T, Terabe K, Aono M. Conductance quantization and synaptic behavior in a ta2o5-based atomic switch. Nanotechnology. 2012;**23**(43):435705

[43] Yang R, Terabe K, Yao Y, Tsuruoka T, Hasegawa T, Gimzewski JK, et al. Synaptic plasticity and memory functions achieved in a wo3- x-based nanoionics device by using the principle of atomic switch operation. Nanotechnology. 2013;**24**(38):384003

[44] Suri M, Bichler O, Querlioz D, Palma G, Vianello E, Vuillaume D, et al. Cbram devices as binary synapses for low-power stochastic neuromorphic systems: Auditory (cochlea) and visual (retina) cognitive processing applications. In: Electron Devices Meeting (IEDM), 2012 IEEE International. Vol. 2012. IEEE. pp. 10-13

[45] Palma G, Suri M, Querlioz D, Vianello E, De Salvo B. Stochastic neuron design using conductive bridge ram. In: Nanoscale Architectures (NANOARCH), 2013 IEEE/ACM International Symposium on. IEEE; 2013. pp. 95-100

[46] Suri M, Parmar V. Exploiting intrinsic variability of filamentary resistive memory for extreme learning machine architectures. Nanotechnology, IEEE Transactions on. 2015;**14**(6):963-968

[47] DeSalvo B, Vianello E, Thomas O, Clermidy F, Bichler O, Gamrat C, et al. Emerging resistive memories for low power embedded applications and neuromorphic systems. In: Circuits and Systems (ISCAS), 2015 IEEE International Symposium on. Vol. 2015. IEEE. pp. 3088-3091

[48] Suri M, Querlioz D, Bichler O, Palma G, Vianello E, Vuillaume D, et al. Bio-inspired stochastic computing using binary cbram synapses. Electron Devices, IEEE Transactions on. 2013;**60**(7):2402-2409

[49] Clermidy F, Heliot R, Valentian A, Gamrat C, Bichler O, Duranton M, et al. Advanced technologies for braininspired computing. In: Design Automation Conference (ASP-DAC), 2014 19th Asia and South Pacific. Vol. 2014. IEEE. pp. 563-569

[50] Roclin D, Bichler O, Gamrat C, Klein J-O. Sneak paths effects in cbram memristive devices arrays for spiking neural networks. In: Proceedings of the 2014 IEEE/ACM International Symposium on Nanoscale Architectures. ACM; 2014. pp. 13-18

[51] DeSalvo B, Vianello E, Garbin D, Bichler O, Perniola L. From memory in our brain to emerging resistive memories in neuromorphic systems. In: Memory Workshop (IMW), 2015 IEEE International. Vol. 2015. IEEE. pp. 1-4

[52] Jang JW, Attarimashalkoubeh B, Prakash A, Hwang H, Jeong YH. "S calable neuron circuit using conductive-bridge ram for pattern reconstructions". IEEE Transactions on Electron Devices. vol. 99. 2016. pp. 1-4

[53] Mahalanabis D, Sivaraj M, Chen W, Shah S, Barnaby H, Kozicki M, et al. Demonstration of spike timing dependent plasticity in cbram devices with silicon neurons. In: Circuits and Systems (ISCAS), 2016 IEEE International Symposium on. Vol. 2016. IEEE. pp. 2314-2317

[54] Yu S, Wong H-S. Modeling the switching dynamics of programmable-metallization-cell (pmc) memory and its application as synapse device for a neuromorphic computation system. In: Electron Devices Meeting (IEDM), 2010 IEEE International. Vol. 2010. IEEE. pp. 22-21

[55] Mahalanabis D, Barnaby H, Gonzalez-Velo Y, Kozicki M, Vrudhula S, Dandamudi P. Incremental resistance programming of programmable metallization cells for use as electronic synapses. Solid-State Electronics. 2014;**100**:39-44

[56] La Barbera S, Vincent A, Vuillaume D, Querlioz D, Alibart F. Short-term to long-term plasticity transition in filamentary switching for memory applications. In: Memristive Systems (MEMRISYS) 2015 International Conference on. IEEE; 2015. pp. 1-2

[57] La Barbera S, Vincent AF, Vuillaume D, Querlioz D, Alibart F. Interplay of multiple synaptic plasticity features in filamentary memristive devices for neuromorphic computing. Scientific Reports. 2016;**6**

[58] La Barbera S, Vuillaume D, Alibart F. Filamentary switching: Synaptic plasticity through device volatility. ACS Nano. 2015;**9**(1):941-949

[59] Covi E, Brivio S, Serb A, Prodromakis T, Fanciulli M, Spiga S. Hfo2-based memristors for neuromorphic applications. In: Circuits and Systems (ISCAS), 2016 IEEE International Symposium on. Vol. 2016. IEEE. pp. 393-396

[60] Gao B, Bi Y, Chen H-Y, Liu R, Huang P, Chen B, et al. Ultra-low-energy three-dimensional oxide-based electronic synapses for implementation of robust highaccuracy neuromorphic computation systems. ACS Nano. 2014;**8**(7):6998-7004

[61] Gao B, Liu L, Kang J. Investigation of the synaptic device based on the resistive switching behavior in hafnium oxide. Progress in Natural Science: Materials International. 2015;**25**(1):47-50

[62] Jha R, Mandal S. Nanoelectronic synaptic devices and materials for brain-inspired computational architectures. In: SPIE NanoScience+ Engineering. International Society for Optics and Photonics; 2014. pp. 91 740S-91 740S

[63] Matveyev Y, Kirtaev R, Fetisova A, Zakharchenko S, Negrov D, Zenkevich A. Crossbar nanoscale hfo 2-based electronic synapses. Nanoscale Research Letters. 2016;**11**(1):1

[64] Yu S, Wu Y, Jeyasingh R, Kuzum D, Wong H-S. An electronic synapse device based on metal oxide resistive switching memory for neuromorphic computation. Electron Devices, IEEE Transactions on. 2011;**58**(8):2729-2737

[65] Matveyev Y, Egorov K, Markeev A, Zenkevich A. Resistive switching and synaptic properties of fully atomic layer deposition grown tin/hfo2/tin devices. Journal of Applied Physics. 2015;**117**(4):044901

[66] Woo J, Moon K, Song J, Kwak M, Park J, Hwang H. Optimized programming scheme enabling linear potentiation in filamentary hfo 2 rram synapse for neuromorphic systems. IEEE Transactions on Electron Devices. 2016;**63**(12):5064-5067

[67] Jia H, Deng N, Pang H. Threshold adaptive transistor realized with rrams for neuromorphic circuits. In: Junction Technology (IWJT), 2014 International Workshop on. Vol. 2014. IEEE. pp. 1-4

[68] Demin V, Emelyanov A, Lapkin D, Erokhin V, Kashkarov P, Kovalchuk M. Neuromorphic elements and systems as the basis for the physical implementation of artificial intelligence technologies. Crystallography Reports. 2016;**61**(6):992-1001

[69] Dongale T, Desai N, Khot K, Mullani N, Pawar P, Tikke R, et al. Effect of surfactants on the data directionality and learning behaviour of al/tio2/fto thin film memristor-based electronic synapse. Journal of Solid State Electrochemistry. 2016;**49**:1-5

[70] Hu M, Wang Y, Qiu Q, Chen Y, Li H. The stochastic modeling of tio 2 memristor and its usage in neuromorphic system design. In: Design Automation Conference (ASP-DAC), 2014 19th Asia and South Pacific. Vol. 2014. IEEE. pp. 831-836

[71] Hu X, Feng G, Li H, Chen Y, Duan S. An adjustable memristor model and its application in small-world neural networks. In: Neural Networks (IJCNN), 2014 International Joint Conference on. IEEE; 2014. pp. 7-14

[72] Park J, Kwak M, Moon K, Woo J, Lee D, Hwang H. Tio x-based rram synapse with 64-levels of conductance and symmetric conductance change by adopting a hybrid pulse scheme for neuromorphic computing. IEEE Electron Device Letters. 2016;**37**(12):1559-1562

[73] O'Kelly CJ, Fairfield JA, McCloskey D, Manning HG, Donegan JF, Boland JJ. Associative enhancement of time correlated response to heterogeneous stimuli in a neuromorphic nanowire device. Advanced Electronic Materials. 2016;**3**:1-6

[74] Chang T, Sheridan P, Lu W. Modeling and implementation of oxide memristors for neuromorphic applications. In: 2012 13th International Workshop on Cellular Nanoscale Networks and their Applications. 2012. pp. 1-3

[75] Du C, Ma W, Chang T, Sheridan P, Lu WD. Biorealistic implementation of synaptic functions with oxide memristors through internal ionic dynamics. Advanced Functional Materials. 2015;**25**(27):4290-4299

[76] Tan Z-H, Yang R, Terabe K, Yin X-B, Zhang X-D, Guo X. Synaptic metaplasticity realized in oxide memristive devices. Advanced Materials. 2015;**28**:377-384

[77] Shi T, Yin X-B, Yang R, Guo X. Pt/wo 3/fto memristive devices with recoverable pseudo-electroforming for time-delay switches in neuromorphic computing. Physical Chemistry Chemical Physics. 2016;**18**(14):9338-9343

[78] Thakoor S, Moopenn A, Daud T, Thakoor A. Solid-state thinfilm memistor

for electronic neural networks. Journal of Applied Physics. 1990;**67**(6):3132-3135

[79] Chang Y-F, Fowler B, Chen Y-C, Zhou F, Pan C-H, Chang T-C, et al. Demonstration of synaptic behaviors and resistive switching characterizations by proton exchange reactions in silicon oxide. Scientific Reports. 2016;**6**

[80] Guo L, Wan Q, Wan C, Zhu L, Shi Y. Short-term memory to long-term memory transition mimicked in izo homojunction synaptic transistors. Electron Device Letters. 2013;**34**(12):1581-1583

[81] Gao L, Wang I-T, Chen P-Y, Vrudhula S, Seo J-S, Cao Y, et al. Fully parallel write/read in resistive synaptic array for accelerating on-chip learning. Nanotechnology. 2015;**26**(45):455204

[82] Wang Y-F, Lin Y-C, Wang I-T, Lin T-P, Hou T-H. Characterization and modeling of nonfilamentary ta/taox/tio2/ti analog synaptic device. Scientific Reports. 2015;**5**

[83] Hu S, Liu Y, Chen T, Liu Z, Yu Q, Deng L, et al. Emulating the ebbinghaus forgetting curve of the human brain with a nio-based memristor. Applied Physics Letters. 2013;**103**(13):133701

[84] Hu SG, Liu Y, Chen T, Liu Z, Yu Q, Deng L, et al. Emulating the paired-pulse facilitation of a biological synapse with a NiOx-based memristor. Applied Physics Letters. 2013;**102**. DOI: 10.1063/1.4804374

[85] Hu S, Liu Y, Liu Z, Chen T, Yu Q, Deng L, et al. Synaptic long-term potentiation realized in pavlov's dog model based on a niox-based memristor. Journal of Applied Physics. 2014;**116**(21): 214502

[86] Yang X, Cai Y, Zhang Z, Yu M, Huang R. An electronic synapse based on tantalum oxide material. In: 2015 15th Non-Volatile Memory Technology Symposium (NVMTS). Vol. 2015. IEEE. pp. 1-4

[87] Wang Z, Yin M, Zhang T, Cai Y, Wang Y, Yang Y, et al. Engineering incremental resistive switching in tao x based memristorsfor brain-inspired computing. Nanoscale. 2016;8:14015-14022

[88] Thomas A, Niehörster S, Fabretti S, Shepheard N, Kuschel O, Küpper K, et al. Tunnel junction based memristors as artificial synapses. Frontiers in Neuroscience. 2015;**9**

[89] Wang C, He W, Tong Y, Zhao R. Investigation and manipulation of different analog behaviors of memristor as electronic synapse for neuromorphic applications. Scientific Reports. 2016;**6**

[90] Wu Y, Yu S, Wong H-S, Chen Y-S, Lee H-Y, Wang S-M, et al. Alox-based resistive switching device with gradual resistance modulation for neuromorphic device application. In: Memory Workshop (IMW), 2012 4th IEEE International. Vol. 2012. IEEE. pp. 1-4

[91] Sarkar B, Lee B, Misra V. Understanding the gradual reset in pt/al2o3/ni rram for synaptic applications. Semiconductor Science and Technology. 2015;**30**(10):105014

[92] Wang L-G, Zhang W, Chen Y, Cao Y-Q, Li A-D, Wu D. Synaptic plasticity and learning behaviors mimicked in single inorganic synapses of pt/hfox/znox/tin memristive system. Nanoscale Research Letters. 2017;**12**(1):65

[93] Moon K, Cha E, Park J, Gi S, Chu M, Baek K, et al. High density neuromorphic system with mo/pr0. 7ca0. 3mno3 synapse and nbo2 imt oscillator neuron. In: 2015 IEEE International Electron

Devices Meeting (IEDM). Vol. 2015. IEEE. pp. 17-16

[94] Jang J-W, Park S, Jeong Y-H, Hwang H. Reram-based synaptic device for neuromorphic computing. In: Circuits and Systems (ISCAS), 2014 IEEE International Symposium on. Vol. 2014. IEEE. pp. 1054-1057

[95] Jang J-W, Park S, Burr GW, Hwang H, Jeong Y-H. Optimization of conductance change in pr 1–x ca x mno 3-based synaptic devices for neuromorphic systems. Electron Device Letters, IEEE. 2015;**36**(5):457-459

[96] Lee D, Park J, Moon K, Jang J, Park S, Chu M, et al. Oxide based nanoscale analog synapse device for neural signal recognition system. In: 2015 IEEE International Electron Devices Meeting (IEDM). Vol. 2015. IEEE. pp. 4-7

[97] Moon K, Cha E, Lee D, Jang J, Park J, Hwang H. Rerambased analog synapse and imt neuron device for neuromorphic system. In: VLSI Technology, Systems and Application (VLSI-TSA), 2016 International Symposium on. Vol. 2016. IEEE. pp. 1-2

[98] Moon K, Cha E, Park J, Gi S, Chu M, Baek K, et al. Analog synapse device with 5-b mlc and improved data retention for neuromorphic system. IEEE Electron Device Letters. 2016;**37**(8):1067-1070

[99] Krzysteczko P, Münchenberger J, Schäfers M, Reiss G, Thomas A. The memristive magnetic tunnel junction as a nanoscopic synapse-neuron system. Advanced Materials. 2012;**24**(6):762-766

[100] Li Y, Zhong Y, Xu L, Zhang J, Xu X, Sun H, et al. Ultrafast synaptic events in a chalcogenide memristor. Scientific Reports. 2013;**3**

[101] Li Y, Zhong Y, Zhang J, Xu L, Wang Q, Sun H, et al. Activity-dependent synaptic plasticity of a chalcogenide electronic synapse for neuromorphic systems. Scientific Reports. 2014;**4**

[102] Tranchant J, Janod E, Corraze B, Stoliar P, Rozenberg M, Besland M-P, et al. Control of resistive switching in am4q8 narrow gap Mott insulators: A first step towards neuromorphic applications. Physica Status Solidi (a). 2015;**212**(2):239-244

[103] Chen Y, Liu G, Wang C, Zhang W, Li R-W, Wang L. Polymer memristor for information storage and neuromorphic applications. Materials Horizons. 2014;**1**(5):489-506

[104] Juarez-Hernandez LJ, Cornella N, Pasquardini L, Battistoni S, Vidalino L, Vanzetti L, et al. Bio-hybrid interfaces to study neuromorphic functionalities: New multidisciplinary evidences of cell viability on poly (anyline)(pani), a semiconductor polymer with memristive properties. Biophysical Chemistry. 2016;**208**:40-47

[105] Li S, Zeng F, Chen C, Liu H, Tang G, Gao S, et al. Synaptic plasticity and learning behaviours mimicked through ag interface movement in an ag/conducting polymer/ta memristive system. Journal of Materials Chemistry C. 2013;**1**(34):5292-5298

[106] Luo W, Yuan F-Y, Wu H, Pan L, Deng N, Zeng F, et al. Synaptic learning behaviors achieved by metal ion migration in a cu/pedot: Pss/ta memristor. In: 2015 15th Non-Volatile Memory Technology Symposium (NVMTS). Vol. 2015. IEEE. pp. 1-4

[107] Luo W, Wu X, Yuan F-Y, Wu H, Pan L, Deng N. Synaptic learning behavior based on a ag/pedot: Pss/ta memristor. In: Next-Generation Electronics (ISNE), 2016 5th International Symposium on. Vol. 2016. IEEE. pp. 1-2

[108] Nawrocki R, Voyles RM, Shaheen SE, et al. Neurons in polymer: Hardware neural units based on polymer memristive devices and polymer transistors. Electron Devices, IEEE Transactions on. 2014;**61**(10):3513-3519

[109] Xiao Z, Huang J. Energy-efficient hybrid perovskite memristors and synaptic devices. Advanced Electronic Materials. 2016;**2**:1-8

[110] Yang X, Wang C, Shang J, Zhang C, Tan H, Yi X, et al. An organic terpyridyl-iron polymer based memristor for synaptic plasticity and learning behavior simulation. RSC Advances. 2016;**6**(30):25179-25184

[111] Zhang C, Tai Y-T, Shang J, Liu G, Wang K-L, Hsu C, et al. Synaptic plasticity and learning behaviours in flexible artificial synapse based on polymer/viologen system. Journal of Materials Chemistry C. 2016;**4**(15):3217-3223

[112] Bennett CH, Chabi D, Cabaret T, Jousselme B, Derycke V, Querlioz D, et al. Supervised learning with organic memristor devices and prospects for neural crossbar arrays. In: Nanoscale Architectures (NANOARCH), 2015 IEEE/ACM International Symposium on. Vol. 2015. IEEE. pp. 181-186

[113] Cabaret T, Fillaud L, Jousselme B, Klein J-O, Derycke V. Electro-grafted organic memristors: Properties and prospects for artificial neural networks based on stdp. In: Nanotechnology (IEEENANO), 2014 IEEE 14th International Conference on. IEEE; 2014. pp. 499-504

[114] Chang C, Zeng F, Li X, Dong W, Lu S, Gao S, et al. Simulation of synaptic short-term plasticity using ba (cf3so3) 2- doped polyethylene oxide electrolyte film. Scientific Reports. 2016;**6**

[115] Erokhin V, Berzina T, Smerieri A, Camorani P, Erokhina S, Fontana MP. Bio-inspired adaptive networks based on organic memristors. Nano Communication Networks. 2010;**1**(2):108-117

[116] Erokhin V. Organic memristive devices: Architecture, properties and applications in neuromorphic networks. In: Electronics, Circuits, and Systems (ICECS), 2013 IEEE 20th International Conference on. IEEE; 2013. pp. 305-308

[117] Erokhina S. Layer-by-layer technique for the fabrication of organic memristors and neuromorphic systems. In: Memristive Systems (MEMRISYS) 2015 International Conference on. IEEE; 2015. pp. 1-2

[118] Kim C-H, Sung S, Yoon M-H. Synaptic organic transistors with a vacuum-deposited charge-trapping nanosheet. Scientific Reports. 2016;**6**

[119] Kong L-A, Sun J, Qian C, Gou G, He Y, Yang J, et al. Iongel gated field-effect transistors with solution-processed oxide semiconductors for bioinspired artificial synapses. Organic Electronics. 2016;**39**:64-70

[120] Lin Y-P, Bennett CH, Cabaret T, Vodenicarevic D, Chabi D, Querlioz D, et al. Physical realization of a supervised learning system built with organic memristive synapses. Scientific Reports. 2016;**6**

[121] Liu G, Wang C, Zhang W, Pan L, Zhang C, Yang X, et al. Organic biomimicking memristor for information storage and processing applications. Advanced Electronic Materials. 2015;**2**:1-8

[122] Nawrocki RA, Voyles RM, Shaheen SE. Simulating hardware neural networks with organic memristors and organic field effect transistors. Intelligent

Engineering Systems through Artificial Neural Networks. 2010;**20**

[123] Wang L, Wang Z, Lin J, Yang J, Xie L, Yi M, et al. Long-term homeostatic properties complementary to hebbian rules in cupc-based multifunctional memristor. Scientific Reports. 2016;**6**

[124] Ishiwara H. Proposal of adaptive-learning neuron circuits with ferroelectric analog-memory weights. Japanese Journal of Applied Physics. 1993;**32**(1S):442

[125] Yoon S-M, Tokumitsu E, Ishiwara H. An electrically modifiable synapse array composed of metal-ferroelectric-semiconductor (mfs) fet's using srbi/sub 2/ta/sub 2/o/sub 9/thin films. Electron Device Letters, IEEE. 1999;**20**(5):229-231

[126] Yoon S-M, Tokumitsu E, Ishiwara H. Adaptive-learning neuron integrated circuits using metal-ferroelectric (SrBi/sub 2/Ta/sub 2/O/sub 9/)-semiconductor (MFS) FET's. IEEE Electron Device Letters. 1999;**20**(10):526-528. DOI: 10.1109/55.791931

[127] Yoon SM, Tokumitsu E, Ishiwara H. Realization of adaptive learning function in a neuron circuit using metal/ferroelectric (SrBi2Ta2O9)/semiconductor field effect transistor (MFSFET). Japanese Journal of Applied Physics. 1999;**38**(4 B):2289-2293. DOI: 10.1143/jjap.38.2289

[128] Yoon SM, Tokumitsu E, Ishiwara H. Ferroelectric neuron integrated circuits using SrBi2Ta2O9-gate FET's and CMOS Schmitt-Trigger oscillators. IEEE Transactions on Electron Devices. 2000;**47**(8):1630-1635. DOI: 10.1109/16.853041

[129] Yoon S-MYS-M et al. Japanese Journal of Applied Physics. 2000;**39**:2119

[130] Kim E, Kim K, Yoon S. Investigation of the ferroelectric switching behavior of p (vdf-trfe)-pmma blended films for synaptic device applications. Journal of Physics D: Applied Physics. 2016;**49**(7):075105

[131] Nishitani Y, Kaneko Y, Ueda M, Morie T, Fujii E. Threeterminal ferroelectric synapse device with concurrent learning function for artificial neural networks. Journal of Applied Physics. 2012;**111**(12):124108

[132] Nishitani Y, Kaneko Y, Ueda M, Fujii E, Tsujimura A. Dynamic observation of brain-like learning in a ferroelectric synapse device. Japanese Journal of Applied Physics. 2013;**52**(4S):04CE06

[133] Yoon S-M, Ishiwara H. Adaptive-learning synaptic devices using ferroelectric-gate field-effect transistors for neuromorphic applications. In: Ferroelectric-Gate Field Effect Transistor Memories. Springer; 2016. pp. 311-333

[134] Nishitani Y, Kaneko Y, Ueda M. Artificial synapses using ferroelectric memristors embedded with cmos circuit for image recognition. In: Device Research Conference (DRC), 2014 72nd Annual. Vol. 2014. IEEE. pp. 297-298

[135] Nishitani Y, Kaneko Y, Ueda M. Supervised learning using spike-timing-dependent plasticity of memristive synapses. IEEE Transactions on Neural Networks and Learning Systems. 2015;**26**(12):2999-3008. DOI: 10.1109/TNNLS.2015.2399491

[136] Wang Z, Zhao W, Kang W, Zhang Y, Klein J-O, Ravelosona D, et al. Compact modelling of ferroelectric tunnel memristor and its use for neuromorphic simulation. Applied Physics Letters. 2014;**104**(5):053505

[137] Tian H, Mi W, Wang X-F, Zhao H, Xie Q-Y, Li C, et al. Graphene dynamic

synapse with modulatable plasticity. Nano Letters. 2015;**15**(12):8013-8019

[138] Wang L, Wang Z, Zhao W, Hu B, Xie L, Yi M, et al. Controllable multiple depression in a grapheme oxide artificial synapse. Advanced Electronic Materials. 2017;**3**(1)

[139] Wan CJ, Liu YH, Feng P, Wang W, Zhu LQ, Liu ZP, et al. Flexible metal oxide/graphene oxide hybrid neuromorphic transistors on flexible conducting graphene substrates. Advanced Materials. 2016;**28**:5878-5885

[140] Wan CJ, Zhu LQ, Liu YH, Feng P, Liu ZP, Cao HL, et al. Proton-conducting graphene oxide-coupled neuron transistors for brain-inspired cognitive systems. Advanced Materials. 2016;**28**(18):3557-3563

[141] Yang Y, Wen J, Guo L, Wan X, Du P, Feng P, et al. Long-term synaptic plasticity emulated in modified graphene oxide electrolyte gated izo-based thin-film transistors. ACS Applied Materials & Interfaces. 2016;**8**(44):30281-30286

[142] Darwish M, Calayir V, Pileggi L, Weldon J. Ultra-Compact Graphene Multigate Variable Resistor for Neuromorphic Computing. IEEE Transactions on Nanotechnology. 2016;**15**:1-1. DOI: 10.1109/TNANO.2016.2525039

[143] Hsu C-C, Parker AC, Joshi J. Dendritic computations, dendritic spiking and dendritic plasticity in nanoelectronic neurons. In: Circuits and Systems (MWSCAS), 2010 53rd IEEE International Midwest Symposium on. Vol. 2010. IEEE. pp. 89-92

[144] Hsu C-C, Parker AC. Border ownership in a nanoneuromorphic circuit using nonlinear dendritic computations. In: Neural Networks (IJCNN), 2014 International Joint Conference on. IEEE; 2014. pp. 3442-3449

[145] Hsu C-C, Parker AC. Dynamic spike threshold and nonlinear dendritic computation for coincidence detection in neuromorphic circuits. In: 2014 36th Annual International Conference of the IEEE Engineering in Medicine and Biology Society. 2014. pp. 461-464

[146] Joshi J, Hsu C, Parker AC, Deshmukh P. A carbon nanotube cortical neuron with excitatory and inhibitory dendritic computations. In: IEEE Xplore/NIH LIfe Science Systems and Applications Workshop. 2009

[147] Parker AC, Joshi J, Hsu C-C, Singh NAD. A carbon nanotube implementation of temporal and spatial dendritic computations. In: Circuits and Systems, 2008. MWSCAS 2008. 51st Midwest Symposium on. Vol. 2008. IEEE. pp. 818-821

[148] Friesz AK, Parker AC, Zhou C, Ryu K, Sanders JM, Wong H-SP, et al. A biomimetic carbon nanotube synapse circuit. In: Biomedical Engineering Society (BMES) Annual Fall Meeting. Springer; Vol. 2(8). 2007. p. 29

[149] Barzegarjalali S, Parker AC. A hybrid neuromorphic circuit demonstrating schizophrenic symptoms. In: Biomedical Circuits and Systems Conference (BioCAS), 2015 IEEE. IEEE; 2015. pp. 1-4

[150] Barzegarjalali S, Parker AC. Neuromorphic circuit modeling directional selectivity in the visual cortex. Annu Int Conf IEEE Eng Med Biol Soc. 2016;**2016**:6130-6133. DOI: 10.1109/EMBC.2016.7592127

[151] Barzegarjalali S, Parker A. A neuromorphic circuit mimicking biological short-term memory. In: Conference Proceedings: ... Annual International Conference of the IEEE Engineering in Medicine and Biology Society. IEEE Engineering in Medicine and Biology Society. Conference.

Vol. 2016. 2016. pp. 1401-1404. DOI: 10.1109/EMBC.2016.7590970

[152] Barzegarjalali S, Parker AC. A bio-inspired electronic mechanism for unsupervised learning using structural plasticity. Future Technologies Conference (FTC). 2016;**2016**:806-815. DOI: 10.1109/FTC.2016.7821696

[153] Gacem K, Retrouvey J-M, Chabi D, Filoramo A, Zhao W, Klein J-O, et al. Neuromorphic function learning with carbon nanotube based synapses. Nanotechnology. 2013;**24**(38):384013

[154] Joshi J, Parker AC, Hsu C-C. A carbon nanotube cortical neuron with spike-timing-dependent plasticity. In: Engineering in Medicine and Biology Society, 2009. EMBC 2009. Annual International Conference of the IEEE. Vol. 2009. IEEE. pp. 1651-1654

[155] Joshi J, Zhang J, Wang C, Hsu CC, Parker AC, Zhou C, Ravishankar U. "A biomimetic fabricated carbon nanotube synapse for prosthetic applications." In: Life Science Systems and Applications Workshop (LiSSA), 2011 IEEE/NIH. IEEE; 2011. pp. 139-142

[156] Kim K, Chen C-L, Truong Q, Shen AM, Chen Y. A carbon nanotube synapse with dynamic logic and learning. Advanced Materials. 2013;**25**(12):1693-1698

[157] Kim K, Tudor A, Chen C-L, Lee D, Shen AM, Chen Y. Bioinspired neuromorphic module based on carbon nanotube/c60/polymer composite. Journal of Composite Materials. 2015;**49**:0021998315573559

[158] Kim S, Yoon J, Kim H-D, Choi S-J. Carbon nanotube synaptic transistor network for pattern recognition. ACS Applied Materials & Interfaces. 2015;7(45):25479-25486

[159] Liao S-Y, Retrouvey J-M, Agnus G, Zhao W, Maneux C, Fr'egon'ese S, et al. Design and modeling of a neuro-inspired learning circuit using nanotube-based memory devices. Circuits and Systems I: Regular Papers, IEEE Transactions on. 2011;**58**(9):2172-2181

[160] Mahvash M, Parker AC. Modeling intrinsic ion-channel and synaptic variability in a cortical neuromorphic circuit. In: Biomedical Circuits and Systems Conference (BioCAS), 2011 IEEE. IEEE; 2011. pp. 69-72

[161] Shen AM, Chen C-L, Kim K, Cho B, Tudor A, Chen Y. Analog neuromorphic module based on carbon nanotube synapses. ACS Nano. 2013;**7**(7):6117-6122

[162] Shen AM, Kim K, Tudor A, Lee D, Chen Y. Doping modulated carbon nanotube synapstors for a spike neuromorphic module. Small. 2015;**11**(13):1571-1579

[163] Yin C, Li Y, Wang J, Wang X, Yang Y, Ren T-L. Carbon nanotube transistor with short-term memory. Tsinghua Science and Technology. 2016;**21**(4): 442-448

[164] Zhao W, Agnus G, Derycke V, FiloramoA,BourgoinJ,GamratC.Nanotube devices based crossbar architecture: Toward neuromorphic computing. Nanotechnology. 2010;**21**(17):175202

[165] Feng P, Xu W, Yang Y, Wan X, Shi Y, Wan Q, et al. Printed neuromorphic devices based on printed carbon nanotube thinfilm transistors. Advanced Functional Materials. 2016;**10**

[166] Chen C, Kim K, Truong Q, Shen A, Li Z, Chen Y. A spiking neuron circuit based on a carbon nanotube transistor. Nanotechnology. 2012;**23**(27):275202

[167] Joshi J, Parker A, Hsu C. A carbon nanotube spiking cortical neuron with

tunable refractory period and spiking duration. In: IEEE Latin American Symp. on Circuits and Systems (LASCAS). 2010

[168] Mahvash M, Parker AC. Synaptic variability in a cortical neuromorphic circuit. Neural Networks and Learning Systems, IEEE Transactions on. 2013;**24**(3):397-409

[169] Najari M, El-Grour T, Jelliti S, Hakami OM, Al-Kamli A, Can N, et al. Simulation of a spiking neuron circuit using carbon nanotube transistors. In: AIP Conference Proceedings. Vol. 1742(1). AIP Publishing; 2016. p. 030013

[170] Kim H, Park J, Kwon M, Lee J, Park B. Silicon-based floating-body synaptic transistor with frequency dependent short- and long-term memories. Electron Device Letters, IEEE. 2016;**99**:1-1

[171] Kim H, Cho S, Sun M-C, Park J, Hwang S, Park B-G. Simulation study on silicon-based floating body synaptic transistor with short-and long-term memory functions and its spike timing-dependent plasticity. Journal of Semiconductor Technology and Science. 2016;**16**(5):657-663

[172] Shao F, Yang Y, Zhu LQ, Feng P, Wan Q. Oxide-based synaptic transistors gated by sol-gel silica electrolytes. ACS Applied Materials & Interfaces. 2016;**5**:3050-3055

[173] Shi J, Ha SD, Zhou Y, Schoofs F, Ramanathan S. A correlated nickelate synaptic transistor. Nature Communications. 2013;**4**

[174] Wan CJ, Zhu LQ, Zhou JM, Shi Y, Wan Q. Memory and learning behaviors mimicked in nanogranular sio 2-based proton conductor gated oxide-based synaptic transistors. Nanoscale. 2013;**5**(21):10194-10199

[175] Wan C, Zhu L, Zhou J, Shi Y, Wan Q. Inorganic proton conducting electrolyte coupled oxide-based dendritic transistors for synaptic electronics. Nanoscale. 2014;**6**. DOI: 10.1039/c3nr05882d

[176] Wan C, Zhu L, Liu Y, Shi Y, Wan Q. Laterally coupled synaptic transistors gated by proton conducting sodium alginate films. Electron Device Letters, IEEE. 2014;**35**(6):672-674

[177] Wan X, Feng P, Wu GD, Shi Y, Wan Q. Simulation of laterally coupled ingazno 4-based electric-double-layer transistors for synaptic electronics. Electron Device Letters, IEEE. 2015;**36**(2):204-206

[178] Wan X, Yang Y, Feng P, Shi Y, Wan Q. Short-term plasticity and synaptic filtering emulated in electrolyte gated igzo transistors. Electron Device Letters, IEEE. 2016;**99**:1-1

[179] Wan C, Liu YH, Zhu LQ, Feng P, Shi Y, Wan Q. Shortterm synaptic plasticity regulation in solution-gated indium-galliumzinc- oxide electric-double-layer transistors. ACS Applied Materials & Interfaces. 2016;**15**:9762-9768

[180] Wang J, Li Y, Yin C, Yang Y, Ren T-L. Long-term depression mimicked in an igzo-based synaptic transistor. IEEE Electron Device Letters. 2016;**38**:191-194

[181] Zhou J, Liu N, Zhu L, Shi Y, Wan Q. Energy-efficient artificial synapses based on flexible igzo electric-double-layer transistors. Electron Device Letters, IEEE. 2015;**36**(2):198-200

[182] Zhu LQ, Xiao H, Liu YH, Wan CJ, Shi Y, Wan Q. Multigate synergic modulation in laterally coupled synaptic transistors. Applied Physics Letters. 2015;**107**(14):143502

[183] Zhu LQ, Wan CJ, Gao PQ, Liu YH, Xiao H, Ye JC, et al. Flexible proton-gated oxide synaptic transistors on si

membrane. ACS Applied Materials & Interfaces. 2016;**8**(33):21770-21775

[184] Zhou J, Wan C, Zhu L, Shi Y, Wan Q. Synaptic behaviors mimicked in flexible oxide-based transistors on plastic substrates. Electron Device Letters, IEEE. 2013;**34**(11):1433-1435

[185] Gkoupidenis P, Schaefer N, Garlan B, Malliaras GG. Neuromorphic functions in pedot: Pss organic electrochemical transistors. Advanced Materials. 2015;**27**(44):7176-7180

[186] Gkoupidenis P, Schaefer N, Strakosas X, Fairfield JA, Malliaras GG. Synaptic plasticity functions in an organic electrochemical transistor. Applied Physics Letters. 2015;**107**(26):263302

[187] Qian C, Sun J, Kong L-A, Gou G, Yang J, He J, et al. Artificial synapses based on in-plane gate organic electrochemical transistors. ACS Applied Materials & Interfaces. 2016;**8**(39):26169-26175

[188] Wan CJ, Zhu LQ, Wan X, Shi Y, Wan Q. Organic/inorganic hybrid synaptic transistors gated by proton conducting methylcellulose films. Applied Physics Letters. 2016;**108**(4):043508

[189] Wood R, Bruce I, Mascher P. Modeling of spiking analog neural circuits with hebbian learning, using amorphous semiconductor thin film transistors with silicon oxide nitride semiconductor split gates. In: Artificial Neural Networks and Machine Learning–ICANN 2012. Vol. 2012. Springer. pp. 89-96

[190] Xu W, Min S-Y, Hwang H, Lee T-W. Organic core-sheath nanowire artificial synapses with femtojoule energy consumption. Science Advances. 2016;**2**(6):e1501326

[191] Alibart F, Pleutin S, Gu'erin D, Novembre C, Lenfant S, Lmimouni K, et al. An organic nanoparticle transistor behaving as a biological spiking synapse. Advanced Functional Materials. 2010;**20**(2):330-337

[192] Alibart F, Pleutin S, Bichler O, Gamrat C, Serrano-Gotarredona T, Linares-Barranco B, et al. A memristive nanoparticle/ organic hybrid synapstor for neuroinspired computing. Advanced Functional Materials. 2012;**22**(3):609-616

[193] Bichler O, Zhao W, Alibart F, Pleutin S, Vuillaume D, Gamrat C. Functional model of a nanoparticle organic memory transistor for use as a spiking synapse. Electron Devices, IEEE Transactions on. 2010;**57**(11):3115-3122

[194] Kwon K-C, Lee J-S, Kim CG, Park J-G. Biological synapse behavior of nanoparticle organic memory field effect transistor fabricated by curing. Applied Physics Express. 2013;**6**(6):067001

[195] Liu YH, Zhu LQ, Feng P, Shi Y, Wan Q. Freestanding artificial synapses based on laterally proton-coupled transistors on chitosan membranes. Advanced Materials. 2015;**27**(37):5599-5604

[196] Wu G, Zhang J, Wan X, Yang Y, Jiang S. Chitosan-based biopolysaccharide proton conductors for synaptic transistors on paper substrates. Journal of Materials Chemistry C. 2014;**2**(31):6249-6255

[197] Wu G, Wan C, Zhou J, Zhu L, Wan Q. Low-voltage protonic/electronic hybrid indium zinc oxide synaptic transistors on paper substrates. Nanotechnology. 2014;**25**(9):094001

[198] Wu G, Feng P, Wan X, Zhu L, Shi Y, Wan Q. Artificial synaptic devices based on natural chicken albumen coupled electricdouble- layer transistors. Scientific Reports. 2016;**6**

[199] Zhou J, Liu Y, Shi Y, Wan Q. Solution-processed chitosangated izo-based transistors for mimicking synaptic plasticity. Electron Device Letters, IEEE. 2014;**35**(2):280-282

[200] Schemmel J, Grübl A, Hartmann S, Kononov A, Mayr C, Meier K, et al. Live demonstration: A scaled-down version of the brainscales wafer-scale neuromorphic system. In: Proceedings of the 2012 IEEE Xplore International Symposium on Circuits and Systems (ISCAS), Seoul, Korea, 20-23 May 2012.

[201] Liu D, Yu H, Chai Y. Low-power computing with neuromorphic engineering. Advanced Intelligent Systems. 2021;**3**:2000150

[202] Balaji A, Adiraju P, Kashyap HJ, Das A, Krichmar JL, Dutt ND, et al. PyCARL: A PyNN interface for hardwaresoftware co-simulation of spiking neural network. arXiv 2020, arXiv:2003.09696

[203] Shi L, Pei J, Deng N, Wang D, Deng L, Wang Y, et al. Development of a neuromorphic computing system. In: Proceedings of the 2015 IEEE International Electron Devices Meeting (IEDM); 7-9 December 2015; Washington, DC, USA.

[204] Chi P, Li S, Xu C, Zhang T, Zhao J, Liu Y, et al. PRIME: A novel processing-in-memory architecture for neural network computation in ReRAM-based main memory. ACM SIGARCH Computer Architecture News. 2016;**44**(27):39

[205] Xia Q, Yang JJ. Memristive crossbar arrays for brain-inspired computing. Nature Materials. 2019;**18**:309-323

[206] Chakraborty I, Jaiswal A, Saha A, Gupta S, Roy K. Pathways to efficient neuromorphic computing with non-volatile memory technologies. Applied Physics Reviews. 2020;7:021308

[207] Islam R, Li H, Chen PY, Wan W, Chen HY, Gao B, et al. Device and materials requirements for neuromorphic computing. Journal of Physics D: Applied Physics. 2019;**52**:113001

[208] Chen A. A review of emerging non-volatile memory (NVM) technologies and applications. Solid-State Electronics. 2016;**125**:25-38

[209] Strenz R. Review and outlook on embedded nvm technologies–from evolution to revolution. In: Proceedings of the 2020 IEEE Xplore International MemoryWorkshop (IMW); 17-20 May 2020; Dresden, Germany

[210] Burr GW, Sebastian A, Vianello E, Waser R, Parkin S. Emerging materials in neuromorphic computing: Guest editorial. APL Materials. 2020;**8**:010401

[211] Sangwan VK, Hersam MC. Neuromorphic nanoelectronic materials. Nature Nanotechnology. 2020;**15**:517-528

[212] Lv Z, Wang Y, Chen J, Wang J, Zhou Y, Han ST. Semiconductor quantum dots for memories and neuromorphic computing systems. Chemical Reviews. 2020;**120**:3941-4006

[213] Qi M, Zhang X, Yang L, Wang Z, Xu H, Liu W, et al. Intensity-modulated LED achieved through integrating p-GaN/n-ZnO heterojunction with multilevel RRAM. Applied Physics Letters. 2018;**113**:223503

[214] Roychowdhury V, Janes D, Bandyopadhyay S, Wang X. Collective computational activity in self-assembled arrays of quantum dots: A novel neuromorphic architecture for

nanoelectronics. IEEE Transactions on Electron Devices. 1996;**43**:1688-1699

[215] Feldmann J, Youngblood N, Wright CD, Bhaskaran H, Pernice WH. All-optical spiking neurosynaptic networks with self-learning capabilities. Nature. 2019;**569**:208-214

[216] Christensen DV et al. Neuromorph. Computer Engineering. 2022;**2**:022501

[217] Bi G-Q, Poo M-M. Synaptic modifications in cultured hippocampal neurons: Dependence on spike timing, synaptic strength, and post synaptic cell type. The Journal of Neuroscience. 1998;**18**(24):10464-10472

[218] Wittenberg GM, Wang SS-H. Malleability of spike-timing-dependent plasticity at the CA3–CA1 synapse. The Journal of Neuroscience. 2006;**26**(24):6610-6617. DOI: 10.1523/JNEUROSCI.5388-05.2006

[219] Abbott LF, Nelson SB. Synaptic plasticity: Taming the beast. Nature Neuroscience. 2000;**3**(Suppl):1178-1183. DOI: 10.1038/81453

[220] Gjorgjieva J, Clopath C, Audet J, Pfister J-P. A triplet spike-timing–dependent plasticity model generalizes the Bienenstock–Cooper–Munro rule to higher-order spatiotemporal correlations. Proceedings of the National Academy of Sciences of the United States of America. 2011;**108**(48):19383-19388. DOI: 10.1073/pnas.1105933108

[221] Rachmuth G, Shouval H-Z, Bear MF, Poon C-S. A biophysically-based neuromorphic model of spike rate- and timing-dependent plasticity. Proceedings of the National Academy of Sciences of the United States of America. 2011;**108**(49):E1266-E1274. DOI: 10.1073/pnas. 1106161108

[222] Indiveri G, Chicca E, Douglas R. A VLSI array of low-power spiking neurons and bistable synapses with spike-timing dependent plasticity. IEEE Transactions on Neural Networks. 2006;**17**(1):211-221. DOI: 10.1109/TNN.2005.860850

[223] Diorio C, Hasler P, Minch BA, Mead CA. A single-transistor silicon synapse. IEEE Transactions on Electron Devices. 1996;**43**(11):1972-1980. DOI: 10.1109/16. 543035

[224] Kuzum D, Jeyasingh RGD, Lee B, Wong H-SP. Nanoelectronic programmable synapses based on phase change materials for brain-inspired computing. Nano Letters. 2012;**12**(5):2179-2186. DOI: 10.1021/nl201040y

[225] Kim S, Ishii M, Lewis S, Perri T, BrightSky M, Kim W, et al. NVM neuromorphic core with 64kcell (256-by-256) phase change memory synaptic array with on-chip neuron circuits for continuous in-situ learning. IEDM Technical Digest. 2015:443-446. DOI: 10.1109/IEDM.2015.7409716

[226] Ambrogio S, Ciocchini N, Laudato M, Milo V, Pirovano A, Fantini P, et al. Unsupervised learning by spike timing dependent plasticity in phase change memory (PCM) synapses. Frontiers in Neuroscience. 2016;**10**:56. DOI: 10.3389/fnins.2016. 00056

[227] Ambrogio S, Balatti S, Nardi F, Facchinetti S, Ielmini D. Spike-timing dependent plasticity in a transistor-selected resistive switching memory. Nanotechnology. 2013;**24**:384012. DOI: 10.1088/0957-4484/24/38/384012

[228] Wang Z-Q, Ambrogio S, Balatti S, Ielmini D. A 2-transistor/1-resistor artificial synapse capable of communication and stochastic learning for neuromorphic systems. Frontiers in Neuroscience. 2015;**8**:438. DOI: 10.3389/fnins.2014.00438

[229] Ambrogio S, Balatti S, Milo V, Carboni R, Wang Z, Calderoni A, et al. Neuromorphic learning and recognition with one-transistor-one-resistor synapses and bistable metal oxide RRAM. IEEE Transactions on Electron Devices. 2016;**63**(4):1508-1515. DOI: 10.1109/TED.2016.2526647

[230] Wang Z, Joshi S, Savel'ev SE, Jiang H, Midya R, Lin P, et al. Memristors with diffusive dynamics as synaptic emulators for neuromorphic computing. Nature Materials. 2017;**16**:101-108. DOI: 10.1038/nmat4756

[231] Ohno T, Hasegawa T, Tsuruoka T, Terabe K, Gimzewski JK, Aono M. Short-term plasticity and long-term potentiation mimicked in single inorganic synapses. Nature Materials. 2011;**10**(8):591-595. DOI: 10.1038/nmat3054

[232] Ielmini D. Brain-inspired computing with resistive switching memory (RRAM): Devices, synapses and neural networks. Microelectronic Engineering. 2018;**190**. DOI: 10.1016/j.mee.2018.01.009

[233] Pedretti G, Milo V, Ambrogio S, Carboni R, Bianchi S, Calderoni A, et al. Memristive neural network for on-line learning and tracking with brain-inspired spike timing dependent plasticity. Scientific Reports. 2017;**7**:5288. DOI: 10.1038/s41598-017-05480-0

[234] Rojas R. Neural Networks: A Systematic Introduction. Springer; 1996

[235] Minsky M, Papert S. Perceptrons. MIT Press; 1969

[236] Sie C. Memory devices using bistable resistivity in amorphous As-Te-Ge films [Ph.D. thesis]. Ames, IA, USA: Iowa State University; 1969

[237] Sie C, Pohm A, Uttecht P, Kao A, Agrawal R. Chalcogenide glass bistable resistivity memory. IEEE MAG-6. 1970;**6**:592

[238] Sie C, Uttecht R, Stevenson H, Griener J, Raghavan K. Electricfield induced filament formation in As-Te-Ge glass. Journal of Non-Crystalline Solids. 1970;**2**:358-370

[239] Suri M, Bichler O, Querlioz D, Cueto O, Perniola L, Sousa V, et al. Phase change memory as synapse for ultra-dense neuromorphic systems: Application to complex visual pattern extraction. In: Proceedings of the IEEE Xplore 2011 International Electron Devices Meeting. 5-7 December 2011; Washington, DC, USA. p. 4

[240] Shelby RM, Burr GW, Boybat I, Di Nolfo C. Non-volatile memory as hardware synapse in neuromorphic computing: A first look at reliability issues. In: Proceedings of the IEEE Xplore International Reliability Physics Symposium; 19-23 April 2015; Monterey, CA, USA. pp. 6A-61A

[241] Yu S. Neuro-inspired computing with emerging nonvolatile memorys. Proceedings of the IEEE. 2018;**106**:260-285

[242] Suri M, Garbin D, Bichler O, Querlioz D, Vuillaume D, Gamrat C, et al. Impact of PCM resistance-drift in neuromorphic systems and drift-mitigation strategy. In: Proceedings of the IEEE Xplore/ACM International Symposium on Nanoscale Architectures (NANOARCH); 15-17 July 2013; Brooklyn, NY, USA. pp. 140-145

[243] Li J, Luan B, Lam C. Resistance drift in phase change memory. In: Proceedings of the IEEE International Reliability Physics Symposium (IRPS); 15-19 April 2012; Anaheim, CA, USA. p. 6C-1

[244] Ielmini D, Lavizzari S, Sharma D, Lacaita AL. Physical interpretation,

modeling and impact on phase change memory (PCM) reliability of resistance drift due to chalcogenide structural relaxation. In: Proceedings of the IEEE International Electron Devices Meeting; 10-12 December 2007; Washington, DC, USA. pp. 939-942

[245] Ielmini D, Sharma D, Lavizzari S, Lacaita AL. Reliability impact of chalcogenide-structure relaxation in phase-change memory (PCM) cells—Part I: Experimental study. IEEE Transactions on Electron Devices. 2009;**56**:1070-1077

[246] Boniardi M, Ielmini D. Physical origin of the resistance drift exponent in amorphous phase change materials. Applied Physics Letters. 2011;**98**:243506

[247] Pirovano A, Lacaita AL, Pellizzer F, Kostylev SA, Benvenuti A, Bez R. Low-field amorphous state resistance and threshold voltage drift in chalcogenide materials. IEEE Transactions on Electron Devices. 2004;**51**:714-719

[248] Available from: https://www.lumenci.com/post/reram

[249] LeCun Y, Bottou L, Bengio Y, Haffner P. Gradient-based learning applied to document recognition. Proceedings of the IEEE. 1998;**86**(11): 2278-2324. DOI: 10.1109/5. 726791

[250] Wu X, Li J, Zhang L, Speight E, Rajamony R, Xie Y. Hybrid cache architecture with disparate memory technologies. ACM Sigarch Computer Architecture News. 2009;**37**:34-45. DOI: 10.1145/1555815.1555761

[251] Peng S, Zhang Y, Wang MX, Zhang YG, Zhao W. Magnetic Tunnel Junctions for Spintronics: Principles and Applications. Springer; 2014. DOI: 10.1002/047134608X.W8231

[252] Caporale N, Dan Y. Spike-timing-dependent plasticity: A Hebbian learning rule. Annual Review of Neuroscience. 2008;**31**:25-46. DOI: 10.1146/annurev.neuro.31.060407.125639

[253] Markram H, Gerstner W, Sjöström PJ. A history of spike-timing-dependent plasticity. Frontiers in Synaptic Neuroscience. 2011;**3**:4. DOI: 10.3389/fnsyn.2011.00004

[254] Dan Y, Poo MM. Spike timing-dependent plasticity: From synapse to perception. Physiological Reviews. 2006;**86**:1033-1048. DOI: 10.1152/physrev.00030.2005

[255] Xu Q, Peng J, Shen J, Tang H, Pan G. Deep CovDenseSNN: A hierarchical event-driven dynamic framework with spiking neurons in noisy environment. Neural Networks. 2020;**121**:512-519. DOI: 10.1016/j.neunet.2019.08.034

[256] Qu L, Zhao Z, Wang L, Wang Y. Efficient and hardware-friendly methods to implement competitive learning for spiking neural networks. Neural Computing and Applications. 2020;**32**(17):13479-13490

[257] Mozafari M, Ganjtabesh M, Nowzari-Dalini A, Thorpe SJ, Masquelier T. Combining STDP and reward-modulated STDP in deep convolutional spiking neural networks for digit recognition. 2018. arXiv:1804.00227

[258] Lee C, Panda P, Srinivasan G, Roy K. Training deep spiking convolutional neural networks with STDP-based unsupervised pre-training followed by supervised fine-tuning. Frontiers in Neuroscience. 2018;**12**:435

[259] Tavanaei A, Kirby Z, Maida AS. Training spiking ConvNets by STDP and gradient descent. In: Proceedings of the International Joint Conference on Neural Networks. Piscataway, NJ: IEEE; 2018. pp. 1-8

[260] Bohte SM, Kok JN, La Poutre H. Error-backpropagation in temporally encoded networks of spiking neurons. Neurocomputing. 2002;**48**(14):17-37. DOI: 10.1016/S0925-2312(01)00658-0

[261] Xu Y, Zeng X, Han L, Yang J. A supervised multi-spike learning algorithm based on gradient descent for spiking neural networks. Neural Networks. 2013;**43**:99-113. DOI: 10.1016/j.neunet.2013.02.003

[262] Wu Y, Deng L, Li G, Zhu J, Shi L. Spatio-temporal backpropagation for training high-performance spiking neural networks. Frontiers in Neuroscience. 2018;**12**:331

[263] Mostafa H. Supervised learning based on temporal coding in spiking neural networks. IEEE Transactions on Neural Networks and Learning Systems. 2017;**29**(7):3227-3235

[264] Zhou S, Chen Y, Ye Q, Li J. Direct training based spiking convolutional neural networks for object recognition. 2019. arXiv:1909.10837

[265] Stromatias E, Soto M, Serrano-Gotarredona T, Linares-Barranco B. An event-driven classifier for spiking neural networks fed with synthetic or dynamic vision sensor data. Frontiers in Neuroscience. 2017;**11**:350. DOI: 10.3389/fnins.2017.00350

[266] Zheng N, Mazumder P. Online supervised learning for hardware-based multilayer spiking neural networks through the modulation of weight-dependent spike-timing-dependent plasticity. IEEE Transactions on Neural Networks and Learning Systems. 2018a;**29**(9):4287-4302. DOI: 10.1109/TNNLS.2017.2761335

[267] Sengupta A, Ye Y, Wang R, Liu C, Roy K. Going deeper in spiking neural networks: VGG and residual architectures. 2018. arXiv:1802.02627

[268] Hu Y, Tang H, Pan G. Spiking deep residual network. 2018. arXiv:1805.01352

[269] Diehl PU, Neil D, Binas J, Cook M, Liu SC, Pfeiffer M. Fast-classifying, high-accuracy spiking deep networks through weight and threshold balancing. In: Proceedings of the 2015 International Joint Conference on Neural Networks. Piscataway, NJ: IEEE; 2015. pp. 1-8

[270] Pfeiffer M, Pfeil T. Deep learning with spiking neurons: Opportunities and challenges. Frontiers in Neuroscience. 2018;**12**:774. DOI: 10.3389/fnins.2018.00774

[271] Rueckauer B, Lungu IA, Hu Y, Pfeiffer M, Liu SC. Conversion of continuous-valued deep networks to efficient event-driven networks for image classification. Frontiers in Neuroscience. 2017;**11**:682. DOI: 10.3389/fnins.2017.00682

[272] Xu Y, Tang H, Xing J, Li H. Spike trains encoding and threshold rescaling method for deep spiking neural networks. In: Proceedings of the 2017 IEEE Symposium Series on Computational Intelligence. Piscataway, NJ: IEEE; 2017. pp. 1-6

[273] Wang Y, Xu Y, Yan R, Tang H. Deep spiking neural networks with binary weights for object recognition. IEEE Transactions on Cognitive and Developmental Systems. 2020;**13**(3):514-523. DOI: 10.1109/TCDS.2020.2971655

[274] Saleh AY, Hameed H, Najib M, Salleh M. A novel hybrid algorithm of differential evolution with evolving spiking neural network for pre-synaptic neurons optimization. International Journal of Advances in Soft Computing and Its Applications. 2014;**6**(1):1-16

[275] Schaffer JD. Evolving spiking neural networks: A novel growth algorithm corrects the teacher. In: Proceedings of the IEEE Symposium on Computational Intelligence for Security and Defense Applications. Piscataway, NJ: IEEE; 2015. pp. 1-8

[276] Vazquez RA. Izhikevich neuron model and its application in pattern recognition. Australian Journal of Intelligent Information Processing Systems. 2010;**11**:35-40

[277] Liu Y, Yenamachintala SS, Li P. Energy-efficient FPGA spiking neural accelerators with supervised and unsupervised spike-timing-dependent-plasticity. ACM Journal on Emerging Technologies in Computing Systems. 2019;**15**(3):1-19

[278] Yusuf ZM, Hamed HNA, Yusuf LM, Isa MA. Evolving spiking neural network (ESNN) and harmony search algorithm (HSA) for parameter optimization. In: Proceedings of the Sixth International Conference on Electrical Engineering and Informatics. Piscataway, NJ: IEEE; 2017. pp. 1-6

[279] Vázquez RA, Garro BA. Training spiking neurons by means of particle swarm optimization. In: Proceedings of the International Conference in Swarm Intelligence. Berlin: Springer; 2011. pp. 242-249

[280] Pavlidis NG, Tasoulis OK, Plagianakos VP, Nikiforidis G, Vrahatis MN. Spiking neural network training using evolutionary algorithms. In: Proceedings of the 2005 IEEE International Joint Conference on Neural Networks. Vol. 4. Piscataway, NJ: IEEE; 2005. pp. 2190-2194 10.1109/IJCNN.2005.1556240

[281] Sherif FF, Ahmed KS. Geographic classification and identification of SARS-CoV2 from related viral sequences. International Journal Biological Biomedical Engineering. 2021;**15**:254-259

[282] Tuchman Y, Mangoma TN, Gkoupidenis P, van de Burgt Y, John RA, Mathews N, et al. Organic neuromorphic devices: Past, present, and future challenges. MRS Bullletin. 2020;**45**(8):619-630. DOI: 10.1557/mrs.2020.196

[283] Sharifshazileh M, Burelo K, Sarnthein J, Indiveri G. An electronic neuromorphic system for real-time detection of high frequency oscillations (HFO) in intracranial EEG. Nature Communications. 2021;**12**(1):3095. DOI: 10.1038/s41467-021-23342-2

[284] Kerman Z, Yu C, An H. Beta oscillation detector design for closed-loop deep brain stimulation of Parkinson's disease with memristive spiking neural networks. In: 2022 23rd International Symposium on Quality Electronic Design (ISQED); Santa Clara, CA, USA. Vol. 2022. pp. 1-6. DOI: 10.1109/ISQED54688.2022.9806207

[285] Ahmed KS. Wheelchair movement control VIA human eye blinks. American Journal of Biomedical Engineering. 2011;**1**(1):27-30

[286] Magour AA, Sayed K, Mohamed WA, El Bahy MM. Locked-in patients' activities enhancement via brain-computer interface system using neural network. Engineering. 2018;**12**

[287] Sherif FF. Discovering Alzheimer Genetic Biomarkers Using Bayesian Networks. 2015. pp. 1-8. DOI: 10.1155/2015/639367

[288] Sherif FF, Ahmed KS. Unsupervised clustering of SARS-CoV-2 using deep convolutional autoencoder. Journal Engineering Application Science. 2022;**69**:72. DOI: 10.1186/s44147-022-00125-0

[289] Khaled Ahmed S, Mohammed Ali R, Fayroz Sherif F. Designing a new fast solution to control isolation rooms in hospitals depending on artificial intelligence decision, Biomedical Signal Processing and Control. 2023;**79**(1)

[290] Greengard S. Neuromorphic chips take shape. Communications of the ACM. 2020;**63**(8):9-11

Chapter 6

A Study of the Comparison between Artificial Neural Networks, Logistic Regression and Similarity Weighted Instance-Based Learning in Modeling and Predicting Trends in Deforestation

Zeynab Moradi and Ali Reza Mikaeili Tabrizi

Abstract

The change in forest cover plays a vital role in ecosystem services, atmospheric carbon balance, and, thus, climate change. In this study, land use maps for the periods 1984 and 2012, derived from Landsat TM satellite imagery, were used. The goal of this study is comparison of three procedures of artificial neural network, logistic regression, and similarity weighted instance-based learning (SIM Weight) to predict spatial trend of forest cover change. The SimWeight considers the nearest instances in the variable space, which are computed based on past changes and the relative importance of the driving variables. The LogReg approach, on the other hand, is a type of generalized linear model that assumes that the current land use pattern reflects the processes of land use in the past. Artificial Neural Network is a nonparametric algorithm that is capable of fitting complex nonlinear functions to find the relations between past changes and their driving variables. Such approaches are expected to produce better fitting between the change potential and their complex relationships with their driving variables. Artificial neural networks in comparison with logistic regression and SimWeight have higher accuracy and less error in modeling and predicting of forest changes.

Keywords: deforestation, neural networks, logistic regression, similarity-weighted instance-based learning in modeling, Gorganroud watershed

1. Introduction

The change in forest cover plays an essential role in ecosystem services, the carbon balance in the atmosphere, and climate change [1, 2]. The forest ecosystems of Gorganrood watershed play a very important role in controlling surface runoff and reducing floods, protecting surface soil and reducing soil erosion, adjusting temperature, and reducing the amount of greenhouse gases.

In consideration of the process of forest destruction in recent decades, it is valuable to examine the changes that have occurred in the fields of natural resources and to interpret the cause and extent of these changes and their impact on other sources [3]. Golestan province forests are very important among the forests in the north of Iran, because of their climatic conditions [4]. Golestan province forests have the highest annual destruction rate among the other northern provinces [5].

So far, many studies have used remote sensing (RS) and geographic information system (GIS) methods to effectively monitor forest cover changes [6–10]. Mahiny and Turner [11] modeled vegetation changes in the watershed of the Boorowa river in Australia by using artificial neural network and logistic regression. The results of their study indicated that artificial neural networks generally achieved better outcomes than logistic regression. Mas et al. [12] in a study that modeled deforestation by GIS and artificial neural network, and the results of their study showed that the rate of forest destruction was higher in areas with gentle slopes, high altitudes, and fertile soils. In addition, their outcomes demonstrated that the intensity of deforestation is greatly reduced by maintaining distance from the road and residential areas. Khoi and Murayama [13] modeled deforestation in an area in northern Vietnam using artificial neural networks and Markov chain models. They found that the destruction is intensive in the borders between forests and agricultural lands, fields near water sources, and areas with lower altitudes. Kumar et al. [14] modeled and predicted forest cover changes in forested areas in India. In order to explain the effects of human interventions in the forest, they used three distance variables, as explanatory variables for forest change (the distance from the edge of the forest, the road, and the city, and the map of the slope classes). The highest regression coefficient ($\beta = -26.892$) was related to the distance from the forest, which present that changes in the forest are more significant near the edge of the forest. Bagheri and Shetaei [15] in the research titled modeling the reduction of forest extent using logistic regression in the Chehel Chai watershed of Golestan province during the years 2016 to 2015, determined that the variables of the slope, distance from the village and the road have an inverse relationship with the amount of destruction. In addition, with the increase in height above the sea level in this area, the amount of destruction increased. Hassanzadeh [16] modeled deforestation in Falred forests by using the multivariate methods and artificial neural networks. According to the estimation error of modeling, artificial neural networks are a better method for modeling such variables than multivariate regression. In a study conducted by Arkhi et al. [17], the forests of North Ilam were simulated using logistic regression. According to the modeling results, it was found that the slope variables, the distance from the population centers and the road, have an opposite relationship with the amount of destruction, and the increase of the height above the sea level in this area resulted in decreasing the amount of destruction. Sardarzadeh et al. [18] predicted the destruction of forests in Chehel Chai watershed of Golestan province using artificial neural networks and Markov chain analysis. The obtained results indicate the destruction of 15.8% of dense forests during the years 1988 to 2010. Gholamalifard et al. [19] conducted a study with the aim of comparing logistic regression and artificial neural networks for modeling the transfer potential of coastal land cover change in Mazandaran province. The results showed that logistic regression has higher accuracy.

2. Materials and methods

2.1 Prediction of land cover changes

In recent years, land cover changes have received special attention. Land cover changes are the result of the complex interaction of several factors such as management, economy, culture, human behavior, and environment [20, 21]. Familiarity with how land cover changes is very important because these changes cause major effects on the environment such as hydrological cycles changing [22], the size and order of natural habitats such as forest areas [23], and species diversity [24] and can overshadow the region's economy [25].

Modeling the spatial pattern of land cover changes provides valuable information for a better understanding of the change process, determining the effective factors, and predicting the areas subject to change. Spatial models of land cover change can be divided into three main groups: Empirical Estimation Models, Dynamic Simulation Models, and Rule-base Simulation Models [26]. Empirical estimation methods use statistical techniques to model the relationship between the change based on the user rule and the factors affecting it.

Knowing the effective processes in change is possible by interpreting the output of statistical models. Empirical estimation methods are one of the most widely used methods for simulating the spatial pattern of land cover and its changes over time due to the simplicity of the structure and the ability to analyze multiple variables [27].

2.2 Land change modeler

The land change modeler is a software for creating sustainable ecological development, which was designed to understand and identify land cover changes and environmental and protection requirements caused by these changes. This software exists as a vertical application in the IDRISI software system [28]. This model is also available as Extension for ArcGIS software. In Land Change Modeler (LCM), tools for the assessment and prediction of land cover change and its implications are organized around major task areas: change analysis, change prediction, and planning interventions. Land change prediction in land change modeler is an empirically driven process that moves in a stepwise fashion from 1) change analysis, 2) transition potential modeling, to 3) change prediction. It is based on the historical change from time 1 to time 2 land cover maps to project future scenarios [29].

2.3 Sensitivity analysis of artificial neural network

The artificial neural network option can model multiple transitions at one time Initially, the dialog for the multilayer perceptron neural network may seem daunting, but most of the parameters presented do not need to be modified (or in fact understood) to make productive use of this very powerful technique.

As launched by LCM, the multilayer perceptron starts training on the samples it has been provided of pixels that have and have not experienced the transitions being modeled. At this point, the MLP is operating in automatic mode whereby it makes its own decisions about the parameters to be used and how they should be changed to better model the data. Automatic mode monitors and modifies the start and end learning rate of a dynamic learning procedure. The dynamic learning procedure starts with an initial learning rate and reduces it progressively over the iterations until the end learning rate is reached when the maximum number of iterations is reached [29].

Figure 1 illustrates a single hidden layer MLP feed-forward neural network. The back-propagation algorithm [30] is the most widely used learning algorithm for an MLP neural network. The learning process consists of two parts: feed-forward and backward pass. The outputs of the Artificial Neural Network (ANN) are calculated in the feed-forward pass process and the output errors are propagated backward to adjust the weights and biases of the ANN.

In the present research, Skill Measure [28] and Accuracy Rate were used to analyze the sensitivity of the multilayer perceptron artificial neural network model. Skill Measure is a statistic to evaluate the ability of the model based on the validation data and measures the skill of the model to predict future changes based on the training data. In fact, this statistic is used to compare the accuracy of the model based on the validation data and the expected accuracy that is supposed to occur randomly.

This statistic considers the ability of the model for each land cover transfer separately and shows the role of variables in the accuracy of the model for predicting change. This statistic is between −1 and − 1. The closer to 1, it indicates the high accuracy of the model in predicting changes, and if it is zero, it indicates the random performance of the model. The expected accuracy and Skill Measure statistics are calculated from Eq. (1) and (2) [29].

$$\mathrm{E(A)} = 1 + (\mathrm{T} + \mathrm{P}) \tag{1}$$

E(A), is the expected accuracy rate. T is the number of considered transitions in the transition potential. P is the number of classes that remain constant in transitions.

$$S(Skill\ Measure) = (A - E(A))/(1 - E(A)) \tag{2}$$

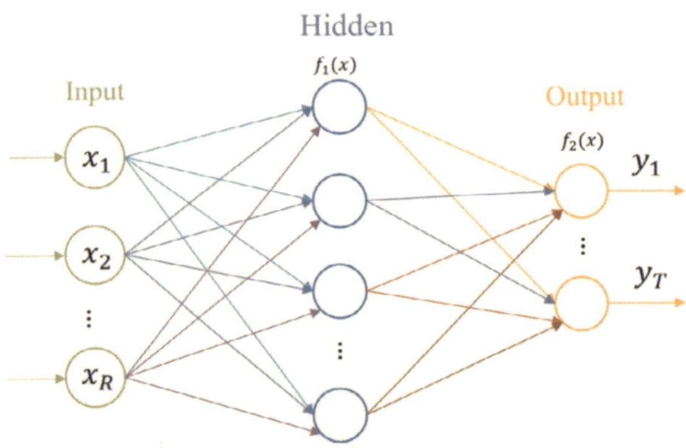

Figure 1.
Example of a single hidden layer MLP feed-forward neural network. X_1, x_2, … , x_R *are input parameters,* y_1, … , y_T *are output targets,* R *is the number of input parameters, and* T *is the number of output targets.* f1 (x) *is the activation function of the hidden layer, and* f_2 (x) *is the activation function of the output layer.*

DOI: http://dx.doi.org/10.5772/intechopen.111615

A is the measured accuracy provided by the model, and E(A) is the expected accuracy.

2.4 Sensitivity analysis of logistic regression

It is one of the experimental models that has been used in many researches in the field of forest area change analysis, urban growth modeling, and agricultural land modeling and has provided very good results [14]. Logistic regression is a probabilistic model that fits between land use change (as a dependent variable) and factors affecting it (as independent variables). Based on this model, the relationship between variables can be explained, the relative importance of index variables can be estimated, and the land cover change probability map can be extracted [31].

In the logistic function, the probability of land use change was defined as a function of explanatory variables, which includes a uniform curve between 0 and 1 [14]. One of the statistical characteristics that are examined in the logistic regression model is Goodness of fit. The goodness of fit is determined based on the difference between the observed values and the predicted values of changes in the value of the dependent variable. The following equation is used to calculate the Goodness of fit (Eq. (3)) [32]:

$$\text{Goodnessoffit} = \sum_{i=1}^{N} (y_i - u_i)/u_i \times (1 - u_i) \tag{3}$$

ui are the observed values for the dependent variable, and y_i are the predicted values for the dependent variable. The basis of Goodness of fit in logistic regression is the probability ratio, which is determined based on two statistical characteristics -2log (L0) and -2log(Likelihood), where L0 is the value of the probability function, provided that all coefficients except the intercept are 0 and Likelihood represents the amount of the probability function for the model. The following two statistical characteristics are defined based on the aforementioned two statistical characteristics (Eq. (4) and (5)) [33]:

$$\text{PseudoRSquare} = 1 - (\log(\text{Likelihood})/\log(\text{L}_0)) \tag{4}$$

Therefore, if the value of R (Pseudo R Square) is equal to 1, it indicates a good Goodness of fit; if the value of this characteristic is 0, it means that there is no relationship between the variables, and if its value is greater than 0.2, it indicates a relatively good Goodness of fit:

$$\text{ChiSquare}(K) = -2(\log(\text{Likelihood}) - \log(\text{L}_0)) \tag{5}$$

Chi-square is also introduced as a statistical characteristic of the probability ratio, which follows the chi-square distribution when the null hypothesis is correct. This statistical characteristic examines hypotheses in which all coefficients are zero except the intercept. The chi-square degree of freedom (K) is equal to the number of independent variables used in land change modeling.

The characteristic of the ROC (Relative Operating Characteristic) is a very suitable statistic for measuring the degree of Goodness of fit in the logistic regression model. The amount of this statistic are between 0 and 1, where 1 indicates proper Goodness of fit and 0.5 indicates random Goodness of fit [14]. Finally, the logistic regression equation is defined as follows (Eq. (6)):

$$\log_e(P) = \beta_0 + \beta_1 X_1 + \beta_2 X_2 + \beta_3 X_3 + \ldots + \beta_n X_n + errorterm \tag{6}$$

β_0: intercept,
$\beta_0, \beta_1, \ldots, \beta_n$: regression coefficients,
$X_1, X_2, \ldots, X_n$: used variables.

2.5 Sensitivity analysis of SimWeight

The similarity-weighted sample-based learning method is a sample-based learning algorithm based on the K-nearest neighbor (KNN) algorithm. This method identifies the relationship between the stimulus variable and the transmission potential prediction for areas that show cases of change. This method is based on the calculation of weighted distances in variable space for known examples of user classes. SimWeight should have two classes (fixed pixel and variable pixel) for each transfer to create transfer potential in the direction of land change modeling.

In this method, the K-nearest neighbor was extracted for each pixel (fixed or variable) and calculates the distance in the variable space from each unknown location to the locations around it (within the range of K) (**Figure 2**). This distance is obtained in the exponential weighting function (e1/di) in order to calculate a continuous level of membership of existing land use classes for each pixel from Eq. (7) [34]:

$$Membership\ of\ Class = \frac{\sum_{i=1}^{c}\left(1.0 - \frac{1}{1+e^{\frac{1}{d_i}}}\right)}{k} \quad (c \le k) \tag{7}$$

where K represents the total number of the closest variable and fixed pixels, c represents the number of variable pixels in the nearest neighbor k, and d is the linear distance of the variable pixel i.

This algorithm, like the K-nearest neighbor method, may be influenced by irrelevant variables [35]. Different stimulus variables may have different importance in

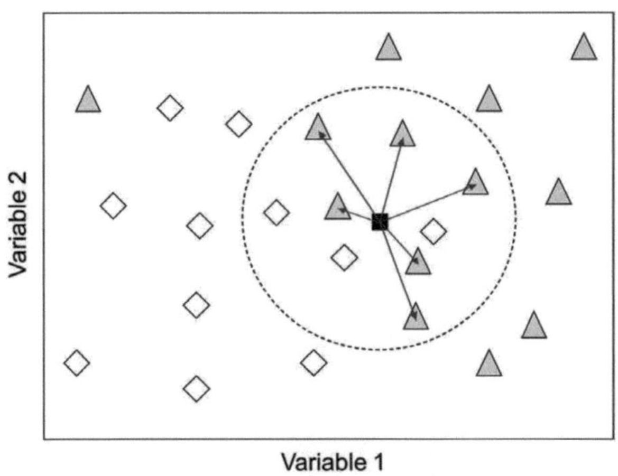

Figure 2.
A representation of the variable space created by two hypothetical variables. In this figure, the triangles represent the variable pixels, and the diamonds represent the fixed pixels. The black square represents the pixel that will be evaluated under the influence of the hypothetical and unknown transmission potential. The dotted circle in the figure represents the range of k (k = 9), which is hypothetically 6 pixels for the changing state and 3 pixels for the fixed state, and also the lines showed are the linear distances from the considered pixel in the space is variable [34].

determining transmission potential. This different importance is due to the relative weight by which each variable is multiplied to determine its ability among different classes of land use. In this method, the weight of the importance of each variable is determined by comparing the standard deviation of the variables that have changed within the considered area to the standard deviation of the variables in the study area (Eq. (8)) [34]:

$$\text{The correlation weight of each variable} = \frac{\textit{the standard deviation of the variable pixels in the change area}}{\textit{the standard deviation of the variables in the study area}} \quad (8)$$

2.6 The location of the study

The Gorganroud river watershed with an area of about 9350 square kilometers with a longitude of 54°2′ to 56°22′ and a latitude of 36°22′ to 37°47′ North is in Golestan province (**Figure 3**). The said river is from Golestan National Park originates from Golidagh Heights, and flows into the Caspian Sea after passing through Gonbadekavous and Aghqala in the west of Khajenafas. This river is located in the southeastern part of the Caspian Sea. Most of its branches are from Alborz mountain and flow from south to north. Among its rivers, we can mention the Madersu, Zaringol, Tilabad, and Chehelchai rivers. The Gorganroud river length is about 300 km, and the direction of the river water flow is from east to west [36].

2.7 Data used

In this study, in order to prepare the forest cover map of 1984 and 2012, the land use maps produced in the Golestan province survey plan have been used. In order to evaluate the accuracy and determine the extent of the forest in 2015, the previous studies [37] by using the TM sensor of the Landsat satellite and also the visual classification of the area through Google Earth images, have been used.

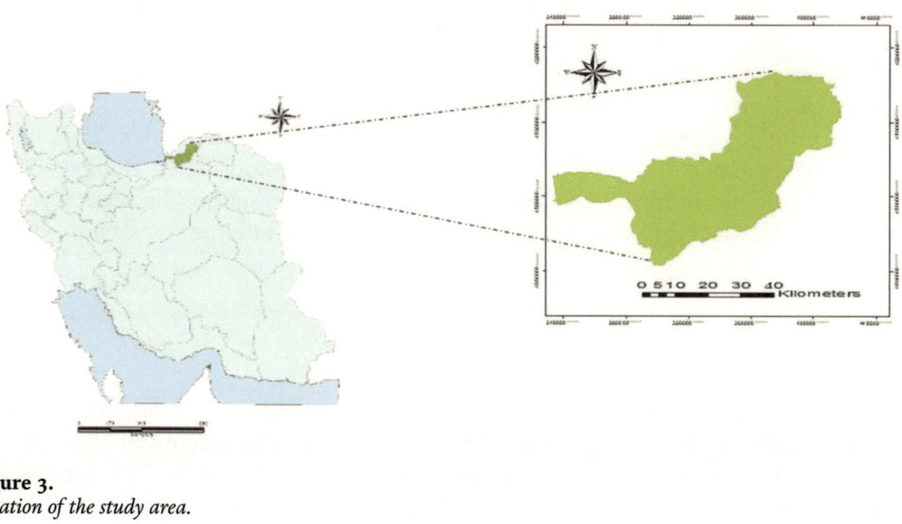

Figure 3.
Location of the study area.

2.8 The transmission potential modeling by using the logistic regression and artificial neural network and similarity weighted instance-based learning (SIM weight)

In this study, by using the forest cover maps of 1984 and 2012, the calibration period is considered, and the transfer potential is modeled by using six variables, and by using the driver variables of the digital model, height, slope, distance from the road, distance from agricultural land, the distance from the edge of the forest and the distance from the village in 1984, the relationship between the change of use from forest to nonforest is determined by using cramer's correlation coefficient.

The probability of changing each user to another user is calculated by using the Markov chain [38]. In this study, changes are predicted for the year 2015 using the hard forecasting model and the calibration period of 1984–2012.

2.9 Validation of the model

For evaluating the accuracy of the modeling, the forest cover map of 2015 and statistics such as the relative performance characteristic curve (ROC), figure of merit [39], and ratio Hits/False Alarms [40] are used. From the ROC/AUC statistic, the range is 0–1 based on the percentage of false positives and true positives. It is used to compare a continuous image of merit with a Boolean image, where a value of 1 indicates complete spatial agreement and a value of 0.5 indicates random agreement. The merit number statistic has a value between zero and one hundred, where the value of one hundred indicates the complete agreement of the predicted map with the ground reality, and the value of zero indicates noncompliance [41]. The closer the figure of merit is to 100. It means that the predicted map has higher accuracy [39]. The figure of merit is obtained from Eq. (9):

$$\text{Figureof merit} = \left(\frac{B}{A+B+C}\right) \tag{9}$$

A: The number of pixels that have changed in reality but remained constant in the prediction (Miss).

B: The number of pixels that have changed in the ground reality, and these changes were correctly predicted by the model (Hits).

C: The number of pixels that have remained constant in the ground reality, but these pixels have changed in the model prediction (False Alarm).

If the ratio of success to error warning in the used model is more than 25%, it can be said that the model has good accuracy in predicting the considered changes [29].

At this stage, according to the relevant statistics, for evaluating the accuracy, the best model with the highest accuracy was selected to continue the research process.

3. Results

The investigation of the changes in the land cover of the Gorganroud watershed showed that during the study period (1984–2012), the largest amount of changes occurred in the field of forest cover destruction (77,214 hectares), and the largest amount of increase is related to agricultural use (28,866 hectares) (**Table 1**).

Class	1984	2012	Changed area Between 1984 and 2012
Residential	3258/4	13,577/3	10,319
Forest	353,128/7	275,914/3	-77,214
Rangeland	178,108/3	197,763/3	19,655
Agriculture	386,290/6	415,156/9	28,866
Water bodies	7253/9	9887/5	2634
Rivers	6825/4	19,821/3	12,996

Table 1.
Changes in all classes of land cover in 1984 and 2012 in terms of hectares.

Figure 4 shows the increase and decrease for each user class during the period of 1984 to 2012 based on the percentage of the study area. The largest increase observed during this period is related to agricultural use (with an increase of 1.2%), and the largest decrease occurred in dense forest use (with an increase of 1.54%).

As **Figure 5** presents, forest destruction during the years 1984 to 2012 mostly occurred in the northeastern part of the area. Severe erosion in the loess soils and destructive floods in this region are among the most important reasons for forest degradation in the northeastern part of this area [42].

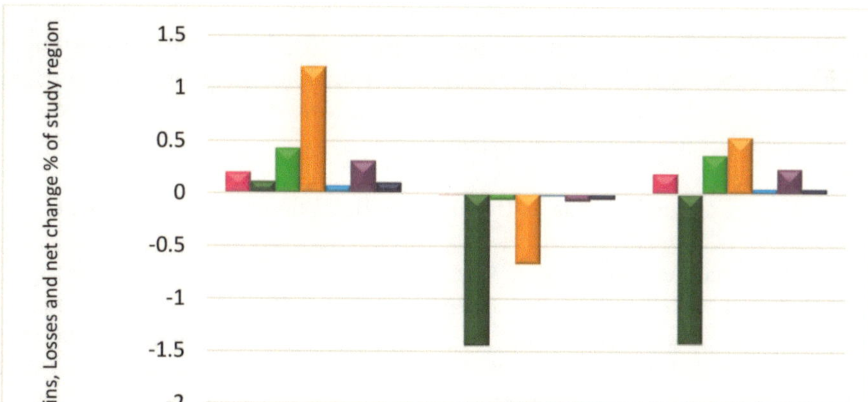

	Gains	Losses	Net change
Residential	0.2	-0.01	0.19
Forest	0.11	-1.45	-1.43
Rangeland	0.43	-0.06	0.37
Agriculture	1.2	-0.67	0.54
Water bodies	0.07	-0.02	0.05
Roads	0.31	-0.07	0.24
Rivers	0.1	-0.05	0.05

Figure 4.
Gains, losses, and net change for each land cover category between 1987 and 2012 in percentage of the study area.

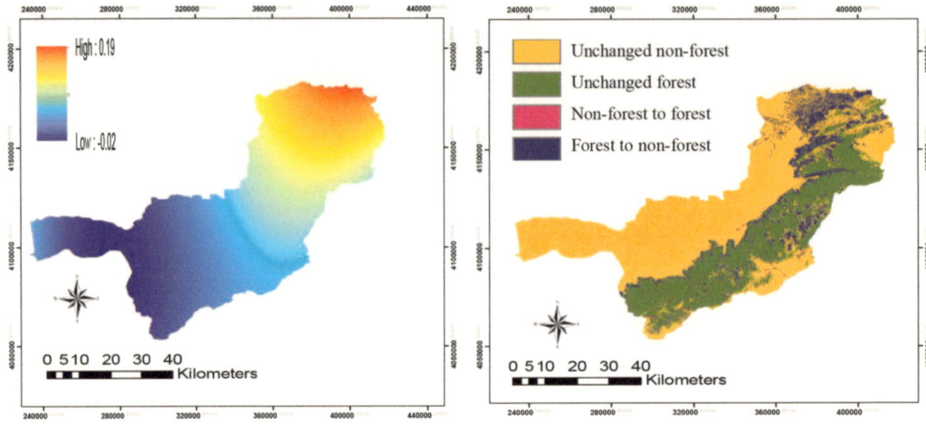

(A) (B)

Figure 5.
Spatial distribution of deforestation. A: Transition of forest cover between 1984 and 2012; B: Spatial trend of changes (deforestation) between 1984 and 2012.

3.1 Modeling of transfer potential and predicting changes by using artificial neural network, logistic regression, and SIM weight model and examination of correctness or integrity

Whit the aim of transfer potential modeling, the connection between motivator factors variants and jungle cover changes by using Crammer coefficient methods has been accomplished for all three modeling (**Table 2**). Cramer's V amounts for the most of variants is upper than 0.15 that demonstrates desirable suitable descripted power of coefficients that been researched [33]. Highest and lowest cramer's V is related to slope and distance from the village respectively.

3.2 Predicting changes for year 2015 using artificial neural network model

Demanding changes has occurred using Markov chain, and predicting changes for the year 2015 has been accomplished (**Figure 6**). Correction and analyze numbers of Skill Measure for examining term had been 84.52% and 0.6850 ordinarily that is declaring suitable and right prediction. The results of correction examining of this model show that the outcome resulted from model with ROC amount, 0.975 has high

Driver variables	Cramer's V
Slope	0.44
Elevation	0.41
Distance from agriculture	0.41
Distance from forest	0.36
Distance from roads	0.22
Distance from village	0.16

Table 2.
Values of the Cramer coefficient for driver variables.

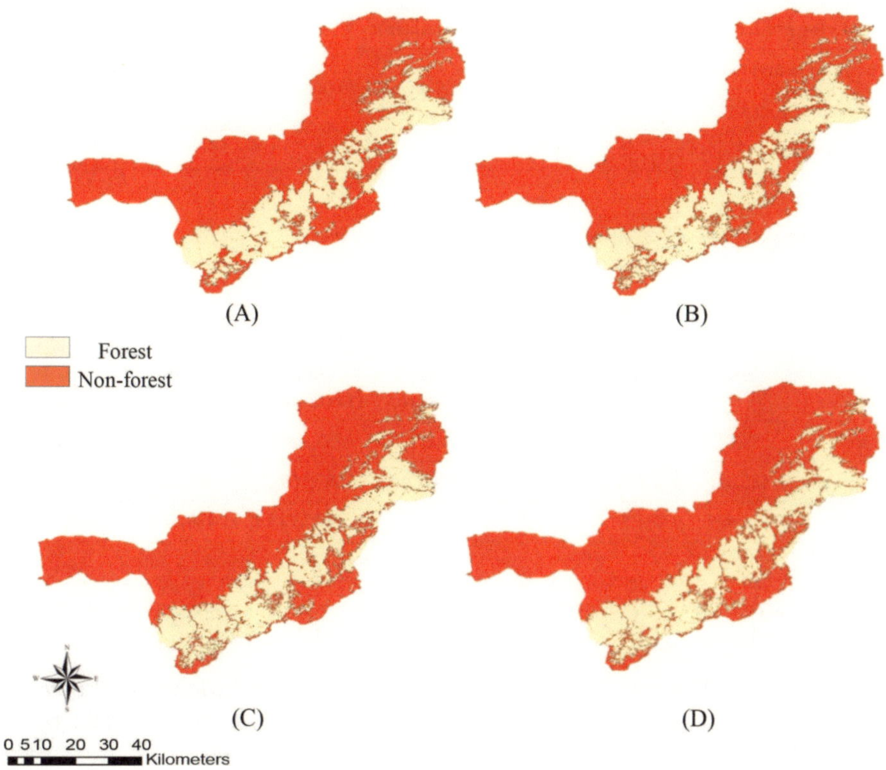

Figure 6.
A: Actual map for 2015; B: Predicted map for 2015 using artificial neural network; C: Predicted map for 2015 using logistic regression; D: Predicted map for 2015 using SIM weight.

harmony with occurred changes. The ratio of Hits/False Alarms for artificial multi-layer perceptron neural network model is 63%, and the ratio of the figure of merit equals 12.

3.3 Predicting changes for year 2015 using logistic regression model

In this case also demanding change has been occurred using Markov chain and changes by prediction for year 2015 has been done (**Figure 6**). Correction examining results logistic regression model shows that outcome resulted from model with ROC amount 0.922 has high harmony with occurred changes. Ratio Hits/False Alarms for logistic regression model is 50% and also amount of figure of merit is 10.

3.4 Predicting changes for year 2015 using SIM weight

In this case, demanding change has also been occurred by using Markov chain and has been done by prediction for year 2015 (**Figure 6**). Correction examining results for SIM Weight Process shows that ROC amount is 0.979. Ratio of Hits/False Alarms for SIM Weight is 52% based on similarity weight model, and also figure of merit is 10.

According to **Table 3**, artificial neural network, in two factors ratio Hits/False Alarms and figure of merit, for examining correction has shown higher quality and, compared to other processes, has more ability and capability for predicting forest

Modeling procedures	Ratio Hits/False Alarms	Figure of merit	ROC
Artificial neural network	63%	12%	0/975
Logistic regression	50%	10%	0/922
SIM Weight	52%	10%	0/979

Table 3.
Comparison between accuracy of the modeling of three approaches was used.

Class	Area (in terms of hectares)						
	1984	2012	2015 (Actual)	2015 (Predicted)	2020 (Predicted)	2025 (Predicted)	2030 (Predicted)
Forest	353,128	275,914	256,659	264,877	253,316	242,169	231,381
Nonforest	585,351	662,565	681,820	673,602	685,163	696,310	707,098
Total	938,479	938,479	938,479	938,479	938,479	938,479	938,479

Table 4.
Areas under forest cover and nonforest during 1984, 2012, 2015, 2020, 2025, and 2030 (actual and predicted).

coverage changes. The amount of ROC has not been observed so much differences between three approaches used. Therefore, artificial neural network model has been chosen as the best model for predicting forest changes prediction.

3.5 Predicting changes of land coverage by using artificial neural network method

Results produced by researching forest cover changes in years 2015 to 2020 shows that in this time 11,561 acres of forest surface (equal to 46 % of region forest surface) been corrupted or ruined. Besides, around 2020 to 2025, approximately 11,147 forest surface (equal to 44 % of the region forest surface) will be declined. Examining forest cover changes between 2025 and 2030 also shows the reduction of 10,788 acres of region forest surface (equal to 42 %'s of region forest surface) (**Table 4**).

4. Discussion

Although many methods are available to model land transition potentials, they are usually not user-friendly and require the specification of many parameters, making the task difficult for decision-makers not familiar with the tools, as well as making the process difficult to interpret. SimWeight is an instance-based learning algorithm based on the logic of the K-nearest neighbor algorithm. The method identifies the relevance of each driver variable and predicts the transition potential of locations given known instances of change. Although computationally simple, the method cannot handle complex nonlinear relations, which is often visible in built-up growth in heterogeneous environments. However, this method can be useful for areas to project infilling or edge-expansion type of built-up growth [43]. SimWeight focused on the distance from the past transitions producing most potential zones near the change areas, LogReg produced suitable areas considering the linear relationship between driving factors and the built-up change.

While the LogReg approach is straightforward and easy to reproduce, it cannot mimic the complex relationship between the variables and the land use pattern in the changing space. This, therefore, indicates the unsuitability of the method for prediction, especially for dynamic and heterogeneous built-up growth. However, this statement might be applicable for only short-time-scale studies and might not be true for applications that incorporate larger time-scale predictions [44].

The Artificial neural networks, as it iterates multiple times to produce the best fit between the transition and driving variables are able to estimate high change potentials for areas of actual change. Artificial neural networks are a sum of the complex of improving methods that can analyze and calculate nonlinear relations after well training and adjusting weights between income and outcome parameters by high currency. Although the SIM weight model and also logistic regression model are not able to calculate nonlinear relations between variants [40]. Artificial neural networks compared to Logistic Regression Model and SIM weight model, do not need a specific formulation for the statement relation between income and outcome data; otherwise, the relation between income and outcome data has been taken by the learning process [45]. At last, using the artificial neural network method to predict jungle cover changes in the future (years 2020, 2025, and 2030) has been discussed.

In relation to the better performance of the artificial neural network, we can mention the following: high processing speed, the ability to learn the pattern, the ability to generalize the pattern after learning, flexibility against unwanted errors, and not causing significant disruption in case of problems in part of the connections due to the distribution of network weights. This discussion declares that artificial neural network has higher power capability for predicting forest changes.

The results of this part demonstrate forest cover changes in the Gorganroud watershed as well and declare that continuing of the current process in recent 30 years, what kind of problems and huge big enormous obtains, and obstacles to the forest of this region will occur. Putting an obvious clear picture of the future in front of Managers and program makers, scheduling Personals can be efficient and effective capable in scheduling for saving and cohabitating forest regions.

5. Conclusion

According to high valued importance of Golestan province forest study and researching and modeling of corruption amount of these forest seems to be urgent and important. In this research of Land Changer Modeler by using Markov chain utilizing three approach logistic regression, artificial neural network, and learning based on similarity weight sample, researching, and examination of forest cover changes has been done. Produced results from evaluating of model showed higher capability and power of land change modeler for predicting forest cover changes that after investigating and comparing dignity correctness of three modeling approaches based on three factors, ratio of Hits/False Alarms and figure of merit and ROC amount, results show a high level of efficiency and potential for artificial neural network and lower errors obtain from this method compared to other two approaches. In fact, the result shows artificial neural networks could correctly predict changes in pixels that have been changed in ground reality (Hit). On the other hand, models mistake pixels that have been changed in ground reality but stayed constant in prediction for artificial neural network model compared to Logistic Regression Model and SIM weight model was lower (Miss). Also, mistakes and bugs resulted from the prediction of pixels that have stayed unchanged but changed in predicting model was also lower in the artificial neural network model compared to the other two approaches (false alarm).

This research's results have coordination with those of Mahiny and Turner who compared artificial neural networks to logistic regression. Also, Bayati et al. [46] compared two models of artificial neural networks and logistic regression in forest surveys. In their study, the artificial neural network model produced better results; In justifying this phenomenon, they stated that the reason for the difference between the better performance of artificial neural networks compared to statistical methods could be found in the ability to estimate and predict artificial neural networks with a small amount of data. This is despite the fact that the performance and accuracy of regression methods depend on the sample size strongly, and the small sample size can be a limitation of statistical models. Therefore, in the designed models of the artificial neural network, the small number of samples has not created a significant limitation.

Author details

Zeynab Moradi* and Ali Reza Mikaeili Tabrizi
Department of Environment, Gorgan University of Agricultural Sciences and Natural Resources, Gorgan, Iran

*Address all correspondence to: zeynab.moradi36@gmail.com

References

[1] Le Quéré C, Raupach MR, Canadell JG, Marland G, Bopp L, Ciais P, et al. Trends in the sources and sinks of carbon dioxide. Nature Geoscience. 2009;**2**:831-836

[2] Van der Werf GR, Morton DC, DeFries RS, Olivier JGJ, Kasibhatla PS, Jackson RB, et al. CO2 emissions from forest loss. Nature Geoscience. 2009;**2**: 737-738

[3] Darvishsefat A, Namiranian M. The Study of Spatial Distribution of Changes in the Northern Forests of Iran. 2004. http://www.GISDevelopment.nat/application/nrm/overview. pp. 1-2

[4] Naad Ali A, Mahini E, Feghhi J, mathematics B. Classification of forest areas of Golestan province with maximum possibility method by using satellite pictures+ETM. 2012 Autumn. Environmental Technology and Science Quarterly. 2001; **Fourteenth Period**(number 3):47-56. (In Persian)

[5] Mirakhoorloo KH, Akhavan, R. Study the changes of upper border of north forests by using satellite data. Forest Study and Iran's Senobar. 2008; **Sixteenth Period**(number 1):139-148. (In Persian)

[6] Kushwaha SPS. Forest type mapping and change detection from satelliteimagery. ISPRS Journal of Photogramming Remote Sensors. 1990; **45**:175-181

[7] FSI. India State of Forest Report 2011. Dehradun: Forest Survey of India, Ministry of Environment and Forests; 2011

[8] Srivastava S, Singh TP, Singh H, Kushwaha SPS, Roy PS. Mappingof large-scale deforestation in Sonitpur district. Assam Current Science. 2002; **82**(12):1479-1484

[9] Nandy S, Kushwaha SPS, Mukhopadhyay S. Monitoring Chilla-Motichurcorridor using geospatial tools. Journal for Nature Conservation. 2007; **15**(4):237-244

[10] Kushwaha SPS, Hazarika R. Assessment of habitat loss in Kameng and Sonitpur elephant reserves. Current Science. 2004;**87**(10):1447-1453

[11] Mahiny AS, Turner BJ. Modeling Past Vegetation Change Through Remote Sensing and GIS: a comparison of neural networks and logistic regression methods. In: Proceedings of the 7th international conference on geocomputation. UK: University of Southampton; 2003. pp. 1-24

[12] Mas JF, Puig H, Palacio JL, Lopez AS. Modelingdeforestation using GIS and artificial neural networks. Environmental Model Software. 2004; **19**:461-471

[13] Khoi D, Murayama Y. Modeling deforestation using a neural network-Maarkov model. Spatial Analysis and Modeling in Geographical Transformation Process. 2011;**100**: 169-190

[14] Kumar R, Nandy S, Agarwal R, Kushwaha SPS. Forestcoverdynamics analysis and prediction modeling using logistic regression model. Ecological Indicators. 2014;**45**:444-455

[15] Bagheri R, Shetayi SH. Modeling the reduction of forest area by using logistic regression. Iran's forest journal. 2010;**3**: 243-252 (In Persian)

[16] Hassanzadeposhtemsari M. Deforestation Modeling by Using Multivariate Statistical Methods and Artificial Neural Network in Felard Forests. MA Thesis of Natural Resources of Shahrekord Collage. 2011. pp. 180 (In Persian)

[17] Arkhi S, Jafarzade EA, Yousefi S. Deforestation modeling by using logistic regression, GIS and remote evaluation. Item: North of Ilam forests. Gography and investment journal. 2012;**29**:31-42 (In Persian)

[18] Sardarzade M, Metkan EA, Sadatinezhad SJ, Ashourloo D. Destruction of forests prediction by using GIS and RS technics and combination of artificial neural networks Markov chain(Chehelchaigolestan province watershed). In: 20th Geomatic National Show. Tehran: Iran mapping organization; 2012 (In Persian)

[19] Gholamalifard M, Jorabianshooshtari SH, Abkar, EA, Naimi B. Logistic regression algorithms and artificial neural network comparison in practical modeling of the change transfer potential in Mazandarn's coastal ereas. Environmental Researches. 2014;**Year 5** (number 1393):167-176. (In Persian)

[20] Wu X, Shen ZY. Analysis of the changes of land use/cover and landscape pattern in the upper reaches of the Yangtze River. Transactions of the CSAE. 2007;**23**(10):86-92

[21] Huang L, Ni L. Object-oriented classification of high resolution satellite image for better accuracy. Spatial Accuracy Assessment in Natural Resources and Environmental Sciences. 2008. pp. 25-27

[22] Sun H, Li YM, Wang XZ, Ni SX. Method and application of landscape ecological evaluation in the typical small watershed's land use. Geomatics and Information Science of Wuhan University. 2003;**28**(2):177-181

[23] Flintrop C, Hohlmann B, Jasper T, Korte C, Podlaha O, Scheele S, et al. Anatomy of pollution: Streams of North Rhine-Wesphalia, Germany. American Journal of Science. 1996;**296**:58-98

[24] Davidson C. Issues in measuring landscape fragmentation. Wildlife Society Bulletin. 1998;**26**:32-37

[25] Burchell RW. Economic and fiscal impacts of alternative land-use patterns. In: The Land Use Decision Making Process: Its Role in a Sustainable Future for Michigan. January 9–10, 1996. Michigan: East Lansing; 1996

[26] He Z, Lo C. Modeling urban growth in Atlanta using logistic regression. Computers, Environment and Urban Systems. 2007;**31**(6):667-688

[27] Zeng YN, Wu GP, Zhan FB, Zhang HH. Modeling Spatial Land Use Pattern Using Auto Logistic Regression The International Archives of the Photogrammetry, Remote Sensing and Spatial Information Science. 2008. pp. 115-119

[28] Eastman J, Luis Solorzano R, Megan F, V. Transition potential Modeling for land-cover change. In: Maguire DJ, Batty M, Goodchild MF, editors. GIS, Spatial Analysis and Modeling. Redlands, CA: ESRI Press; 2005. pp. 357-385

[29] Eastman JR. IDRISI Andes, Guide to GIS and Image Processing. Worcester, MA: Clark Labs, Clark University; 2006

[30] Li S, Yang B, Qi F. Accelerate global sensitivity analysis using artificial neural network algorithm: Case studies for

combustion kinetic model. Combustion and Flame. 2016;**168**:53-64

[31] ZareGarizi A, Sheikh VB, Saadoddin A, Salman Mahini A. The application of logistic regression method in modeling the spatial pattern of the probability of vegetation change (a case study of Chehel chai watershed in Golestan province). Geographical Space Scientific-Research Quarterly. 2011; **12**(37):55-68 (In Persian)

[32] Eastman JR. IDRISI Taiga, Guide to GIS and Remote Processing. Worcester: Clark University; 2009

[33] DiBari JN. Evaluation of five landscape-level metrics for measuring the effects of urbanization on landscape structure: The case of Tucson, Arizona, USA. Landscape and Urban Planning. 2006;**79**:308-313

[34] Sangermano F, Eastman JR, Zhu H. Similarity weighted instance-based learning for the generation of transition potentials in land use change Modeling. Transactions in GIS. 2010; **14**(5):569-580

[35] Tahir TA, Bouridane A, Kurugollu F. Simultaneous feature selection and feature weighting using hybrid Tabu search/k-neareast neighbor classifier. Pattern Recognition Letters. 2007;**218**: 438-446

[36] Khakpour A, Mehrdadi N, Noori RA, Soroush M. Evaluation of the quality of the Gorganroud River based on field studies. In: Third Environmental Engineering Conference. Tehran: University of Tehran; 2009 (In Persian)

[37] Management and planning organization of Golestan province. Executive document of Golestan province. Tehran: National Program and Budget Organization; 2012

[38] Haibo Y, Longjiang D, Hengliang G, Jie Z. Tai'an land use analysis and prediction based on RS and Markov model. Procedia Environmental Sciences. 2011;**10**(C):2625-2630

[39] Kim OS. An assessment of deforestation models for reducing emissions from deforestation and forest degradation (REDD). Transactions in GIS. 2010;**14**(5):631-654

[40] Eastman JR. IDRISI Help System. Accessed in IDRISI Selva 17.02. Worcester, MA: Clark Labs, Clark University; 2012

[41] Pontius RG, Schneider LC. Land-cover change model validation by an ROC method for the Ipswich watershed, Massachusetts, USA. Agriculture, Ecosystems & Environment. 2001;**85**(1): 239-248

[42] Ajami M, Khormali F. Pedogenic and micro-morphological evidence of degradation of Losi forestland in eastern Golestan province. Journal of of Agricultural Science and Technology, Water and Watersheds Science. 2012; **16**(61):153-141 (In Persian)

[43] Mas JF, Kolb M, Paegelow M, Olmedo MTC, Houet T. Inductive pattern-based land use/cover change models: A comparison of four software packages. Environmental Modelling & Software. 2014;**51**:94-111

[44] Overmars KP, Verburg PH, Veldkamp TA. Comparison of a deductive and an inductive approach to specify land suitability in a spatially explicit land use model. Land Use Policy. 2007;**24**:584-599

[45] Schaap MG, Bouten W. Modeling water retention curves of sandy soils using neural networks. Water Resources Research. 1996;**32**:3033-3040

[46] Bayati H, Najafi A, Abdulmaleki P. Comparison between artificial neural network (ANN) and regression analysis in tree felling time estimation. Iranian Forest and Spruce Research Quarterly. 2012;**20**(4):559-607 (in persian)

Section 4

Temporal Neural Network and Spiking Neural Network

Chapter 7

Spiking Neural Encoding and Hardware Implementations for Neuromorphic Computing

Honghao Zheng and Yang (Cindy) Yi

Abstract

Due to the high requirements of the computational power of modern data-intensive applications, the traditional von Neumann structure and neuromorphic computing structure started to play complementary roles in the area of computing. Thus, neuromorphic computing architectures have attracted much attention with high data capacity and power efficiency. In this chapter, the basic concept of neuromorphic computing is discussed, including spiking codes and neurons. The spiking encoder can transfer analog signals to spike signals, thus avoiding using power-consuming analog-to-digital converters. Comparisons of training accuracy and robustness of neural codes are carried out, and the circuit implementations of the spiking temporal encoders are briefly introduced. The encoding schemes are evaluated on the PyTorch platform with the most common datasets, such as Modified National Institute of Standards and Technology (MNIST), Canadian Institute for Advanced Research, 10 classes (CIFAR-10), and The Street View House Numbers (SVHN). From the result, the multiplexing temporal code has shown high data capacity, robustness, and low training error. It achieves at least 6.4% more accuracy than other state-of-the-art works using other encoding schemes.

Keywords: analog/mixed-signal integrated circuit (IC) design, neuromorphic computing, neural spike encoding, multiplexing temporal encoder, gamma alignment

1. Introduction

Since the late 1980s, researchers have been paying attention to neuromorphic computing [1]. Researchers have noticed that by mimicking the biological neural systems with either software or hardware implementations, computing systems' data capacity and power efficiency can be greatly improved. With the limitations of conventional von Neumann structures appealing to the progress of application requirements, the self-training mechanism of neuromorphic computing structures has attracted more and more attention, both from the industrial and academic areas. For instance, neuromorphic computing systems can more efficiently process data-intensive tasks like speech processing and image recognition [2–4] due to the nature of parallel computing. Besides that, the comparison between the conventional structure and neuromorphic chips has shown that the power consumed by neuromorphic computing systems is also lower than traditional von Neumann structures and it is achieved by the parallelism and distributed processing as well as event-driven processing nature of neuromorphic computing. For example, the IBM TrueNorth chip is very power efficient when processing recognition applications [5]. It only consumes less than 3 watts of power. Compared with conventional central processing units (CPUs) or graphics processing units (GPUs), it has shown great improvements. Although GPU has more small specialized cores than CPU, it still consumes tens or even hundreds of watts for the same tasks. Therefore, neuromorphic computing systems, especially those realized by application-specific integrated circuits (ASICs), have demonstrated their superiority in both learning capability and power efficiency aspects.

Among all the types of artificial neural networks (ANNs), one special type is the so-called spiking neural network (SNN) [6–9]. Inspired by the signal transformation in biological neural networks, researchers realized that information could be transmitted in the form of spikes in neural systems [10]. A neuron's function is to receive stimulus and output impulses. It is formed of four main parts, dendrites, soma, axons, and synapses. A dendrite receives the stimulus and transmits it to soma. Soma is the central computing unit in the biological neural network and generates an output signal when the input exceeds a threshold voltage. Each neuron has a specific threshold voltage, and when the output signal is generated, the soma fires a spike to the axon, which transmits the output signal to the synapse. Finally, the impulse will be conveyed through the synapse to the subsequent neurons. With such property, neurons in the network can stay silent unless triggered by coming spikes. In this way, the operating power consumption can be greatly saved. To convert information to the form of spikes, several encoding schemes have been investigated [11, 12]. This research started a few decades ago. These encoding schemes can be categorized into two main types, rate encoding and temporal encoding [13]. The rate encoding scheme represents that information is converted to the number of spikes in one spike train [14]. The spike rate in this encoding scheme means the number of spikes in one encoding window. The larger the input is, the higher the spike rate will be in the corresponding encoding window. This encoding scheme is very straightforward to implement and is one of the most commonly used encoding schemes. On the other hand, temporal encoding takes account of the number of spikes in the encoding window and utilizes the temporal property of the spike train [15]. With the different temporal properties used, the temporal encoding scheme can be more deeply divided into several types, like the Time-to-First-Spike (TTFS) [16], the Interspike Interval (ISI) [17], and the Phase of firing encoding [18].

With the advancement of neuroscience, researchers noticed a special encoding scheme in biological neural systems. This encoding scheme can integrate multiple encoding schemes that operate in different time scales [19]. This kind of encoding scheme is called multiplexing encoding [20]. For instance, ISI encoding can be integrated with phase encoding, forming the multiplexing ISI-phase encoding. Compared with just one encoding scheme, multiplexing encoding has various advantages, including high data capacity and high robustness, especially in noisy environments. The advantages and disadvantages of these encoding schemes are summarized in **Table 1**. For example, rate encoding is not only easier to implement than other schemes but it also has lower data capacity. Temporal encoding schemes not only have higher data capacity than rate encoding but also have lower robustness compared with multiplexing encoding.

Encoding scheme	Rate	Temporal	Multiplexing
Advantage	Straightforward Easy to implement	Higher data capacity	Highest data capacity High robustness
Disadvantage	Low data capacity	Low robustness against noise	High complexity High power and area cost

Table 1.
Advantages and disadvantages of different encoding schemes.

A literature review of those encoding schemes is also carried out. This literature review provides a concise overview of key studies and advancements in the area of spike neural encoding.

In Rolls and Treves [21], the authors have carried out a quantitative analysis of information in the realm of neural encoding. The researchers noticed the existence of the firing rate encoding in the short time window. In the aspect of quantitative analysis, more information is encoded by the rate encoding scheme rather than temporal encoding. In rate code, neurons have been found to be able to take synaptic weighted sums of the inputs for training purposes.

In Auge et al. [22], the authors have summarized the theoretical foundation as well as the applications of encoding schemes. It includes both the rate encoding and the temporal encoding schemes. They concluded that the rate encoding has high robustness since it does not rely on the precise firing timing of spikes. They also noticed that temporal codes have been shown to have higher information capacity, faster reaction times, and higher transmission speeds.

In Kayser et al. [23], researchers have verified the hypothesis that different codes might be employed concurrently and provide complementary stimulus information. They also quantified the information encoded in the auditory cortex of animals and found that multiplexing those codes together will provide a much higher information level. What's more, the authors also found that the multiplexing codes with phase of firing code are very much robust to sensory noise added to the stimulus.

In this chapter, a deeper discussion of these mentioned encoding schemes will be presented in Section 2. Section 3 will discuss the ASIC implementations of these encoding schemes and their simulation results. Lastly, the training results of these encoding schemes working with some common datasets and the hardware testbench of the multiplexing temporal encoder will both be illustrated in Section 4.

2. Neural encoding schemes

As mentioned in Section 1, the neural encoding schemes represent the different ways the input signals get converted to spike signals in spiking neural networks. Researchers have put much effort into finding various encoding schemes that utilize different properties of spike trains in the SNN [24]. The most straightforward and, thus, the first discovered encoding scheme is rate encoding [21]. It uses the number of spikes in one spike train to typify the input information. **Figure 1(a)** shows that the input stimulus is transferred to the firing rate during the sampling window. Therefore, as long as the numbers of spikes are the same, two different spike trains still stand for the same input signal. The simplicity of this code leads to its common use in nowadays applications. For instance, the Intel Loihi chip, which is a neuromorphic

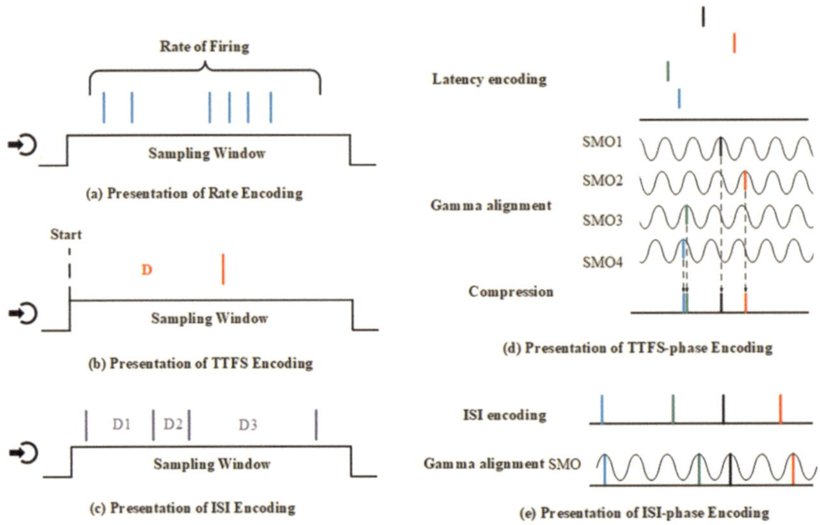

Figure 1.
Examples of encoding scheme. (a) Presentation of rate encoding. Input stimulus is transferred to the firing rate in the encoding window. (b) Presentation of TTFS encoding. Input stimulus is converted to the time difference between the onset of the window and the first spike. (c) Presentation of ISI encoding. Input information is transferred to the time intervals of spikes.

research test chip designed by Intel Labs, uses an asynchronous SNN to implement adaptive self-modifying event-driven fine-grained parallel computations used to implement learning and inference with high efficiency. It utilizes rate encoding in its neural network and has been evaluated through lots of applications such as adaptive robot arm control and drone motor control [25, 26]. It only consumes less than 1 watt of power while maintaining a good operating speed. Moreover, the Tianjic chip also implements rate encoders in the neural network and achieves high accuracy for pattern recognition applications [27, 28]. However, this simple encoding scheme has its disadvantages. First, rate encoding has a relatively lower data capacity than other encoding schemes since it only utilizes the number of spikes in a spike train but ignores the temporal patterns of the spike train. Second, the low data capacity also caused the low robustness of the encoding scheme against noises and errors. Since one spike train only represents one input information, any mistakes of the spike train will lead to an inaccurate result.

To overcome these drawbacks, other encoding schemes that can use other properties of spike trains are proposed for the spike encoding process. Temporal patterns, which mean the different timings of spikes in the spike train, are the most used aspects for encoding [29]. Therefore, a large category of neural code is called temporal encoding, which employs both the spike number and the temporal pattern of the spike train for stimulus. Among these temporal codes, three are the most widely used, the TTFS code, the ISI code, and the phase of firing code.

The time of the first spike encoding, also known as latency encoding, is the most basic temporal encoding scheme [22, 30]. Just as literal, the TTFS encoding converts the input information to the time difference between the onset of the sampling window and the first spike. Since the only useful spike is the first one, for energy efficiency, normally there is only one spike in the spike train for TTFS encoding, as demonstrated in **Figure 1(b)**. Since the onset of sampling windows is often defined by external references, the precision of the encoding process is very much dependent on the accuracy of external signals. Any variation in the external source could affect the performance of the encoder [31]. Another shortcoming of the TTFS encoding scheme is that its robustness is low. With the property that only one spike is effective, even only one mistake in the TTFS-encoded spike train could cause enormous errors in the final output of the encoding process. Thus, the function of the TTFS encoder is not robust against even minor noise or error.

Another neural code is proposed to overcome the disadvantages of the TTFS code, the ISI code. As demonstrated in **Figure 1(c)**, instead of being converted to the time difference between the onset and the first spike, the input stimuli are converted to time intervals of spikes [15]. Unlike latency encoding, ISI encoding utilizes the spikes as the internal reference frame for each other, thus avoiding the dependence on the external references. As discussed in the previous paragraph, one main drawback of latency encoding is its relatively low data capacity. However, since the ISI code has multiple spikes in one sampling window, it can transfer more information than latency encoding. There are two types of ISI encoders, the parallel structure and the iteration structure, introduced in Zhao et al. [30]. The parallel encoder, the simpler type, could convey information faster but maintains fewer spikes in one encoding window. On the contrary, the iteration encoder generates more spikes in the sampling window but also takes more time. Both structures' spike numbers relate to the number of neurons in the encoder. The parallel structure has this relation:

$$N_S = N, \quad (1)$$

where N and N_S are the number of spikes in one sampling window and the number of neurons in the encoder. The iteration structure has an exponential relation:

$$N_S = 2^{N-1}. \quad (2)$$

From Eqs. 1 and 2, we can notice that the iteration structure has more spikes when the encoder has more than two neurons. Thus, when looking for a high data capacity and high robustness, the iteration encoder is a promising candidate [17].

Besides relying on the number of spikes and intervals between spikes, information can also be conveyed as relative position on internal reference frames. The internal reference frames are called the subthreshold membrane oscillation (SMO). The SMO can replace the external reference frame and thus overcome the precision issue. What's more, with the help of SMO, the phase of firing encoding scheme can be implemented in neuromorphic computing systems [32–34]. In the phase encoding, input signals are transferred to the phase of SMO. When the SMO arrives at this phase, one spike will be fired. The math model of the SMO can be written as follows:

$$SMO_i = A \cdot \cos(\omega T + \phi_i), \quad (3)$$

where A, ω, and ϕ present the amplitude of the SMO, the angular velocity of the signal, and the start phase of the sine SMO signal, respectively.

To further improve the performance of neuromorphic computing systems, another encoding mechanism has been investigated by researchers. First found in biological neural systems, the multiplexing encoding schemes combine multiple neural codes, especially the ones with different time scales, to have higher data capacity [23]. Each independent encoding scheme can carry a certain amount of information. For instance, in ISI-phase encoding, the ISI encoding scheme and phase of firing encoding

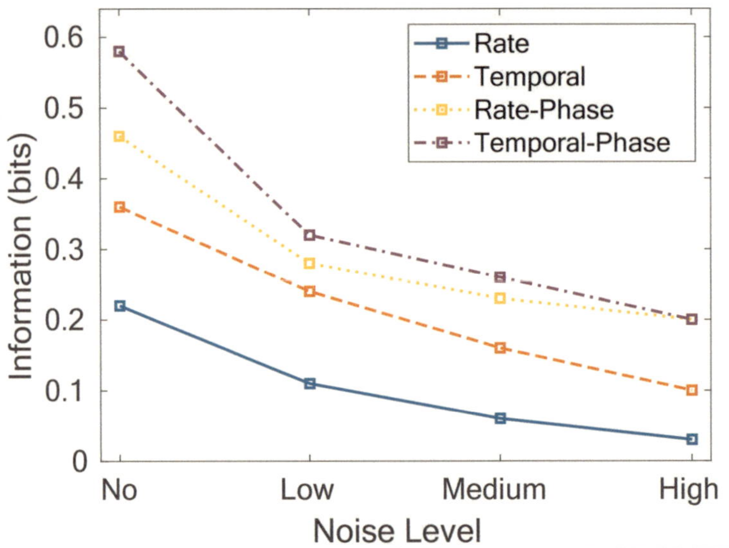

Figure 2.
Information in codes for different noise levels. The blue line indicates the information carried in the rate code with different noise levels. The orange line represents the information level of the temporal encoding scheme. Similarly, the yellow line means the information in the rate and phase of firing multiplexed encoding scheme, and the purple line demonstrates the information carried in the temporal and phase of firing multiplexed code.

scheme carry their own information. After the multiplexing process, the two parts of information are combined and transferred within one sampling time window. Therefore, with multiplexing encoding, the same amount of information can be conveyed within a shorter sampling window and thus increase the data transmission rate [35].

The multiplexing encoding schemes are more robust than the other ones. Experiments have been conducted to quantify the data density in the different neural codes with different levels of input sensory noise [23], as shown in **Figure 2**. From the figure, it is noticeable that although the information from all the encoding schemes decreases with the increase in noise level, the multiplexing encoding schemes always keep the highest data density. Moreover, we can also notice that temporal encoding keeps more information than rate encoding; thus, temporal-phase encoding also has higher data capacity than rate-phase encoding. With the result discussed above, it is proved that the multiplexing encoding scheme is more robust against noisy environments and the high data capacity in all noise levels helps with data transferring from noisy input to spike signals.

The multiplexing encoding requires two separate steps to transfer the input signals to multiplexing encoded spikes [36]. The first one is the encoding process that transferred the analog inputs to differently encoded spikes. For example, if rate encoding is utilized, the encoded spikes are in the form of spike trains with different numbers of spikes. For TTFS encoding, the encoded spike is normally in the form of a single spike. As for the ISI code, the spikes are spike trains with the same number of spikes but different temporal patterns.

After the encoding process, the spikes need to be shifted to meet the phase encoding mechanism [37, 38]. This step is called the gamma alignment step. In this step, the already generated spikes are shifted to the next immediate local maximum of SMOs. The relationship between the original spikes and the shifted spikes can be expressed as

$$P'_\tau = P_t, \tag{4}$$

where t is the timing of the original spike and τ is the next immediate local maximum.

As depicted in **Figure 1(d)**, the TTFS-encoded spikes are processed by the gamma alignment step and become the TTFS-phase encoded spikes. In this figure, the spikes are divided into four different channels. Each channel has its corresponding SMO. The SMOs have the same amplitude and angular velocity, and their phases follow this relationship:

$$\phi_i = \phi_0 + (i-1)\frac{2\pi}{N}, \tag{5}$$

where i means the ith channel and N represents the total number of channels. As for the ISI-phase encoding scheme, since only one channel in **Figure 1(e)** exists, the ISI-coded spikes are shifted to their immediately following local maximums of the same SMO [39].

3. Circuit implementations of the neural encoders

To utilize the various encoding schemes in the application-specific integrated circuit (ASIC) of neural network systems, the circuit implementations of these neural encoders need to be investigated. This section will discuss the circuit implementations of these different encoders and their simulation results with analog sinusoidal current input.

3.1 Rate encoder

The schematic of the rate encoder is demonstrated in **Figure 3(a)**. After the clock signal CLK resets the voltage across the membrane capacitor C1 with the switch transistor M8, an encoding window begins. The voltage across the membrane capacitor C1 increases when the input current is injected. When the membrane voltage exceeds the reference voltage Vref, a spike will be fired through the buffer. The fired spike will also trigger the switch transistor M7 to bring the membrane voltage back to

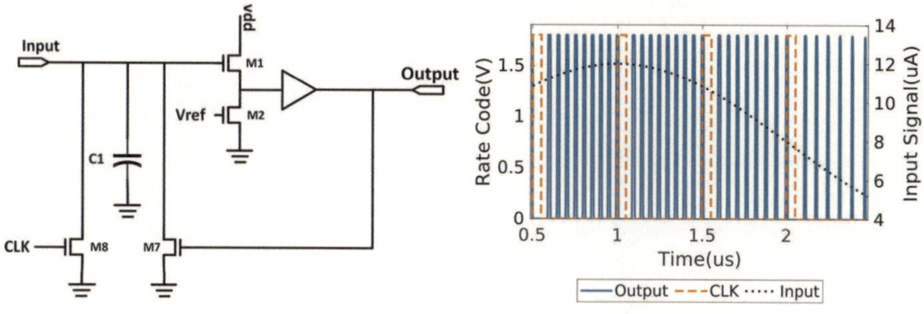

(a) Circuit schematic of rate encoder

(b) Simulation result of rate encoder

Figure 3.
Circuit schematic and simulation result of rate encoder.

the ground so that the integration process will start over. Thus, the relation between the input current and the spike numbers can be written as:

$$N = \frac{P}{T} = \frac{P}{\frac{C_m V_{ref}}{I_{in}}} = \frac{P I_{in}}{C_m V_{ref}}, \tag{6}$$

where N, P, and T represent the spike number, encoding window period, and integration time of one spike. I_{in}, C_m, and V_{ref} mean the input current, membrane capacitance, and the reference voltage. The formula shows that the number of spikes in the sampling window and the input current have a linear relationship. From **Figure 3(b)**, a similar relationship can be observed. When the input current is high, there are more spikes in the sampling window; when the input current is relatively smaller, there are fewer spikes in the sampling window.

3.2 TTFS encoder

The schematic of the TTFS encoder is depicted in **Figure 4(a)** [40]. The charge integration mechanism starts after the CLK signal resets the membrane voltage with switch transistor M11. Along with the voltage across the membrane capacitor C1 increasing, the voltage at the source of the transistor M1 will increase at a rate controlled by Vref. When the source voltage of M1 exceeds the threshold voltage of the inverter composing M3 and M4, the output will be digitally high. Almost immediately after the output becomes digital high, the four-transistor clock-controlled inverter also gives a digital high feedback signal to the switch M11 so that the membrane voltage goes back to the ground. Thus, at the output of the encoder, there will only be a spike instead of a square wave at digital high. Moreover, until the next CLK signal, the feedback signal will always be at the high voltage, so there will only be one spike in one sampling window. The time difference between the onset and the spike can be written as:

$$T = \frac{C_m V_{ref}}{I_{in}}. \tag{7}$$

Figure 4(b) shows that the time difference is inversely proportional to the input current. The larger the input is, the closer the spike is to the CLK signal.

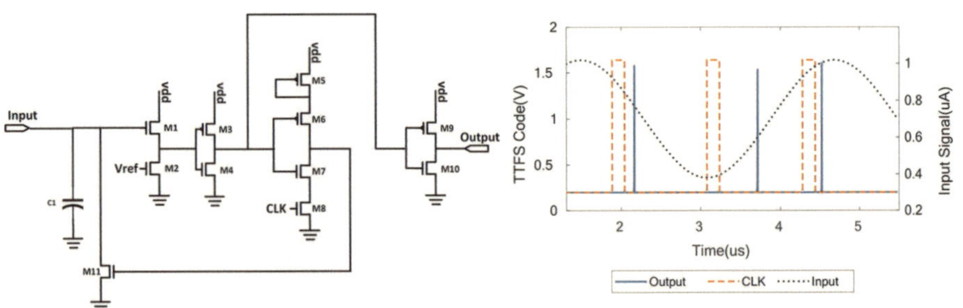

(a) Circuit schematic of TTFS encoder

(b) Simulation result of TTFS encoder

Figure 4.
Circuit schematic and simulation result of TTFS encoder.

3.3 ISI encoder

As mentioned in Section 2, the ISI encoder has two different structures. In this section, we will talk in detail about the parallel structure. Although more neurons in the ISI encoder lead to more spikes in one encoding window, it also leads to higher power consumption and a larger design area. Thus, the two-neuron parallel structure ISI encoder circuit will be discussed in this section. The schematic of the encoder is demonstrated in **Figure 5(a)**. The two neurons utilize the same CLK signal and have the same encoding window. The input currents of the two neurons are the same, so the charge integration rates of the neurons are the same. The only difference between the neurons is that they have different reference voltages so that the neurons will fire spike at different times. Afterward, an OR gate is implemented to integrate the two spikes into a two-spike train. With that, the input information is converted to the time intervals of spikes, which can be expressed as:

$$D = T_2 - T_1 = \frac{C_m\left(V_{ref2} - V_{ref1}\right)}{I_{in}}. \tag{8}$$

The simulation result of the ISI encoder is illustrated in **Figure 5(b)**. It is noticeable that when the input is smaller, the time interval of the spike is larger and vice versa. Thus, this encoder has fulfilled the mathematical relation of the information conversion.

3.4 TTFS-phase encoder

As discussed in Section 2, the TTFS-phase encoding scheme shifts the TTFS-encoded spikes to the immediate local maximum of their corresponding SMOs. Since there is only one channel in our design, as shown in **Figure 6(a)**, the TTFS-phase encoder utilizes only one SMO [37]. To carry out the spike-shifting process, a gamma alignment block is implemented. Inside the gamma alignment block, a peak detector captures and holds the coming spike. The spike voltage will be held across the capacitor with a diode-connected transistor. After that, when the local maximum of the SMO arrives, a spike will be fired by the AND gate and outputted after being stabilized by a buffer. Meanwhile, the spike will trigger the switch transistor and bring the captured voltage back to the ground until the next spike comes.

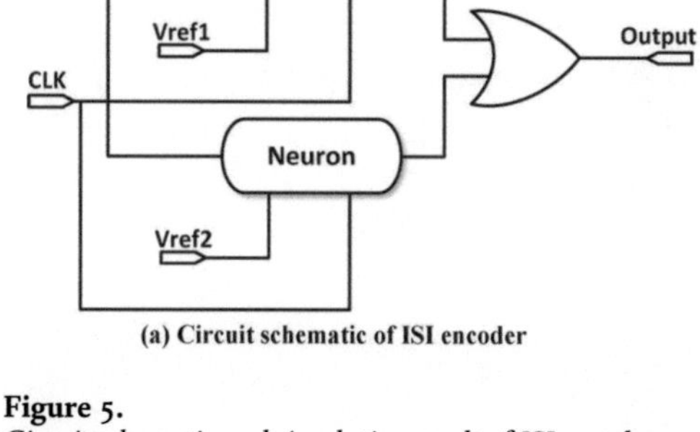

(a) Circuit schematic of ISI encoder

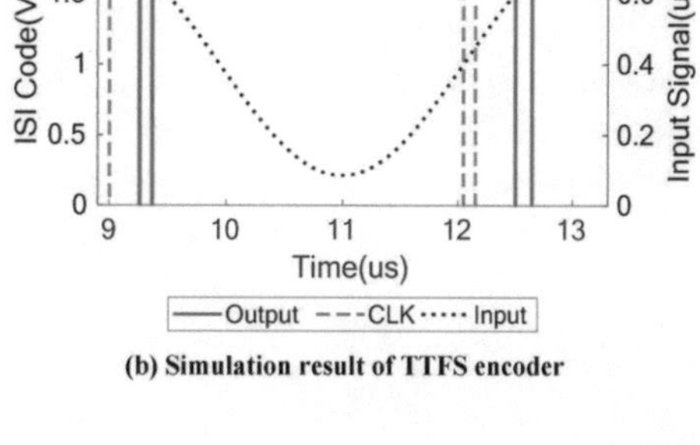

(b) Simulation result of TTFS encoder

Figure 5.
Circuit schematic and simulation result of ISI encoder.

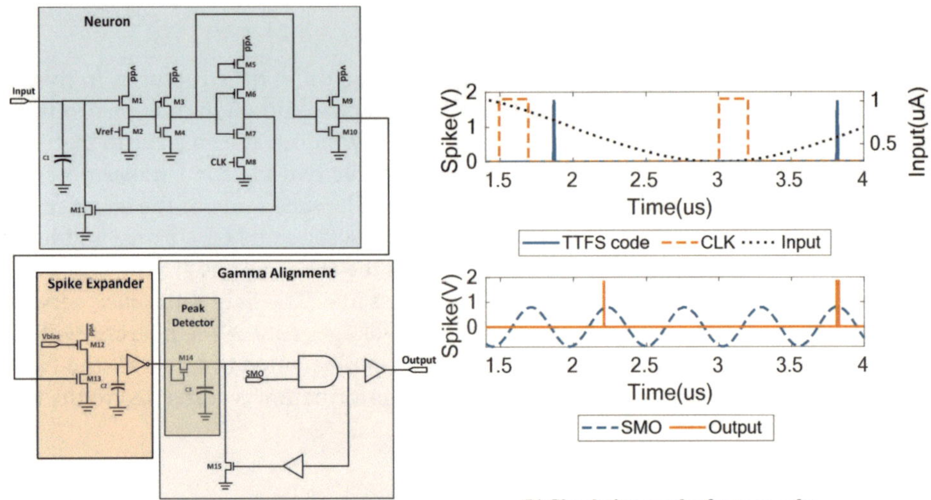

(a) Circuit schematic of TTFS-phase encoder

(b) Simulation result of rate encoder

Figure 6.
Circuit schematic and simulation result of TTFS-phase encoder.

The peak detector can only detect spikes that last longer than 10 nanoseconds (ns). However, the TTFS neuron can only output 1 ns spikes. Thus, a spike expander, as demonstrated in **Figure 6(a)**, is designed to extend the width of spikes. With Vbias controlling the charging rate on the capacitor, the width of spikes can be adjusted without changing the capacitor and thus save a lot of design area. **Figure 6(b)** illustrates the signal flows in the TTFS-phase encoder. The top panel of the figure represents the TTFS encoding function, while the bottom panel depicts the gamma alignment process. After being encoded by the TTFS neuron, the current signal is converted to spikes, and in the gamma alignment block, the TTFS spikes are moved to the next local maximum of the SMO.

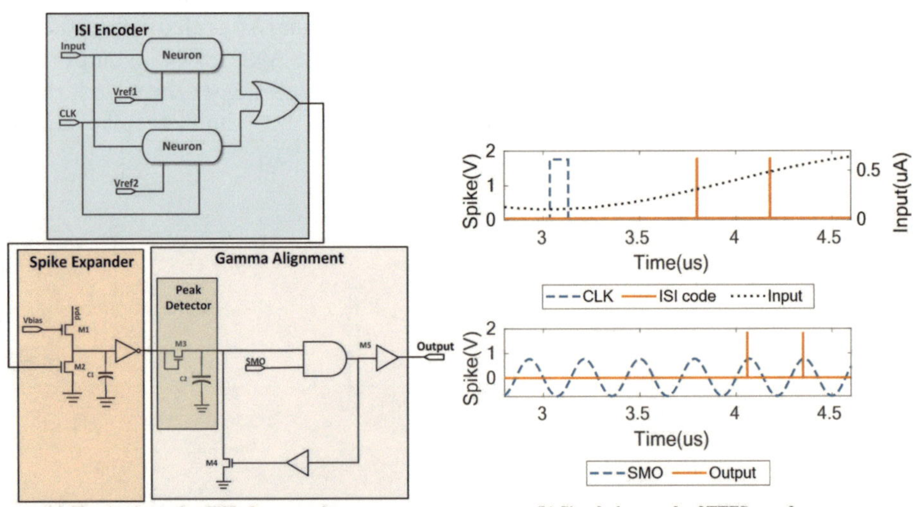

(a) Circuit schematic of ISI-phase encoder

(b) Simulation result of TTFS encoder

Figure 7.
Circuit schematic and simulation result of ISI-phase encoder.

3.5 ISI-phase encoder

Similar to the TTFS-phase encoder, the ISI-phase encoder is designed with an ISI encoder with one spike expander and one gamma alignment block, as shown in **Figure 7**. With the spike expander, the output spikes of the ISI encoder have a width of over 10 ns and thus can be captured by the peak detector. To shift the expanded spikes to the local maximum of the SMO, the gamma alignment block is implemented for the spike train in one sampling window. With the two neurons in the ISI encoder, there are two spikes in one encoding window, and thus the SMO frequency should be higher; otherwise, the two spikes in the same encoding window will possibly be moved to the same local maximum to leave only one spike in the sampling window.

4. Encoder performance analysis

4.1 Performance comparison of encoding schemes

We have implemented SNNs in Python to compare the performance of these various encoding schemes, especially the classification accuracy for popular datasets. For instance, the MNIST dataset [41], the CIFAR-10 dataset [42], and the SVHN dataset [43] are utilized in the comparison for the accuracy of the above-mentioned encoding schemes. In all, 60,000 data points served as the training samples, and 10,000 were testing samples in the MNIST dataset, 50,000 samples were training ones, and 10,000 samples were testing ones. As for the SVHN samples, 73,257 are for training and 26,032 are for testing. The process of verifying encoding scheme performance can be divided into three steps. The first step is to design encoders to convert the datasets to spikes. The second is to build neural networks with these spike encoders and tune the neural networks according to the encoding schemes and datasets to get the desired output. The last step is to run the simulation and compare the accuracy achieved by the various codes.

First, the rate encoder needs to be designed. Due to the nature of the rate encoding that the spiking number in the sampling window has a linear proportional relation with the input signal amplitude, the rate encoding is implemented in Python that outputs spike numbers within the range of 0 to 16. A larger input pixel value leads to a greater number of spikes in one spike train.

Second, the TTFS encoder is implemented to realize the inversely proportional relation of the input pixel value and the first spike time. One more thing to notice is that only the pixel values larger than the threshold can be transferred to spikes. This property maps the neuron functionality more closely. Similarly, the ISI encoder is implemented with multiple TTFS encoders with different thresholds. With those encoders, spikes with different timings are outputted. Those spikes have time intervals also inversely proportional to the input value. For both the TTFS and ISI encoders, the input pixel values are first linearly assigned to the range of 0 to 8 and then transferred to spike trains.

As for the multiplexing neural encoders, the corresponding encoders, the TTFS and ISI encoders, need to be added with the gamma alignment process. Realizing gamma alignment in Python means updating the spike times of the TTFS and ISI encoder outputs with the numbers of an arithmetic progression. The various frequencies of the SMO can be achieved by changing the common difference of the arithmetic progression. Thus, the TTFS-phase encoder and ISI-phase encoder are achieved with different SMO frequencies.

The datasets utilized in this experiment are transferred into different architectures with the help of the TTFS-phase and ISI-phase encoders. Certain training neural networks need to be designed and tuned accordingly to verify the training accuracy of these encoding schemes. These networks are implemented with a PyTorch package for spiking neural networks, SpykeTorch [44]. The two parameters are the threshold of neurons and the size of each layer. The threshold values of neurons are 15 for the first layer and 10 for the rest of the layers. For the MNIST dataset, a 3-layer convolutional neural network is implemented. Due to the datasets' complexity, a deeper neural network is implemented for the CIFAR-10 and SVHN datasets to get the desired results. This network is a 10-layer convolutional neural network. The neurons in both networks are all leaky integrate-and-fire (LIF) neuron models. The neuron model's parameters are: initial $V_m = 0$, $E_L = 0$, $C_m = 100pF$, $R_m = 10kOhm$, $V_{reset} = 0$, and $\tau_m = 1ms$. What's more, the synapses in the SNNs simply take into account the weights and output of neurons and provide excitation current. They do not give any specific influence themselves. The spike-timing-dependent plasticity (STDP) is the training algorithm used in these neural networks since the spiking neural network cannot be trained with backpropagation and is most commonly trained with the STDP training rule. As demonstrated in **Figure 8**, the two convolutional SNNs contain a decision-maker layer to provide reward/punishment signal as part of the reinforcement STDP learning rule. The neurons between the convolutional layers are connected in the relation of N:1. N equals the size of the kernel in the former convolutional layer. Neurons in convolutional layers are organized into local receptive fields that slide across the input data, and they share weights to efficiently learn and capture local patterns.

This simulation is performed with a 12 GB NVIDIA Tesla K80 GPU and 13 GB RAM on the Google Colab platform. For the sake of fairness, the encoding schemes are compared between the other state-of-the-art works and this work, as shown in **Table 2**. For conciseness, the results without reference are from this work. This work

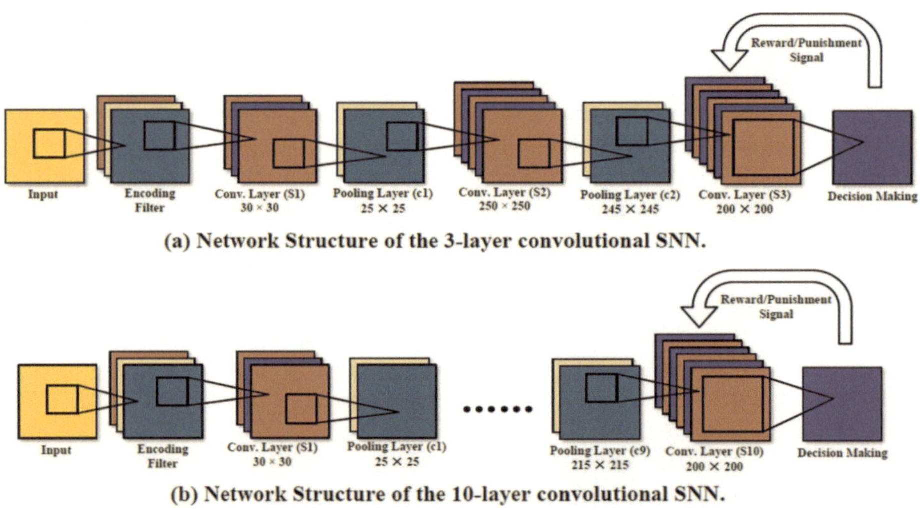

Figure 8.
(a) Network structure of the 3-layer convolutional SNN. (b) Network structure of the 10-layer convolutional SNN.

Encoder type	Rate	TTFS	ISI	TTFS-phase	ISI-phase
MNIST	[45] – 83.0%	[45] – 85.0%	[45] – 90.0%	91.8%	93.8%
CIFAR-10	[46] – 79.7%	[47] – 77.4%	[48] – 83.7%	77.9%	83.8%
SVHN	[49] – 75.0%	[50] – 82.1%	82.8%	82.5%	86.4%

Table 2.
Performance comparison of code-level encoders with the MNIST, CIFAR-10, and SVHN datasets.

achieves 91.8% of accuracy and 93.78% of accuracy for the MNIST dataset. Compared with other works, the multiplexing encoding schemes classify the dataset with at most 10.78% higher accuracy. The ISI-phase encoding gets 83.83% of accuracy for the CIFAR-10 dataset, while the other works at most get 83.71% of accuracy. It classifies the dataset at most 6.4% more accurately than other neural codes. When it comes to the SVHN dataset, the multiplexing encoding has got even more desired results, especially ISI-phase encoding. It has achieved 86.4% of testing accuracy while the rate encoder only gets to 75% of accuracy. The 11.4% of accuracy difference has proved the superiority of the multiplexing encoding schemes. With these comparison results, we notice that the multiplexing encoding, although more complex than other encoding schemes since it requires one more processing step, yields the highest training accuracy for the commonly used datasets. If considering the complexity of the state-of--the-art networks is often higher than that given in this work, the multiplexing encoding can have even higher accuracy for image classification applications. Thus, the multiplexing encoding schemes have the capability to convert datasets into a more classifiable structure and get better training performance for the whole system.

5. Conclusions

In this chapter, we discussed different encoding schemes and the advantages and disadvantages of each spiking neural code. It shows that the rate encoding is straightforward but has low data capacity, the temporal codes have higher data capacity but are not robust against noise, and the multiplexing encoding schemes not only have both high data capacity and high robustness but also have high complexity and thus great power and area cost. Moreover, the mechanisms of the neural encoding schemes are also explained. The input signals are converted to different properties of spike trains in sampling windows. For instance, the signals are converted to the time intervals between the spikes for the ISI encoding, and the inputs are transferred to the spike number for the rate encoding. To utilize these encoding schemes in the analog circuit neural systems, the circuit implementations of these schemes are introduced as long as the mathematical models of the neural codes are employed. To the best of our knowledge, the ISI, TTFS-phase, and ISI-phase encoders proposed by our group are the first IC implementations. We have also built neural networks with different encoders to compare their performance when working with some commonly used image classification datasets. For fairness, we compared our group's performance of multiplexing encoders with those of the state-of-the-art works. For MNIST, the multiplexing encoder achieves 10.78% higher accuracy. For CIFAR-10, the ISI-phase encoder can classify the images 6.4% more accurately. The ISI-phase encoder gets 11.4% of higher accuracy than other works for SVHN. These comparison results have shown that although the multiplexing encoding may require more power and area, it has the potential to achieve better training performance for the whole system.

As for future work, our group is going to implement spike neural networks as well as those above-mentioned encoding schemes in the Neural Simulation Tool (NEST) simulator [51]. With such a simulator, a more detailed and more realistic simulation of SNN can be carried out and there will be more convincing evidence that multiplexing encoding schemes achieve higher data capacity and robustness than rate or temporal codes alone. What's more, we will also investigate the various training algorithms for SNNs, including STDP, spike-based backpropagation, and ANN-SNN conversion. We will dig into the cooperation of different encoding schemes with various training algorithms and find out the most suitable one for the multiplexing encoder. The hardware implementation difficulties of the training algorithms will also be considered as part of the tradeoff.

Author details

Honghao Zheng* and Yang (Cindy) Yi
The Bradley Department of Electrical and Computer Engineering, Virginia Tech, USA

*Address all correspondence to: zhenghh@vt.edu

References

[1] Mead C. Neuromorphic electronic systems. Proceedings of the IEEE. 1990; **78**(10):1629-1636

[2] Bai K, Yi Y. Opening the "black box" of silicon chip design in neuromorphic computing [Internet]. In: Bio-Inspired Technology. London, UK: IntechOpen; 2019. DOI: 10.5772/intechopen.83832

[3] Bai K, Yi Y. DFR: An energy-efficient analog delay feedback reservoir computing system for brain-inspired computing. ACM Journal on Emerging Technologies in Computing Systems (JETC). 2018;**14**(4):1-22

[4] Hamedani K, Zhou Z, Bai K, Liu L. The novel applications of deep reservoir computing in cyber-security and wireless communication [Internet]. In: Intelligent System and Computing. London, UK: IntechOpen; 2020. DOI: 10.5772/intechopen.89328

[5] Akopyan F, Sawada J, Cassidy A, Alvarez-Icaza R, Arthur J, Merolla P, et al. Truenorth: Design and tool flow of a 65 mw 1 million neuron programmable neurosynaptic chip. IEEE Transactions on Computer-Aided Design of Integrated Circuits and Systems. 2015; **34**(10):1537-1557

[6] Hamedani K, Liu L, Hu S, Ashdown J, Wu J, Yi Y. Detecting dynamic attacks in smart grids using reservoir computing: A spiking delayed feedback reservoir based approach. IEEE Transactions on Emerging Topics in Computational Intelligence. 2019;**4**(3): 253-264

[7] Bai K, Li J, Hamedani K, Yi Y. Enabling an new era of brain-inspired computing: Energy-efficient spiking neural network with ring topology. In: 2018 55th ACM/ESDA/IEEE Design Automation Conference (DAC). New York, NY: IEEE; 24 Jun 2018. pp. 1-6

[8] Bai K, Bradley YY. A path to energy-efficient spiking delayed feedback reservoir computing system for brain-inspired neuromorphic processors. In: 2018 19th International Symposium on Quality Electronic Design (ISQED). Santa Clara, CA: IEEE; 13 Mar 2018. pp. 322-328

[9] Hamedani K, Liu L, Liu S, He H, Yi Y. Deep spiking delayed feedback reservoirs and its application in spectrum sensing of MIMO-OFDM dynamic spectrum sharing. Proceedings of the AAAI Conference on Artificial Intelligence. 2020;**34**(02):1292-1299

[10] Ghosh-Dastidar S, Adeli H. Spiking neural networks. International Journal of Neural Systems. 2009;**19**(04): 295-308

[11] Zhao C, Li J, Yi Y. Making neural encoding robust and energy efficient: An advanced analog temporal encoder for brain-inspired computing systems. In: 2016 IEEE/ACM International Conference on Computer-Aided Design (ICCAD). Austin, TX: ACM; 7 Nov 2016. pp. 1-6

[12] Zhao C, Danesh W, Wysocki BT, Yi Y. Neuromorphic encoding system design with chaos based CMOS analog neuron. In: 2015 IEEE Symposium on Computational Intelligence for Security and Defense Applications (CISDA). Verona, NY: IEEE; 26 May 2015. pp. 1-6

[13] Zhao C, Hamedani K, Li J, Yi Y. Analog spike-timing-dependent resistive crossbar design for brain inspired computing. IEEE Journal on Emerging and Selected Topics in Circuits and Systems. 2017;**8**(1):38-50

[14] Cullen KE. The neural encoding of self-motion. Current Opinion in Neurobiology. 2011;**21**(4):587-595

[15] Zhao C, Wysocki BT, Liu Y, Thiem CD, McDonald NR, Yi Y. Spike-time-dependent encoding for neuromorphic processors. ACM Journal on Emerging Technologies in Computing Systems (JETC). 2015; **12**(3):1-21

[16] Nomura O, Sakemi Y, Hosomi T, Morie T. Robustness of spiking neural networks based on time-to-first-spike encoding against adversarial attacks. IEEE Transactions on Circuits and Systems II: Express Briefs. 2022;**69**(9): 3640-3644

[17] Zhao C, Yi Y, Li J, Fu X, Liu L. Interspike-interval-based analog spike-time-dependent encoder for neuromorphic processors. IEEE Transactions on Very Large Scale Integration (VLSI) Systems. 2017;**25**(8): 2193-2205

[18] Montemurro MA, Rasch MJ, Murayama Y, Logothetis NK, Panzeri S. Phase-of-firing coding of natural visual stimuli in primary visual cortex. Current Biology. 2008;**18**(5):375-380

[19] Panzeri S, Brunel N, Logothetis NK, Kayser C. Sensory neural codes using multiplexed temporal scales. Trends in Neurosciences. 2010;**33**(3):111-120

[20] Lankarany M, Al-Basha D, Ratté S, Prescott SA. Differentially synchronized spiking enables multiplexed neural coding. National Academy of Sciences of the United States of America. 2019; **116**(20):10097-10102

[21] Rolls ET, Treves A. The neuronal encoding of information in the brain. Progress in Neurobiology. 2011;**95**(3): 448-490

[22] Auge D, Hille J, Mueller E, Knoll A. A survey of encoding techniques for signal processing in spiking neural networks. Neural Processing Letters. 2021;**53**(6):4693-4710

[23] Kayser C, Montemurro MA, Logothetis NK, Panzeri S. Spike-phase coding boosts and stabilizes information carried by spatial and temporal spike patterns. Neuron. 2009;**61**(4):597-608

[24] Yi Y. Analog Integrated Circuit Design for Spike Time Dependent Encoder and Reservoir in Reservoir Computing Processors. Lawrence, United States: University of Kansas Center for Research, Inc.; 1 Jan 2018

[25] Davies M, Srinivasa N, Lin TH, Chinya G, Cao Y, Choday SH, et al. Loihi: A neuromorphic manycore processor with on-chip learning. IEEE Micro. 2018;**38**(1):82-99

[26] Davies M, Wild A, Orchard G, Sandamirskaya Y, Guerra GA, Joshi P, et al. Advancing neuromorphic computing with loihi: A survey of results and outlook. Proceedings of the IEEE. 2021;**109**(5):911-934

[27] Deng L, Wang G, Li G, Li S, Liang L, Zhu M, et al. Tianjic: A unified and scalable chip bridging spike-based and continuous neural computation. IEEE Journal of Solid-State Circuits. 2020; **55**(8):2228-2246

[28] Pei J, Deng L, Song S, Zhao M, Zhang Y, Wu S, et al. Towards artificial general intelligence with hybrid Tianjic chip architecture. Nature. 2019; **572**(7767):106-111

[29] Zhao C, An Q, Bai K, Wysocki B, Thiem C, Liu L, et al. Energy efficient temporal spatial information processing circuits based on stdp and spike iteration. IEEE Transactions on Circuits

and Systems II: Express Briefs. 2019; **67**(10):1715-1719

[30] Zhao C, Wysocki BT, Thiem CD, McDonald NR, Li J, Liu L, et al. Energy efficient spiking temporal encoder design for neuromorphic computing systems. IEEE Transactions on Multi-Scale Computing Systems. 2016;**2**(4): 265-276

[31] Rueckauer B, Liu SC. Conversion of analog to spiking neural networks using sparse temporal coding. In: 2018 IEEE International Symposium on Circuits and Systems (ISCAS). Florence, Italy: IEEE; 27 May 2018. pp. 1-5

[32] Paraskevopoulou SE, Constandinou TG. A sub-1μW neural spike-peak detection and spike-count rate encoding circuit. In: 2011 IEEE Biomedical Circuits and Systems Conference (BioCAS). San Diego, CA: IEEE; 10 Nov 2011. pp. 29-32

[33] Masquelier T, Hugues E, Deco G, Thorpe SJ. Oscillations, phase-of-firing coding, and spike timing-dependent plasticity: an efficient learning scheme. Journal of Neuroscience. 2009;**29**(43): 13484-13493

[34] Cattani A, Einevoll GT, Panzeri S. Phase-of-firing code. arXiv preprint arXiv:1504.03954. Apr 15 2015

[35] Akam T, Kullmann DM. Oscillatory multiplexing of population codes for selective communication in the mammalian brain. Nature Reviews Neuroscience. 2014;**15**(2):111-122

[36] Nadasdy Z. Information encoding and reconstruction from the phase of action potentials. Frontiers in systems neuroscience. 2009;**3**:6

[37] Zheng H, Mohammadi N, Bai K, Yi Y. Low-power analog and mixed-signal ic design of multiplexing neural encoder in neuromorphic computing. In: 2021 22nd International Symposium on Quality Electronic Design (ISQED). Santa Clara, CA: IEEE; 7 Apr 2021. pp. 154-159

[38] Arriandiaga A, Portillo E, Espinosa-Ramos JI, Kasabov NK. Pulsewidth Modulation-Based Algorithm for Spike Phase Encoding and Decoding of Time-Dependent Analog Data. IEEE Transactions on Neural Networks and Learning Systems. 2019;**31**(10): 3920-3931

[39] Zheng H, Anderson J, Yi Y. Approaching the area of neuromorphic computing circuit and system design. In: 2021 12th International Green and Sustainable Computing Conference (IGSC). Pullman, WA: IEEE; 18 Oct 2021. pp. 1-8

[40] Bai K, An Q, Liu L, Yi Y. A training-efficient hybrid-structured deep neural network with reconfigurable memristive synapses. IEEE Transactions on Very Large Scale Integration (VLSI) Systems. 2019;**28**(1):62-75

[41] Deng L. The mnist database of handwritten digit images for machine learning research [best of the web]. IEEE Signal Processing Magazine. 2012;**29**(6): 141-142

[42] Krizhevsky A, Hinton G. Learning multiple layers of features from tiny images

[43] Netzer Y, Wang T, Coates A, Bissacco A, Wu B, Ng AY. Reading digits in natural images with unsupervised feature learning

[44] Mozafari M, Ganjtabesh M, Nowzari-Dalini A, Masquelier T. Spyketorch: Efficient simulation of convolutional spiking neural networks

with at most one spike per neuron. Frontiers in Neuroscience. 2019;**13**:625

[45] Nowshin F, Yi Y. Memristor-based deep spiking neural network with a computing-in-memory architecture. In: 2022 23rd International Symposium on Quality Electronic Design (ISQED). Santa Clara, CA: IEEE; 6 Apr 2022. pp. 1-6

[46] Nguyen VT, Trinh QK, Zhang R, Nakashima Y. STT-BSNN: An In-Memory Deep Binary Spiking Neural Network Based on STT-MRAM. IEEE Access. 2021;**9**:151373-151385

[47] Cao Y, Chen Y, Khosla D. Spiking deep convolutional neural networks for energy-efficient object recognition. International Journal of Computer Vision. 2015;**113**(1):54-66

[48] Park S, Kim S, Choe H, Yoon S. Fast and efficient information transmission with burst spikes in deep spiking neural networks. In: 2019 56th ACM/IEEE Design Automation Conference (DAC). Las Vegas, NV: IEEE; 2 Jun 2019. pp. 1-6

[49] Wang Z, Liu J, Ma Y, Chen B, Zheng N, Ren P. Perturbation of spike timing benefits neural network performance on similarity search. IEEE Transactions on Neural Networks and Learning Systems. Sep 2022;**33**(9): 4361-4372

[50] Ma C, Yan R, Yu Z, Yu Q. Deep spike learning with local classifiers. IEEE Transactions on Cybernetics. May 2023; **53**(5):3363-3375

[51] Plesser HE, Diesmann M, Gewaltig MO, Morrison A. NEST: the Neural Simulation Tool. In: Jaeger, D., Jung, R. (eds) Encyclopedia of Computational Neuroscience. New York, NY: Springer; 2015. DOI: 10.1007/978-1-4614-6675-8_258

Chapter 8

Cortical Columns Computing Systems: Microarchitecture Model, Functional Building Blocks, and Design Tools

John Paul Shen and Harideep Nair

Abstract

Reverse-engineering the human brain has been a grand challenge for researchers in machine learning, experimental neuroscience, and computer architecture. Current deep neural networks (DNNs), motivated by the same challenge, have achieved remarkable results in Machine Learning applications. However, despite their original inspiration from the brain, DNNs have largely moved away from biological plausibility, resorting to intensive statistical processing on huge amounts of data. This has led to exponentially increasing demand on hardware compute resources that is quickly becoming economically and technologically unsustainable. Recent neuroscience research has led to a new theory on human intelligence, that suggests Cortical Columns (CCs) as the fundamental processing units in the neocortex that encapsulate intelligence. Each CC has the potential to learn models of complete objects through continuous predict-sense-update loops. This leads to the overarching question: *Can we build Cortical Columns Computing Systems (C3S) that possess brain-like capabilities as well as brain-like efficiency?* This chapter presents ongoing research in the Neuromorphic Computer Architecture Lab (NCAL) at Carnegie Mellon University (CMU) focusing on addressing this question. Our initial findings indicate that designing truly intelligent and extremely energy-efficient C3S-based sensory processing units, using off-the-shelf digital CMOS technology and tools, is quite feasible and very promising, and certainly warrants further research exploration.

Keywords: neuromorphic computing, neocortical columns, spiking temporal neural networks, online continual learning, intelligent sensory processing

1. Introduction

The field of deep learning (DL) has seen extraordinary progress over the last decade and has established itself as the de facto standard technology for sensory processing tasks such as visual object recognition/detection, audio/time-series signal processing, natural language processing, etc. However, this progress has thrived with heavy dependence on increasing hardware resources, and power and energy consumption. These hardware resources such as CPUs, GPUs and specialized accelerators primarily employ Turing computation model and von-Neumann computer architecture. Such computation model and architecture were originally developed for numerical and symbolic processing tasks that were difficult for humans. In contrast, human neocortex is highly efficient in sensory processing and pattern recognition and capable of online continual learning. Current deep learning is attempting to replicate human-like sensory processing capability using conventional Turing-von Neumann computing machines.

It has been observed by Hans Moravec and others that these two computation models and systems are quite different and distinct; this observation is known as the "Moravec's Paradox" [1–3]. The current overarching grand challenge is: Can we design much more energy-efficient computing hardware systems for human-like sensory processing with continuous learning capability by mimicking the architecture, organization, and operation of the human neocortex? This grand challenge is not new and has been around for decades. In 1990, Carver Mead first coined the term "neuromorphic computing" [4] for this grand challenge. But the same notion and aspiration can be traced back to Frank Rosenblatt's "perceptrons" from the late 1950's [5]. This chapter describes our particular approach and strategy in pursuing this grand challenge as part of the current resurgence of interest in neuromorphic computing [6].

Our research builds on the seminal works by James E. Smith on temporal neural networks (TNNs) [7, 8] and is strongly influenced by Jeff Hawkins' recent book "A Thousand Brains: A New Theory of Intelligence" [9]. Unlike convolutional neural networks (CNNs), TNNs are temporal spiking neural networks that embrace a strong adherence to biological plausibility [10]. TNNs encode and process information in temporal form mimicking the brain's neocortical sensory signal processing. We developed a microarchitecture model for implementing highly efficient TNN designs [11, 12] and demonstrated state-of-the-art clustering performance on a wide variety of time-series signals [13].

Hawkins' new theory on intelligence [9], informed by extensive neuroscience research, suggests Cortical Columns (CCs) as the key compute units within the human neocortex. The neocortex gains its intelligence through CC's ability to model sensory information in structured Reference Frames (RFs), and continuously update its models as the sensor interacts with the environment. Each CC is computationally powerful and can learn any specific task. There is great synergy between CCs and TNNs.

Our current research focuses on extending the TNN design and implementation framework to incorporate CC attributes. Unlike artificial neural networks that separate training and inference, CCs store and process information in RFs, support online continuous learning, and can dynamically adapt to sensory input changes. Sensory processing units built based on such CCs, can be truly "intelligent" as per Hawkins' definition, and can enable contextualization and personalization of applications and services for supporting edge AI on diverse mobile and wearable devices.

This chapter presents a framework for designing and implementing *Cortical Columns Computing Systems (C3S)*. This framework consists of three major components: (1) a microarchitecture model for designing and implementing cortical columns and CC-based computing systems using off-the-shelf digital CMOS technology; (2) a suite of specialized functional building blocks to implement application-specific CC-based processing units with significant improvements on Power-Performance-Area (PPA) efficiency; and (3) a PyTorch-based software simulator tool for rapid design space exploration targeting specific applications; and a design synthesis tool for direct CMOS implementation of special-purpose C3S sensory processing units in the form of chiplets, potentially supporting the latest Universal Chiplet Interconnect Express (UCIe) [14] standard. This chapter presents the progress we have made on C3S and our future research directions.

2. Background and motivation

2.1 Unsustainable trend for DL compute

Deep neural networks (DNNs) [15–18] have advanced state-of-the-art in a plethora of applications, particularly those mimicking human sensory processing tasks such as image recognition, object detection, time-series signal (e.g., speech) processing etc. However, due to exponentially growing models performing high-dimensional tensor processing and global gradient backpropagation, their compute requirements are scaling unsustainably. Specifically, a study by OpenAI in 2018 [19] illustrated that DNN training compute is doubling every 3.4 months. Comparing this with hardware driven by Moore's Law doubling every 24 months, the gap between compute demands and hardware supply is increasing at an exponential rate of 8x per year or 500x every 3 years (see **Figure 1**). There is also growing evidence that the current trajectory of deep learning compute scaling is economically, technologically and environmentally unsustainable [20–22]. On a more optimistic note, some other recent studies have claimed that the rate of DNN model explosion has slowed down between 2018 and 2020 [23, 24], and that the trend is expected to plateau and shrink in the future [25].

There are significant ongoing efforts to deal with this DNN compute complexity explosion. Reducing data value precision through quantization and DNN model sizes

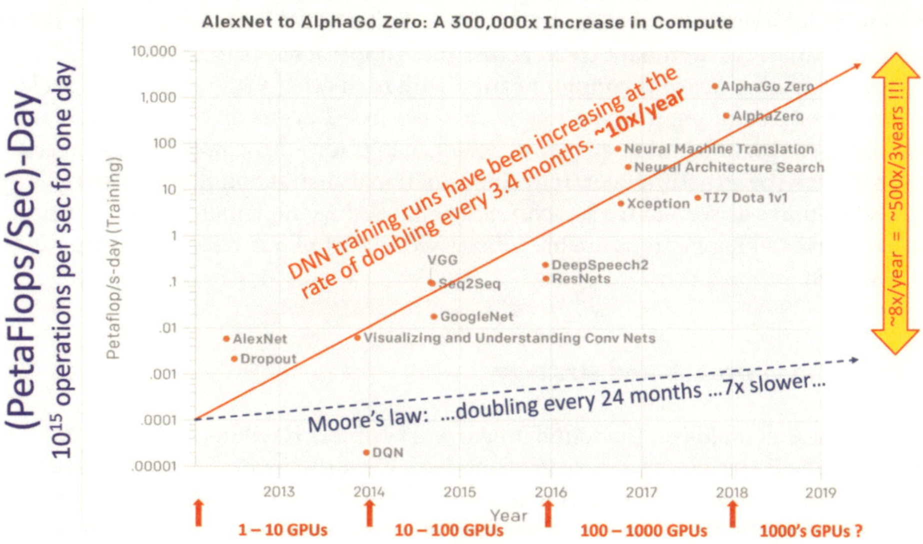

Figure 1.
Figure taken from [19] with annotations added. There is a significant trending gap between DNN computation demand and hardware resource supply provided by Moore's law. This demand-supply gap is increasing at the rate of 8x/year. Thus, to sustain this trend, every year the amount of hardware resources would need to increase by 8× or the training time will need to increase by 8×. A promising candidate solution is to revisit and develop computing paradigms based on the brain's neocortex, which is an existence proof for a highly efficient computing machine for intelligent sensory processing.

through pruning can help mitigate the computation complexity [26–31]. More efficient ways of using the computation hardware infrastructure, including static or dynamic exploitation of value sparsity to avoid unnecessary computation, are also being developed [32–38]. These are all valuable efforts to mitigate the complexity explosion and to help sustain the continuation of the current productive trends. However, there is also the need to concurrently explore other potentially promising alternative paradigms and approaches.

2.2 Right time to revisit neuromorphic computing

All current commercial accelerators for Artificial Intelligence (AI) computation employ the same computing paradigm that emerged more than 70 years ago based on the Turing computation model and the von Neumann stored-program computer architecture. However, these systems were not originally developed for targeting human-type sensory processing workloads that constitute majority of modern AI compute. This underscores the potential need to explore a much more complexity- and energy-efficient computing model and architecture for AI computing. One approach is to revisit biology and examine how to mimic not just the functional behaviors, but also the structural organization of biological neural networks. Such an approach can potentially enable real intelligent computing with significantly less computation complexity and much better energy efficiency. One of the last major efforts taking such "neuromorphic computing" approach was by Carver Mead and his PhD students at CalTech back in the late 1980's utilizing analog VLSI circuits to model neural computation [4, 39, 40]. Both experimental neuroscience research and silicon fabrication technology have made tremendous advancements in the past 40 years, making this a good time to revisit the neuromorphic computing approach [6].

In recent years, several neuromorphic chips have been introduced both from academia and industry, including analog [41], digital [42–48] and mixed-signal [49] implementations. A dominant trend across these approaches is to implement a large number of spiking neurons communicating with each other via event-driven packets that encapsulate spiking information. Such spike-based event-driven computations have been shown to be more energy-efficient than traditional compute. However, they still lack the structural hierarchy and functional abstraction in the form of cortical columns as seen in the neocortex. Digital CMOS implementation of cortical columns and spiking neurons are both discussed as part of our research strategy in the next section.

3. Research approach and strategy

Our research builds on the foundational works of Jeff Hawkins and James E. Smith on reverse-engineering the neocortex from both a neuroscience theorist's perspective and a computer architect's perspective, respectively. These two independent perspectives are uncannily synergistic, and we fully embrace and integrate both of these perspectives in formulating our research approach.

3.1 Hawkins' new theory of neocortical computation

In 2021, Jeff Hawkins published a fascinating book entitled *"A Thousand Brains: A New Theory of Intelligence"* [9]. In this book, accompanied by additional publications from others at Numenta [50–53], Hawkins proposes a new theory of neocortical computation informed by extensive neuroscience research. Hawkins suggests Cortical Columns (CCs) as the key compute units within the human neocortex, which contains about 150,000 of such CCs (see **Figure 2**). The neocortex gains its intelligence through CC's ability to model sensory information in structured Reference Frames (RFs) and continuously update its models as the sensor interacts with the environment.

Each CC is computationally powerful and capable of modeling complete objects. As illustrated in **Figure 2**, complete models of an object exist in multiple CCs across different sensory modalities (e.g., vision, hearing, touch) and across different hierarchy layers at different scales. Multiple CCs, across sensory modalities and layer hierarchies, can communicate and reach consensus via a voting process. This is in contrast to the traditional strict hierarchical view, involving increasing complexity and sophistication with ascending hierarchical layers. Hawkins suggests the increasing level of intelligence demonstrated by higher-functioning mammals is mainly due to the increase in the total number of CCs and not necessarily due to increased complexity or sophistication of the CCs. Furthermore, in contrast to current DNNs that separate training and inference, CCs that store and process information in RFs can support online, concurrent, and continuous learning and can dynamically adapt to sensory input changes.

Recent discussions in the DL community seem to align with the basic ideas of Hawkins' theory. Continually learning structured models of the world via interactions with the environment can help overcome the brittleness and catastrophic forgetting associated with DNNs [54]. With the ability to continually learn and adapt to new

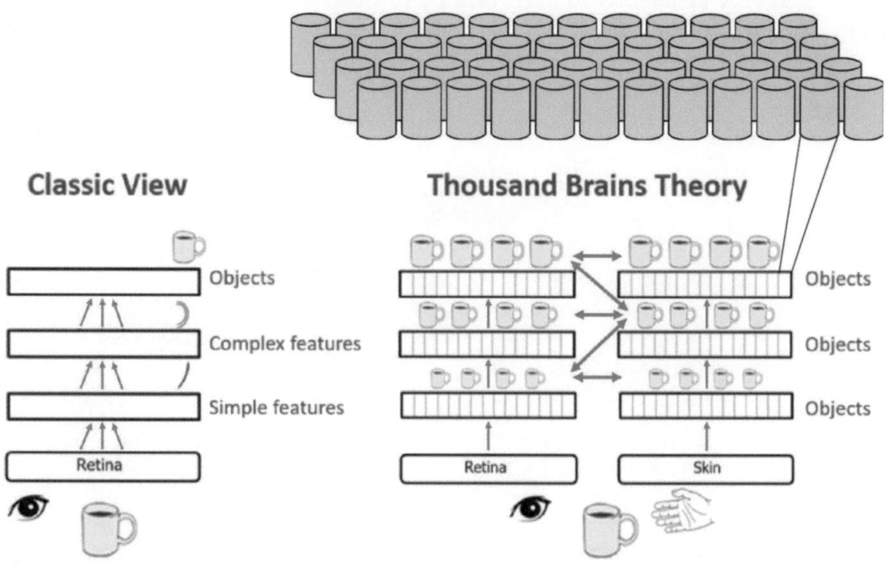

Figure 2.
Figure taken from [51, 53]. Left: Traditional hierarchical view where complexity of features recognized increases up the hierarchy with complete objects detected only at the top. Right: Hawkins' view where neocortex contains about 150 K cortical columns (CCs) and multiple CCs across different sensory modalities and different hierarchy layers can all learn complete models of objects and can communicate to reach a consensus on the output. For example, a coffee cup can be quickly recognized through touch and vision, wherein two sets of CCs with modality-specific models of the coffee cup at different scales all vote together to determine the object.

inputs from the environment, the need to achieve near-perfect accuracy through extensive offline training process can be alleviated. This can potentially reduce the offline training cost and complexity.

3.2 Smith's biologically plausible neural networks

We leverage and build on another significant body of work by James E. Smith on reverse-architecting the brain with the goal of replicating it using off-the-shelf digital CMOS silicon [7, 8, 10, 55]. He proposes a new category of spiking neural networks called temporal neural networks (TNNs) [7] that are architected to mimic the key attributes of biological neocortex (see **Figure 3**). Unlike DNNs performing real-valued tensor-based computations supported by global back propagation for stochastic gradient descent, TNNs employ spiking neurons that encode and process inputs as timings of events or spikes, and can learn using local biologically plausible algorithm called spike timing dependent plasticity (STDP). TNNs also differ from most other spiking neural networks (SNNs) that encode values based on the rate of spikes as opposed to spike timing. **Figure 3** shows this taxonomy. Consequently TNNs, unlike most other ANNs, are more truly "neuromorphic" due to their strong adherence to biological plausibility.

Smith has also developed a TNN-based architecture for the cortical columns and demonstrated the capability for doing unsupervised learning with rapid convergence, and the capability to do online, concurrent, and continuous learning [10, 55]. Formulating the underlying mathematical basis for implementing TNNs, Smith has proposed a new algebra called space-time (temporal) algebra [8]. Based on this new temporal algebra (instead of Boolean algebra), temporal functions can be implemented very efficiently in hardware by re-purposing the current digital logic gates to make use of time as a "free" resource for both encoding and processing of information.

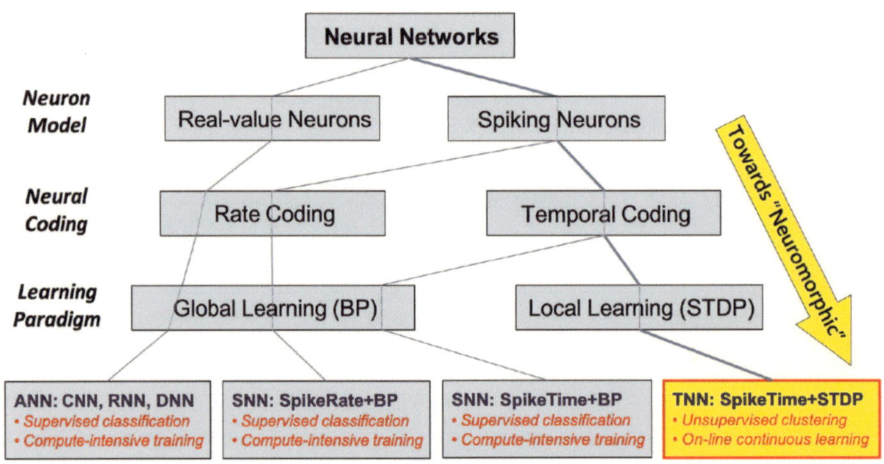

Figure 3.
Neural network taxonomy contrasting neocortex-inspired temporal neural networks (TNNs) with other artificial neural networks (ANNs). In contrast to other ANNs including deep neural networks (DNNs), TNNs incorporate attributes with strong adherence to biological plausibility, including spiking neuron model, temporal coding of inputs, and local simple spike timing dependent plasticity (STDP) learning rules.

3.3 Cortical columns computing systems

Our research builds on these two bodies of prior works to explore the potential of creating brain-inspired and brain-like computing fabric, which we call *Cortical Columns Computing Systems (C3S)*. **Figure 4** compares our C3S approach against conventional computers and biological neocortex. Our goal is to leverage the best attributes of the other two paradigms to create a new computing paradigm. **Figure 4** shows the linkage of comparable levels of abstractions across these three distinct computing system paradigms. Our research spans all four levels of abstractions including sensory processing applications, a processor-level microarchitecture model, RTL-level functional units, and gate-level building blocks. In order to target and impact mass market computing, our research must be able to leverage current digital CMOS technology and design tools. One of our goals is to develop an end-to-end framework for designing and implementing Cortical Columns Computing Systems (C3S). This will require the development of novel design exploration and design synthesis tools. Our target applications include diverse sensory signal processing units capable of energy-efficient and edge-native AI inference and online continuous learning.

As illustrated in **Figure 5**, this framework consists of three key components: (1) a microarchitecture model that can facilitate RTL implementation of Cortical Columns (CCs) and Reference Frames (RFs) employed in C3S designs; (2) a suite of highly optimized functional units and building blocks implemented in System Verilog to support efficient application-specific implementations of C3S designs; and (3) a software tool suite consisting of a PyTorch simulator for rapid design space exploration of C3S designs and a design synthesis flow for mapping PyTorch functional models to corresponding C3S hardware. Specific applications of current interest include visual object recognition, anomaly detection on time-series signals (e.g., ECG), edge-native always-on keyword spotting, and multi-modal human activity recognition (HAR) [56].

ABSTRACTIONS	CONVENTIONAL COMPUTERS	CORTICAL COLUMNS COMPUTING SYSTEMS	BIOLOGICAL COMPUTING SYSTEMS
LEVEL 1: SYSTEM	Computing Systems	NEUROMORPHIC COMPUTING: SPECIALIZED SILICON NEOCORTEX (Applications & System Architecture)	Cerebral Neocortex
LEVEL 2: SUBSYSTEMS	CPU, GPU, TPU, NPU, Mem, I/O	DIVERSE SENSORY PROCESSING UNITS: CORTICAL COLUMNS, REFERENCE FRAMES, VOTE (Processor-Level Microarchitecture)	Neocortical Lobes/Regions
LEVEL 3: FUNCTIONAL UNITS	ALU, FPU, LSU, ARF, ROB, LSQ	HIGHER-LEVEL PATTERN CLUSTERING: MINICOLUMNS, INPUT ENCODE, OUTPUT DECODE (RTL-Level Functional Units Implementation)	Neocortical Columns
LEVEL 4: BUILDING BLOCKS	Adders, Decoders, Multiplexers, etc.	LOWER-LEVEL FEATURE EXTRACTION: NEURON, DENDRITE, SYNAPSE, STDP, WTA (Gate-Level Building Blocks Implementation)	Neurons Excitatory/Inhibitory

Figure 4.
Comparison of cortical columns computing systems (C3S) against conventional computers and biological computing systems in terms of levels of computing abstraction. Our research strategy is highlighted in the red box and spans the four abstraction layers: Applications and system architecture, processor-level microarchitecture, RTL-level functional units, and gate-level building blocks.

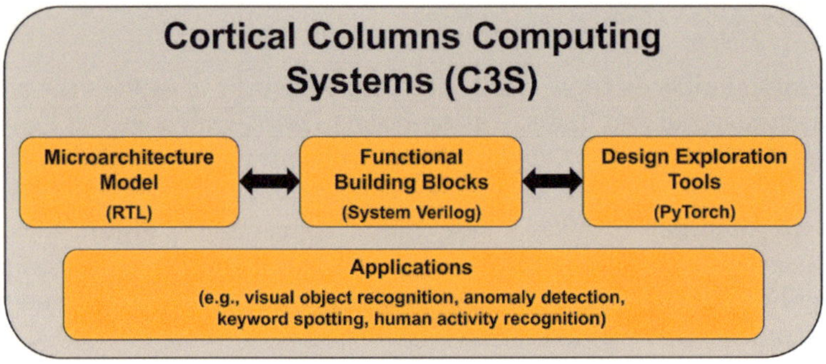

Figure 5.
Proposed framework for implementing C3S designs consists of three key components: Microarchitecture model [11], functional building blocks [12] and design exploration tools. Some applications of interest are visual object recognition, anomaly detection [13], keyword spotting, and human activity recognition [56]. These framework components currently support TNN design and implementation and key relevant publications are cited here. Our ongoing research aims to extend this framework to support more general C3S designs.

4. C3S microarchitecture model

4.1 Mini-columns and TNNs (completed work)

We have developed a microarchitecture model for building highly efficient TNN designs [11]. In this model, values are encoded as timings of events or spikes. Spikes are implemented as logic pulses whose timings are calibrated using unit hardware clock. We refer to this as "direct" implementation as hardware clock itself defines the time unit for temporal processing and spikes directly arrive at precisely timed clock cycles. Here, spike timings are not encoded as binary values propagated in the form of packets. The proposed TNN microarchitecture model consists of the following key modules: (1) a counter-based synapse that performs temporal processing of input spikes to generate specialized responses and unsupervised/ supervised STDP learning of synaptic weight; (2) a multi-synapse neuron that accumulates the synaptic responses and fires an output spike when the accumulated response crosses a threshold; and (3) a multi-neuron TNN column (referred henceforth as "mini-column" to distinguish from cortical column) that applies winner-take-all (WTA) inhibition across its neurons. **Figure 6** illustrates the implementation of a mini-column.

A mini-column with p synaptic inputs feeding q neurons via a pxq synaptic crossbar is referred to as a *(pxq) mini-column* in **Figure 6**. The figure illustrates the actual implementation of the major components of a (pxq) mini-column, including the synapses, neuron body, STDP local learning, and WTA lateral inhibition. In our prior work [11], we have taken the RTL designs of various configurations of the mini-column through both the synthesis and physical design tools, and obtained PPA (power, performance, area) results by scaling both p and q. We observed that power and area scale linearly with the total number of synapses (pxq), whereas performance limited by the critical path delay scales logarithmically with the number of synaptic inputs (p). We also derived a set of characteristic equations from the gate-level designs that provide qualitative estimation of gate-level PPA complexity for arbitrary mini-column designs.

DOI: http://dx.doi.org/10.5772/intechopen.110252

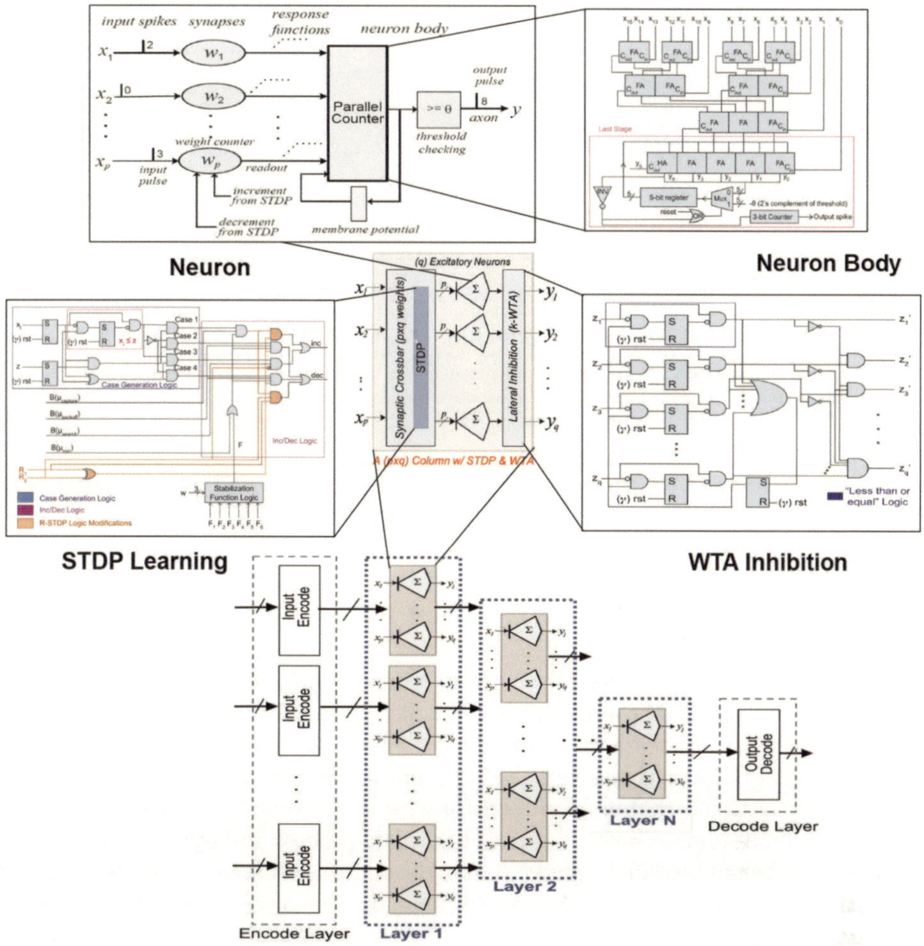

Figure 6.
Microarchitecture model for implementing multi-column and multi-layer TNNs [11]. This figure highlights details of a generic TNN column (or mini-column). Each mini-column consists of p inputs feeding a stack of q neurons via a pxq synaptic crossbar. Each cross point in the pxq crossbar stores a weight value that is updated based on spike timing dependent plasticity (STDP) updating rules. Each neuron performs the weighted sum of its inputs. Each mini-column is also supported by winner-take-all (WTA) inhibition across its neuron outputs that selects the winner neuron. Gate-level implementations of these mini-column components are illustrated.

Multiple mini-columns can be grouped and organized into a hierarchy of layers to form multi-layer temporal neural networks (TNNs), as shown at the bottom of **Figure 6**. Multiple multi-neuron mini-columns are stacked to form a single layer and multiple multi-column layers are cascaded to form a multi-layer TNN. A TNN is typically bookended by input-encode and output-decode layers, and is effectively a feed-forward network. The mini-column is the fundamental building block for TNNs and can learn to distinguish distinct input patterns. Hence, a single mini-column can be viewed as a fully functioning TNN. **Table 1** in Section 6 demonstrates the efficacy of single mini-column designs in performing unsupervised clustering with minimal PPA complexity.

UCR Mini-column Design (pxq)	UCR Benchmark Name	Synapse Count	Power (μW)	Comp. Time (ns)	Area (mm^2)
65x2	SonyAIBORobotSurface2	130	0.78	13.86	0.001
96x2	ECG200	192	1.12	14.49	0.001
152x2	Wafer	304	1.76	15.96	0.002
343x2	ToeSegmentation2	686	3.95	17.32	0.005
637x2	Lightning2	1274	7.33	18.75	0.011
470x5	Beef	2350	14.62	22.08	0.018
270x25	WordSynonyms	6750	39.00	17.51	0.054

All mini-columns are specifically trained for their corresponding benchmarks and achieve competitive performance relative to state-of-the-art. Even the largest mini-column design only consumes 39 μ W.

Table 1.
Design space exploration for UCR time-series clustering: Using TNN7 macros, PPA [12] for seven sample TNN prototype designs (mini-columns) across diverse synapse counts and application benchmarks from [13].

4.2 Cortical columns with RFs (ongoing work)

The current phase of our research aims to leverage the high synergy between cortical columns and TNN mini-columns. We plan to extend the TNN microarchitecture model to incorporate CCs consisting of RFs, making it generic enough to accomplish integration across diverse sensory modalities (see **Figure** 7).

In contrast to TNNs that employ feed-forward processing, CCs store structured information about inputs in Ref. Frames (RFs), and process and update the stored information through feedback connections. This continuous feedback loop of predict-sense-update effectively introduces an additional dimension of "memory" that remembers past observations and patterns. This "sequential" behavior is missing in feed-forward TNNs. Hence, TNN mini-columns combined with feedback mechanism implementing the predict-sense-update loop can be used to build cortical columns. As shown in **Figure** 7, multiple CCs targeting different sensory modalities can interact with each other and seek consensus on the output via voting within and across sensory modalities. Each CC, irrespective of its sensory modality, broadly implements two components: 1) *a Reference Frame* that maintains a "map" of the sensory information, and 2) an *Agent* that achieves goal-oriented behavior based on information from the Reference Frame and the input signals. Agent comprises of two TNN-type mini-columns performing unsupervised clustering and supervised classification. Reference Frame involves three functionalities which can be mapped to three types of mini-columns: *Where* Column, *What* Column and *Output* Column. In the context of visual object recognition, the three types of mini-columns together create models of objects by tracking locations of features on the object. The *Output* Column is envisioned to be very similar to the TNN mini-column discussed previously and determines the object identity. The *Where* and *What* Columns take the outputs from the *Output* Column as feedback information. The Where Column generates location of sensor on the object based on this feedback and the latest movement information from the agent. The *What* Column predicts object features based on the result from the *Where* Column and updates its model based on the actual sensory input and the feedback from the *Output* Column. In order to apply this model to other time-series applications, a key

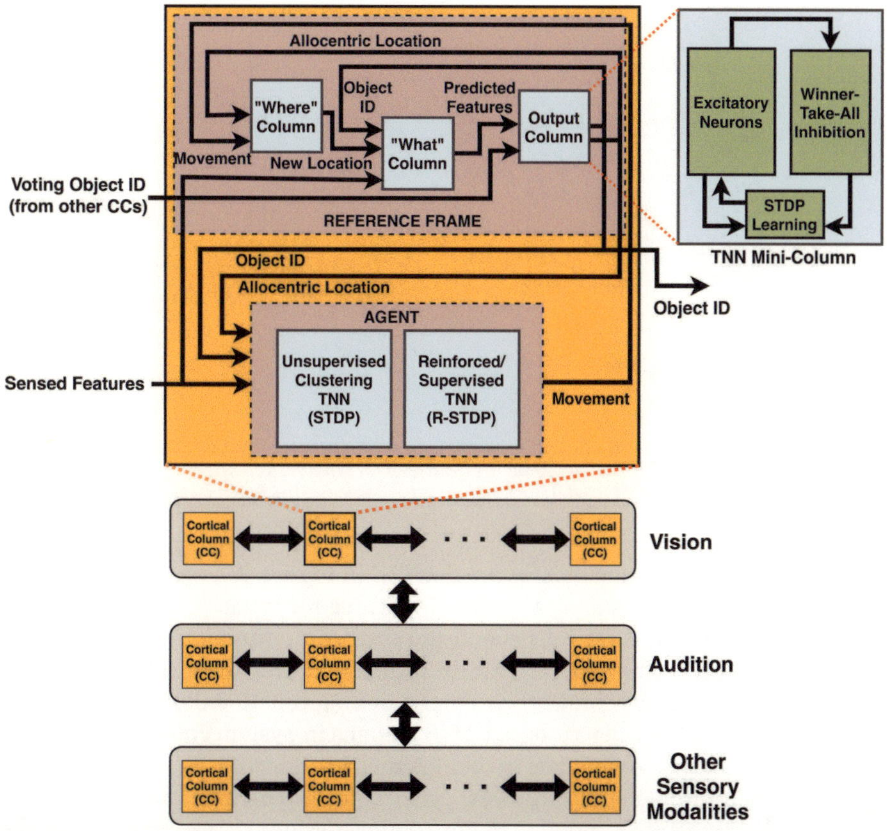

Figure 7.
Cortical columns computing system (C3S) architecture consisting of multiple CCs targeting multiple sensory modalities interacting with each other to form a consensus on the output via voting. Each CC broadly consists of five TNN-style mini-columns: Where, What and Output mini-columns that together implement the Reference Frame (RF), and unsupervised and supervised mini-columns comprising the agent. For visual object recognition, the respective functionalities of the three RF mini-columns are: derive locations of sensor on the object, map features to locations, and derive the object ID based on the feature map. In contrast to feedforward TNNs, each CC learns through feedback from output and possesses a form of "memory" in the learning process.

component that needs to be investigated is pre-processing of the diverse sensory signals to extract abstract "location-feature" pairs from raw sensory signals. Developing a detailed parameterized design template of a generic Cortical Column (CC) containing the five mini-column types is largely ongoing research.

5. C3S functional building blocks

5.1 Custom macro cells for mini-column designs (completed work)

From the TNN microarchitecture model, we identified the key fundamental building blocks and implemented them as a suite of custom standard cell macros in 7 nm CMOS, called *TNN7* [12]. Using these custom macros, we can further optimize the TNN designs and enhance their scalability. These building blocks and the

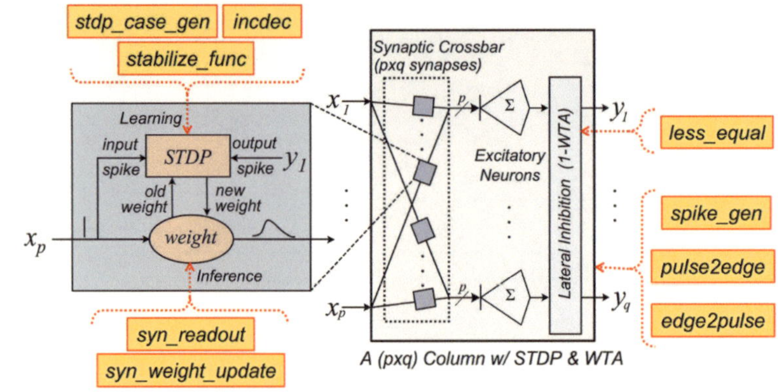

Figure 8.
Functional building blocks of a mini-column (q neurons with p synapses each) and associated proposed TNN7 macros (highlighted in yellow) [12]. Two macros are developed for synaptic feedforward processing, three for synaptic STDP learning, one for WTA inhibition and three for generic utility functions such as spike encoding.

corresponding proposed nine macros are detailed in **Figure 8**. Two macros are implemented for synaptic temporal processing, three for synaptic STDP learning, one for WTA inhibition, and three for general-purpose utility functions such as spike encoding. These macros have been optimized such that they incur minimal numbers of gates and transistors to achieve their corresponding functionalities. After implementing the custom macro-based TNN designs in SystemVerilog, we observed significant improvements in all PPA metrics. Specifically, we achieved 14, 16, 28, and 45% improvements in power, performance, area, and energy-delay product (EDP) respectively, relative to our original designs that simply use off-the-shelf standard cells. Post-layout designs are also considerably simplified using *TNN7* (see **Figures 9a** and **b**). Further, instantiating the *TNN7* macros during logic synthesis reduces the netlist generation runtime considerably (by more than 3x typically).

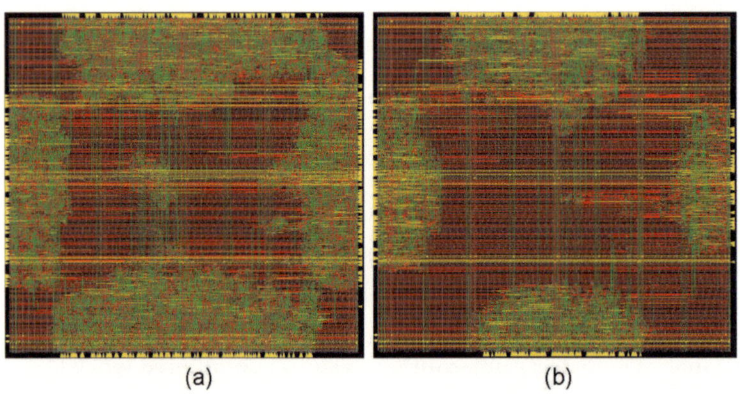

Figure 9.
Layout comparison for a sample mini-column [12]: Original design using off-the-shelf standard cell library vs. custom design using enhanced cell library with TNN7 macros. TNN7 design incurs much lesser wiring and logic complexity, and delivers 14, 16, 28, and 45% improvements in power, performance, area, and energy-delay product respectively. Note: TNN7 macros are developed using standard toolchain and design flow.

5.2 Custom macro cells for C3S designs (ongoing work)

In our follow-on work, we plan to examine decomposition of the extended C3S microarchitecture model into its fundamental building blocks, and implement them as custom macro cells. These new macro cells can be added to our earlier *TNN7* library to create a new extended library with custom macros for supporting C3S designs. We envision distinct specialized macros for "storage", "predict", "update" and "voting" functionalities that are generic enough to build *Where*, *What* and *Output* mini-columns at arbitrary scales and modalities. A key goal here is to have a rich set of macros that support diverse sensory modalities to enable implementation of potentially new types of mini-columns in the future. Currently, content-addressable memories (CAMs) seem to be a promising candidate as building blocks for implementing mini-columns for Reference Frames. In order to effectively predict relevant information based on stored knowledge in RFs, we believe these CAMs would need to support fuzzy matching; not strict exact matching as done conventionally. Implementing fuzziness in an RF can help recover similar features that can aid rapid learning of new patterns and features. This is largely ongoing and future work.

6. C3S design tools

6.1 TNN simulation tool and example applications (completed work)

Along with the hardware framework consisting of microarchitecture model and *TNN7* custom macros, we also developed a software tool, called *TNNSim*, that follows the above TNN microarchitecture model and allows rapid application-specific design exploration (see **Figure 10**). This tool is based on PyTorch [57], which is a popular deep learning framework widely used in academia. Using PyTorch enables easy integration of standard DL benchmarks and fast parallel processing using its native computational libraries. The *TNNSim* toolflow and its major libraries of functions are illustrated in **Figure 10**. Combining design exploration on *TNNSim* with *TNN7* implementation, we demonstrate TNN designs for the following two classes of applications. **Tables 1** and **2** summarize ten TNN design points for these applications.

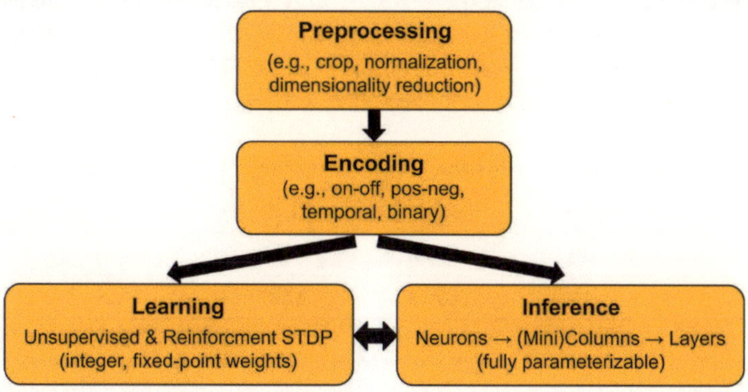

Figure 10.
TNNSim flow consisting of libraries implementing the four major functionalities: Preprocessing, encoding, inference and continual learning. Each functionality is implemented as a separate class and can be instantiated in a modular fashion to design arbitrary TNNs. TNNSim has enabled online learning demonstration and application performance exploration in our prior works [11–13].

6.1.1 UCR time-series clustering benchmark suite

Using *TNNSim*, we designed 36 single-(mini) column TNNs of various sizes to perform unsupervised clustering on 36 UCR time-series datasets [58]. These single-column TNNs can achieve state-of-the-art performance surpassing most of the contemporary comparable algorithms [13]. **Table 1** provides synapse count and PPA for 7 of the 36 benchmark-specific designs to demonstrate the efficacy of single mini-column designs across various configurations and various benchmark applications. The seven benchmarks and their input signal types are briefly described: (1) *SonyAIBORobotSurface2* - accelerometer signals to detect two types of walking surfaces; (2) *ECG200* - ECG signals to detect normal heartbeat vs. myocardial infarction; (3) *Wafer* - fabrication process control sensor signals to detect normal vs. abnormal silicon wafers; (4) *ToeSegmentation2* - motion sensors to detect normal vs. abnormal walking; (5) *Lightning2* - power-density series derived from optical and RF sensor spectogram to detect lightning; (6) *Beef* - food spectograph to detect varying levels of adulteration; and (7) *WordSynonyms* - 1D series from word outlines to detect 25 different words.

The smallest mini-column (130 synapses) for *SonyAIBORobotSurface2* and the largest mini-column (6750 synapses) for *WordSynonyms* perform very efficient unsupervised clustering within only 1 μW and 40 μW power, respectively [12]. PPA metrics scale with synapse count as expected. Note that mini-column for *Beef* has the highest input synapse count p and therefore incurs the largest computation time, as delay depends on p. These benchmarks demonstrate even relatively small single mini-column designs can perform practically useful clustering with minimal hardware complexity and power consumption.

6.1.2 MNIST handwritten digit recognition benchmark

Here, we demonstrate much larger multi-column multi-layer TNN designs with each layer composed of multiple TNN mini-columns, trained to recognize 10 classes of handwritten digits from 0 through 9 [59]. Three multi-layer TNNs are presented here. They include 2-layer, 3-layer and 4-layer designs containing 389 K, 1.3 M and 3 M total synaptic counts respectively, as shown in **Table 2**. With increasing number of layers, these three TNNs can achieve 93, 97 and 99% accuracy on the MNIST dataset, while consuming only about 2, 8 and 18 mW of power, respectively [12]. This demonstrates the huge energy-efficiency potential of TNNs. Using *TNNSim*, we also illustrate online incremental learning capability wherein a TNN design learns a new

MNIST TNN Design	Error Rate	Synapse Count	Power (mW)	Comp. Time (ns)	Area (mm^2)
2-Layer	7%	389 K	2.25	41.38	3.09
3-Layer	3%	1310 K	7.57	66.16	10.42
4-Layer	1%	3096 K	17.89	91.58	24.63

2-layer, 3-layer and 4-layer TNN designs achieve progressively decreasing error rates (7%, 3% and 1%) while consuming increasing power. The 4-layer design incurs just 18 mW power in delivering state-of-the-art accuracy of 99%. Note the synapse counts of these multi-column multi-layer TNN designs are significantly higher than that of the single mini-column TNN designs in ***Table 1****.*

Table 2.
Design space exploration for MNIST: Using TNN7 macros, PPA for three multi-layer TNN prototype designs from [10].

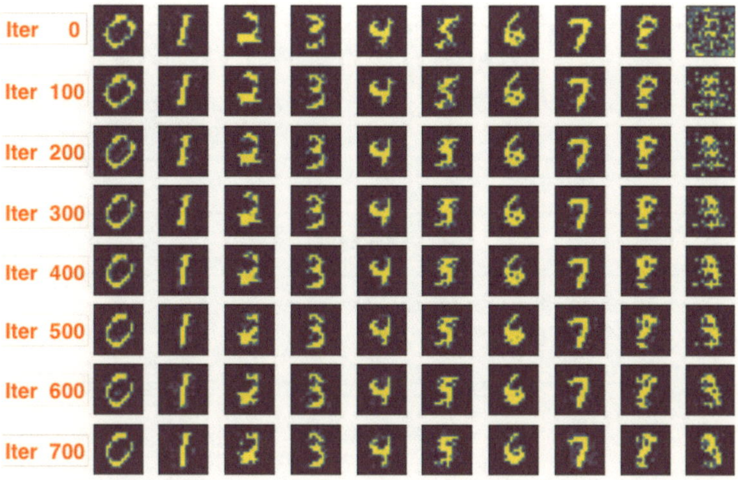

Figure 11.
Online incremental learning: A single TNN mini-column trained only on the digits 0–8 learns a new previously unseen digit '9' within 500 samples (that consist about 50 samples of digit '9') in an unsupervised manner [11]. This demonstrates the online learning capability of TNNs, which can quickly learn new classes from relatively few examples without forgetting previously learned information.

previously unseen pattern using unsupervised STDP. **Figure 11** shows the converged weights of a single mini-column that is first trained using only digits 0–8 (iteration 0). When introducing the previously unseen digit '9', within around 500 new examples (iteration 500) that contain about 50 examples of '9', it learns the new digit '9' without forgetting previously learned digits in an unsupervised fashion [11]. This demonstrates the ability for this mini-column to perform online continual learning.

6.2 Proposed C3S simulation & synthesis tools (ongoing work)

Building on the previously described three components, the overarching goal of our current research is to develop an end-to-end framework that can automatically translate application-specific C3S models in software to highly customized hardware designs. It can be utilized as an architecture and design synthesis framework for implementing C3S processing units. We envision this as a complete design framework spanning applications, architecture, microarchitecture, and custom macro suite, to generate application-specific C3S processing units for diverse sensory processing applications. As shown in **Figure 12**, the framework consists of two main components.

C3S-Sim (Application and Architectural Design Exploration): This is the software simulator framework, consisting of (1) a PyTorch simulator (extended from *TNNSim* to incorporate C3S functional modeling), and (2) a cycle-accurate architectural simulator in C++ to mimic hardware and derive accurate latency performance information for C3S designs.

C3S-Syn (Micro-architectural Design and Implementation): This is the hardware implementation framework that takes in the PyTorch C3S functional models and generates C3S hardware designs. It is envisioned to include (1) PyVerilog conversion for automated RTL generation from PyTorch, (2) automated RTL-to-GDSII flow that leverages C3S-specialized custom macro cells and generates application-specific post-layout netlist and PPA results for a specific C3S design.

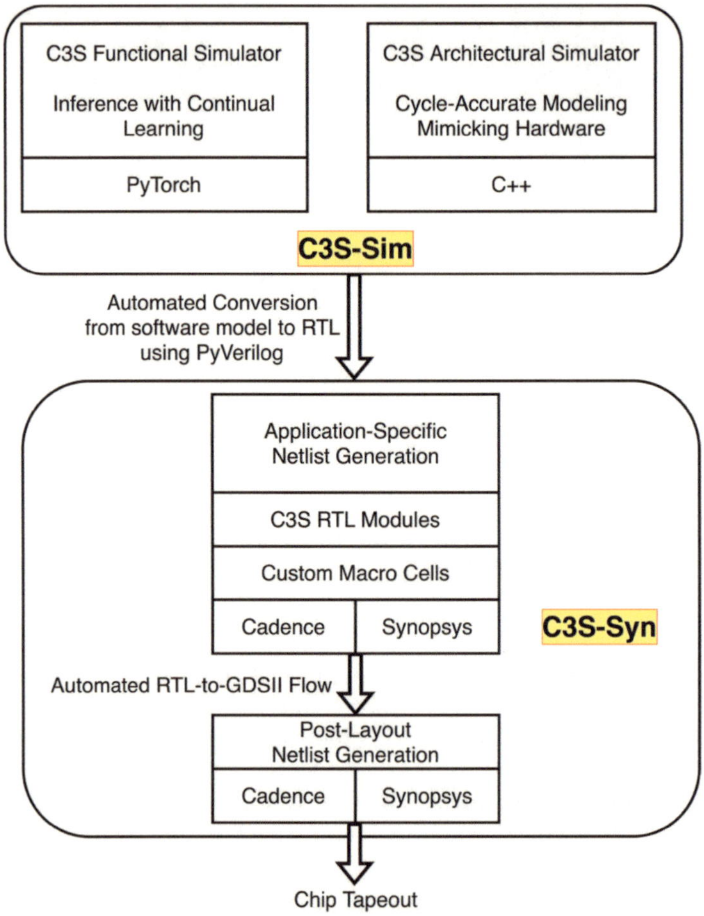

Figure 12.
Envisioned end-to-end cortical columns computing system (C3S) design framework consisting of C3S-Sim for application exploration and C3S-Syn for microarchitectural implementation. C3S-Sim consists of a PyTorch tool (extension of TNNSim) to design application-specific C3S functional models and a cycle-accurate architectural simulator for hardware performance estimation. C3S-Syn incorporates the extended microarchitecture model and functional building blocks for C3S implementation, with an automated design flow to translate PyTorch functional models to application-specific hardware designs.

6.3 Targeted applications for C3S designs (future work)

Our targeted applications for C3S designs are influenced by the emergence of several dominant industry trends: (1) *Increasing AI Compute Demand*: The computation required for supporting Deep Learning (DL) is increasing exponentially at the staggering rate of doubling every 3.4 months [19]. Many believe that this trend is not technologically nor economically sustainable [20–22]. (2) *Migration Towards Edge AI*: There is strong efficiency, security, and privacy motivations for migrating AI computation from the centralized cloud infrastructure to distributed personal edge devices. (3) *Emerging Ubiquity of Wearables*: Mobile devices, e.g. smartphones, have replaced PCs as the technology, innovation, and profitability driver for mass market computing, due to their more stringent form factor and energy efficiency requirements. As wearable devices become more ubiquitous, we believe they will in turn become the new technology and innovation driver for mass market computing [60]. (4) *Interests in Neuromorphic Chips*: Orders of magnitude improvement on energy efficiency is necessary to support autonomous intelligent wearable devices. We believe to achieve such level of improvement will require adopting the neuromorphic approach in creating new computing devices that mimic the organization and operation of biological neural networks of the neocortex.

Hence, our C3S research focuses on targeting the design and implementation of *Neuromorphic Intelligent Sensory Processing (NISP) Chiplets* with truly brain-like sensory processing capability as well as brain-like energy efficiency. This new genre of C3S based sensory processing units exhibit the following attributes: (1) **Neuromorphic**: based on reverse architecting the neocortex; (2) **Intelligent**: capable of edge-native autonomous and continuous online learning; (3) **Sensory Processing**: support near-sensor on-device processing of diverse modalities of sensory signals; (4) **Chiplets**: target chiplet implementations (potentially UCIe-compliant [14]) in standard digital CMOS, for ease of integration into diverse edge AI platforms on mobile, wearable, and IoT devices.

7. Summary and conclusion

Figure 13 summarizes our overall research vision and provides a broader context for our work. Current deep learning approach adopts the path on the left, employing high dimensional tensor processing and hardware accelerators with massive array of MAC units. Our approach takes the path on the right, leveraging spike timing processing and hardware fabric inspired by cortical columns. Our initial focus was on feed-forward mini-columns based on TNN principles. Our current focus broadens to

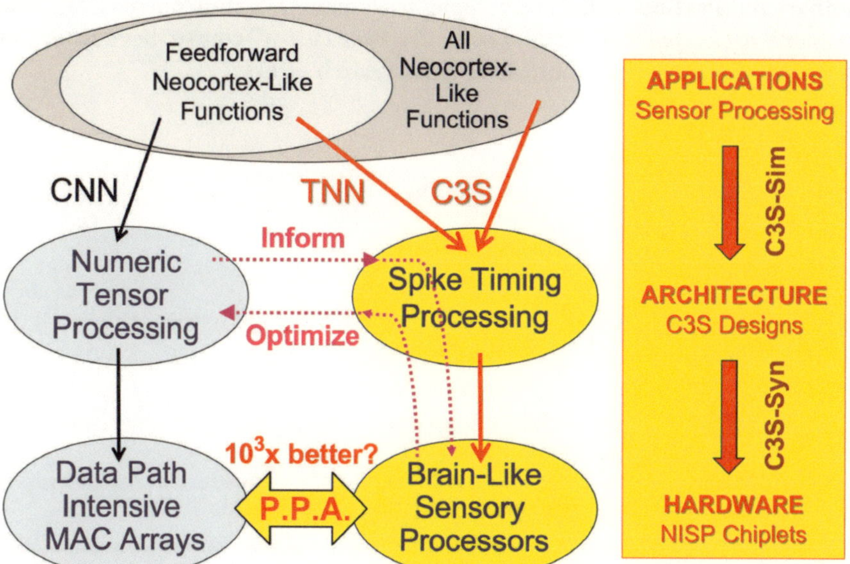

Figure 13.
Summary of envisioned research in a broader context. Left path corresponds to the current deep learning approach that mainly employs numeric tensor processing deployed on hardware accelerators consisting of large MAC arrays. Right path illustrates our research approach that employs spike timing processing in the design and implementation of special-purpose processing units inspired by cortical columns, or Neuromorphic Intelligent Sensory Processing (NISP) units. Interesting cross-over ideas between both paths should be explored.

all cortical columns with feedback in the form of reference frames. We are developing a framework and software tool suite for designing and implementing application-specific, highly energy-efficient sensory processing units, or Neuromorphic Intelligent Sensory Processing (NISP) units, with continuous online on-device learning capability. Our goal is to achieve potentially up to three orders of magnitude improvements on power and energy efficiency, relative to current deep learning accelerators. We also anticipate there could be interesting cross-over ideas between the current deep learning path and our new C3S path that should be explored.

This chapter presents a neuromorphic computer architecture and design approach that focuses on implementing neocortical computing fabric using digital off-the-shelf CMOS technology. This work builds on the foundational works of Jeff Hawkins' "*A Thousand Brains Theory*" and James E. Smith's *biologically plausible neural networks*. This research effort in NCAL at CMU aims to build Cortical Columns Computing Systems (C3S) that exhibit brain-like capabilities and brain-like efficiency. We hope our work serves as one step towards the holy grail of building a silicon neocortex. We hope to generate broad interest through this chapter that can lead to a vibrant research community pursuing this line of research. We believe there are tremendous opportunities for novel innovations that can have significant industry impact.

Acknowledgements

Our research effort has been inspired and guided by the foundational work of **James E. Smith** (Emeritus Professor, University of Wisconsin, and Adjunct Professor, Carnegie Mellon University). We are indebted to Jim's work and guidance. Two other PhD students in NCAL (Neuromorphic Computer Architecture Lab), **Shreyas Chaudhari** and **Prabhu Vellaisamy**, along with many MS students in CMU's 18-743 "*Neuromorphic Computer Architecture and Processor Design*" course, have made and are still making significant contributions to this research effort.

Author details

John Paul Shen* and Harideep Nair
Carnegie Mellon University, Pittsburgh, USA

*Address all correspondence to: jpshen@cmu.edu

References

[1] Moravec H. Mind Children: The Future of Robot and Human Intelligence. Cambridge, MA, USA: Harvard University Press; 1988

[2] Minsky M. Society of Mind. New York, NY, USA: Simon and Schuster; 1988

[3] Brooks RA. Intelligence without representation. Artificial Intelligence. 1991;**47**(1–3):139-159

[4] Mead C. Neuromorphic electronic systems. Proceedings of the IEEE. 1990; **78**(10):1629-1636

[5] Rosenblatt F. The perceptron: A probabilistic model for information storage and organization in the brain. Psychological Review. 1958;**65**(6): 386

[6] Schuman C, D, Potok TE, Patton RM, Birdwell DJ, Dean ME, Rose GS, et al. A survey of neuromorphic computing and neural networks in hardware. arXiv preprint arXiv:1705.06963. 2017

[7] Smith JE. Space-time computing with temporal neural networks. Synthesis Lectures on Computer Architecture. 2017;**12**(2):i-215

[8] Smith J. Space-time algebra: A model for neocortical computation. In: 2018 ACM/IEEE 45th Annual International Symposium on Computer Architecture (ISCA). Los Angeles, CA, USA: IEEE; 2018. pp. 289-300

[9] Hawkins J. A Thousand Brains: A New Theory of Intelligence. London, United Kingdom: Hachette UK; 2021

[10] Smith JE. A temporal neural network architecture for online learning. arXiv preprint arXiv:2011.13844. 2020

[11] Nair H, Shen JP, Smith JE. A microarchitecture implementation framework for online learning with temporal neural networks. In: 2021 IEEE Computer Society Annual Symposium on VLSI (ISVLSI). Tampa, FL, USA: IEEE; 2021. pp. 266-271

[12] Nair H, Vellaisamy P, Bhasuthkar S, Shen JP. Tnn7: A custom macro suite for implementing highly optimized designs of neuromorphic tnns. In: 2022 IEEE Computer Society Annual Symposium on VLSI (ISVLSI). Pafos, Cyprus: IEEE; 2022. pp. 152-157

[13] Chaudhari S, Nair H, Moura JMF, Shen JP. Unsupervised clustering of time series signals using neuromorphic energy-efficient temporal neural networks. In: ICASSP 2021–2021 IEEE International Conference on Acoustics, Speech and Signal Processing (ICASSP). Toronto, Canada: IEEE; 2021. pp. 7873-7877

[14] Sharma DD, Pasdast G, Qian Z, Aygun K. Universal chiplet interconnect express (ucie): An open industry standard for innovations with chiplets at package level. IEEE Transactions on Components, Packaging and Manufacturing Technology. 2022;**12**(9): 1423-1431

[15] LeCun Y, Bengio Y, Hinton G. Deep learning. Nature. 2015;**521**(7553): 436-444

[16] Pouyanfar S, Sadiq S, Yan Y, Tian H, Tao Y, Reyes MP, et al. A survey on deep learning: Algorithms, techniques, and applications. ACM Computing Surveys (CSUR). 2018;**51**(5):1-36

[17] Alom MZ, Taha TM, Yakopcic C, Westberg S, Sidike P, Nasrin MS, et al. A

state-of-the-art survey on deep learning theory and architectures. Electronics. 2019;**8**(3):292

[18] Dong S, Wang P, Abbas K. A survey on deep learning and its applications. Computer Science Review. 2021;**40**: 100379

[19] OpenAI. Ai and compute. 2018. Available from: https://openai.com/blog/ai-and-compute/.

[20] Thompson NC, Greenewald K, Lee K, Manso GF. The computational limits of deep learning. arXiv preprint arXiv:2007.05558. 2020

[21] Lohn A, Musser M. Ai and Compute: How Much longer Can Computing Power Drive Artificial Intelligence Progress. Washington, D.C., USA: Center for Security and Emerging Technology (CSET); 2022

[22] Numenta Inc. Ai Is Harming our Planet: Addressing ai's Staggering Energy Cost. CA, USA: Numenta Inc.; 2022

[23] Alex Lyzhov. "ai and compute" trend isn't predictive of what is happening (blog post). 2021. Available from: https://www.alignmentforum.org.

[24] Sevilla J, Heim L, Ho A, Besiroglu T, Hobbhahn M, Villalobos P. Compute trends across three eras of machine learning. arXiv preprint arXiv: 2202.05924. 2022

[25] Patterson D, Gonzalez J, Hölzle U, Le Q, Liang C, Munguia L-M, et al. The carbon footprint of machine learning training will plateau, then shrink. Computer. 2022;**55**(7):18-28

[26] Han S, Mao H, Dally WJ. Deep compression: Compressing deep neural networks with pruning, trained quantization and huffman coding. arXiv preprint arXiv:1510.00149. 2015

[27] Li F, Zhang B, Liu B. Ternary weight networks. arXiv preprint arXiv: 1605.04711. 2016

[28] Hubara I, Courbariaux M, Soudry D, El-Yaniv R, Bengio Y. Quantized neural networks: Training neural networks with low precision weights and activations. The Journal of Machine Learning Research. 2017;**18**(1):6869-6898

[29] He Y, Zhang X, Sun J. Channel pruning for accelerating very deep neural networks. In: Proceedings of the IEEE International Conference on Computer Vision. Venice, Italy: ICCV; 2017. pp. 1389-1397

[30] Wu H, Judd P, Zhang X, Isaev M, Micikevicius P. Integer quantization for deep learning inference: Principles and empirical evaluation. arXiv preprint arXiv:2004.09602. 2020

[31] Sun X, Wang N, Chen C-Y, Ni J, Agrawal A, Cui X, et al. Vijayalakshmi Viji Srinivasan, and Kailash Gopalakrishnan. Ultra-low precision 4-bit training of deep neural networks. Advances in Neural Information Processing Systems. 2020;**33**:1796-1807

[32] Wen W, Chunpeng W, Wang Y, Chen Y, Li H. Learning structured sparsity in deep neural networks. Advances in Neural Information Processing Systems. 2016;**29**:1-9

[33] Gale T, Elsen E, Hooker S. The state of sparsity in deep neural networks. arXiv preprint arXiv:1902.09574. 2019

[34] Hoefler T, Alistarh D, Ben-Nun T, Dryden N, Peste A. Sparsity in deep learning: Pruning and growth for efficient inference and training in neural

networks. Journal of Machine Learning Research. 2021;**22**(241):1-124

[35] Zhang S, Zidong D, Zhang L, Lan H, Liu S, Li L, et al. Cambricon-x: An accelerator for sparse neural networks. In: In 2016 49th Annual IEEE/ACM International Symposium on Microarchitecture (MICRO). Taipei, Taiwan: IEEE; 2016. pp. 1-12

[36] Parashar A, Rhu M, Mukkara A, Puglielli A, Venkatesan R, Khailany B, et al. Scnn: An accelerator for compressed-sparse convolutional neural networks. ACM SIGARCH computer architecture news. 2017;**45**(2): 27-40

[37] Gondimalla A, Chesnut N, Thottethodi M, Vijaykumar TN. Sparten: A sparse tensor accelerator for convolutional neural networks. In: Proceedings of the 52nd Annual IEEE/ACM International Symposium on Microarchitecture. 2019. pp. 151-165

[38] Yang D, Ghasemazar A, Ren X, Golub M, Lemieux G, Lis M. Procrustes: A dataflow and accelerator for sparse deep neural network training. In: 2020 53rd Annual IEEE/ACM International Symposium on Microarchitecture (MICRO). IEEE; 2020. pp. 711-724

[39] Mead C, Ismail M. Analog VLSI Implementation of Neural Systems. Vol. 80. New York, NY, USA: Springer, Science & Business Media; 1989

[40] Douglas R, Mahowald M, Mead C. Neuromorphic analogue vlsi. Annual Review of Neuroscience. 1995;**18**: 255-281

[41] Schemmel J, Fieres J, Meier K. Wafer-scale integration of analog neural networks. In: 2008 IEEE International Joint Conference on Neural Networks (IEEE World Congress on Computational Intelligence). Hong Kong: IEEE; 2008. pp. 431-438

[42] Furber SB, Lester DR, Plana LA, Garside JD, Painkras E, Temple S, et al. Overview of the spinnaker system architecture. IEEE Transactions on Computers. 2012;**62**(12):2454-2467

[43] Frenkel C, Lefebvre M, Legat J-D, Bol D. A 0.086-mm ^{2} 12.7-pj/sop 64k-synapse 256-neuron online-learning digital spiking neuromorphic processor in 28-nm cmos. IEEE Transactions on Biomedical Circuits and Systems. 2018; **13**(1):145-158

[44] Frenkel C, Legat J-D, Bol D. Morphic: A 65-nm 738k-synapse/mm^2 quad-core binary-weight digital neuromorphic processor with stochastic spike-driven online learning. IEEE Transactions on Biomedical Circuits and Systems. 2019;**13**(5):999-1010

[45] Stuijt J, Sifalakis M, Yousefzadeh A, Corradi F. μ brain: An event-driven and fully synthesizable architecture for spiking neural networks. Frontiers in Neuroscience. 2021;**15**:538

[46] Merolla PA, Arthur JV, Alvarez-Icaza R, Cassidy AS, Sawada J, Akopyan F, et al. A million spiking-neuron integrated circuit with a scalable communication network and interface. Science. 2014;**345**(6197): 668-673

[47] Davies M, Srinivasa N, Lin T-H, Chinya G, Cao Y, Choday SH, et al. Loihi: A neuromorphic manycore processor with on-chip learning. IEEE Micro. 2018;**38**(1):82-99

[48] BrainChip Holdings Ltd. Akida neuromorphic system-on-chip. Available from: https://brainchip.com/akida-neural-processor-soc/

[49] Benjamin BV, Gao P, McQuinn E, Choudhary S, Chandrasekaran AR, Bussat J-M, et al. Neurogrid: A mixed-analog-digital multichip system for large-scale neural simulations. Proceedings of the IEEE. 2014;**102**(5):699-716

[50] Hawkins J, Ahmad S, Cui Y. A theory of how columns in the neocortex enable learning the structure of the world. Frontiers in Neural Circuits. 2017; **11**:81

[51] Hawkins J, Lewis M, Klukas M, Purdy S, Ahmad S. A framework for intelligence and cortical function based on grid cells in the neocortex. Frontiers in Neural Circuits. 2019;**12**:121

[52] Lewis M, Purdy S, Ahmad S, Hawkins J. Locations in the neocortex: A theory of sensorimotor object recognition using cortical grid cells. Frontiers in Neural Circuits. 2019;**13**:22

[53] Kjell Jørgen Hole and Subutai Ahmad. A thousand brains: Toward biologically constrained ai. SN Applied Sciences. 2021;**3**(8):1-14

[54] Heaven D. Deep trouble for deep learning. Nature. 2019;**574**(7777): 163-166

[55] Smith JE. A macrocolumn architecture implemented with temporal (spiking) neurons. arXiv preprint arXiv: 2207.05081. 2022

[56] Nair H, Tan C, Zeng M, Mengshoel OJ, Shen JP. Attrinet: Learning mid-level features for human activity recognition with deep belief networks. In: Adjunct Proceedings of the 2019 ACM International Joint Conference on Pervasive and Ubiquitous Computing and Proceedings of the 2019 ACM International Symposium on Wearable Computers. 2019. pp. 510-517

[57] Paszke A, Gross S, Massa F, Lerer A, Bradbury J, Chanan G, et al. Pytorch: An imperative style, high-performance deep learning library. Advances in Neural Information Processing Systems. 2019; **32**:8026-8037

[58] Dau HA, Bagnall A, Kamgar K, Yeh C-CM, Zhu Y, Gharghabi S, et al. The ucr time series archive. IEEE/CAA Journal of Automatica Sinica. 2019;**6**(6): 1293-1305

[59] Yann LeCun, Corinna Cortes, Christopher Burges JC. The mnist database of handwritten digits. Availalbe from: http://yann.lecun.com/exdb/mnist/.

[60] John Dian F, Vahidnia R, Rahmati A. Wearables and the internet of things (iot), applications, opportunities, and challenges: A survey. IEEE Access. 2020; **8**:69200-69211

Section 5

Deep Learning, Sparse Learning, and Digital Twins

Chapter 9

Artificial Intelligence Approaches for Studying the *pp* Interactions at High Energy Using Adaptive Neuro-Fuzzy Interface System

Doaa Mahmoud Habashy, Mahmoud Yaseen El-Bakry, El-Sayed Ahmed El-Dahshan and Hanem Ibrahim Lebda

Abstract

Adaptive Neuro-Fuzzy Inference System (ANFIS), a popular machine learning model, is introduced in this chapter. ANFIS has a long development history and good agreement on scientific accomplishments. The value of ANFIS has grown dramatically along with the great interest in deep learning. We will examine how machine learning and ANFIS are related. Different methods can be used to implement machine learning models. ANFIS is a Fuzzy Inference System (FIS) that works within the context of adaptive networks. It merges the ideas of Artificial Neural Networks (ANNs) and Fuzzy Logic (FL) into a single framework. This framework can learn to estimate nonlinear functions and operates as a universal estimator. This chapter aimed to investigate the behavior of $\boldsymbol{D}$ mesons ratios production cross section ($\boldsymbol{D^+/D^0}$, $\boldsymbol{D^{*+}/D^0}$, $\boldsymbol{D_s^+/D^0}$, ***and*** $\boldsymbol{D_s^+/D^+}$), differential production cross section of prompt ($\boldsymbol{D^0}, \boldsymbol{D^+}, \boldsymbol{D^{*+}}$ ***and*** $\boldsymbol{D_s^+}$ mesons) as a function of $\boldsymbol{P_T}$ in pp collisions at ($\sqrt{s}$ = 5.02 and 7 TeV) and predict the behavior for others. The ANFIS model was created through a series of trial-and-error experiments. The ANFIS-based model simulation results perfectly fit the experimental data. When tested with non-training data points, the ANFIS prediction capabilities performed well. The ANFIS offers extensive procedures for high-energy physics modeling.

Keywords: adaptive neuro-fuzzy interface system, artificial neural network, deep learning, machine learning, artificial intelligence

1. Introduction

Artificial Intelligence (AI) has been quickly assimilated into our daily lives during the past 10 years. It drives a lot of our behaviors, including how we use social media and definitely the future of the planet. So to be employable in the expanding sector, it is essential to understand the primary distinctions between Artificial Intelligence (AI), Machine Learning (ML,) and Deep Learning (DL) [1–5]. We will define AI, ML, and

DL in the following sections, as well as emphasize the key distinctions between the two latter subsets (see **Figure 1**).

The term "Artificial Intelligence" was popularized in 1956 at the Dartmouth conference, which was organized by John McCarthy, the father of AI. It is the development of intelligent devices and systems that are capable of performing tasks which would normally require human intelligence. It is defined as a branch of computer science concerned with the simulation of intelligent behaviors in computers. AI systems possess the ability to do a variety of activities similar to human intellect, including planning, learning, manipulation, and problem-solving. Artificial narrow intelligence and artificial general intelligence are the two most commonly types of artificial intelligence. Narrow Artificial Intelligence is systems or computers with artificial limited intelligence, commonly referred to as weak AI, that are capable of carrying out solitary tasks. However, they are unable to perform tasks that are not part of their intended function and capability. The kinds of AI machines or programs that we use on a daily basis are instances of limited or weak AI. Examples that are often used include facial recognition software, email spam filters, and Google Translate. General Artificial Intelligence is a more sophisticated machine that has all of a human's talents, including emotional intelligence and creativity, which is referred to as strong AI. However, it is a difficult challenge that has not yet succeeded in replicating human intelligence, emotions, and the capacity to respond in unanticipated circumstances. The only way we can see this type of AI in action is in science fiction films. Machine Learning (ML) is a branch of AI [6]. It focuses on developing algorithms that can take in the provided data, learn from it, and make judgments based on the patterns found in it. When an unpleasant or wrong choice is made, these intelligent systems will need human intervention. ML is used to teach machines how to manage data more efficiently. We may be unable to perceive the obtained information from the data after viewing it. In that case, we employ ML. With an abundance of datasets available, the demand for machine learning is increasing. Many industries use ML to extract data. The goal of ML is to learn from data. Deep Learning (DL) is a further division of ML. Without requiring human input, it processes data using a number of layers of algorithms and artificial neural network to arrive at an accurate conclusion. DL was first introduced in the 1980s, but it has achieved popular success since 2006. Due to the enormous quantity of data needed to train a DL network, significant computational power and time are required. The time required to train the network could eventually be cut in half, from weeks to hours, as cloud computing and Graphics Processing Units

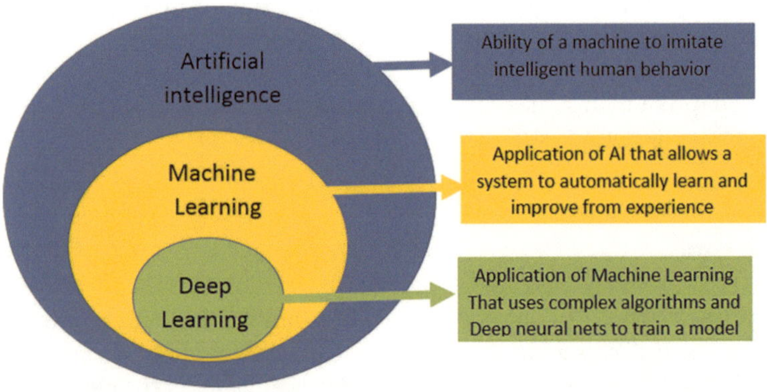

Figure 1.
Overview of artificial intelligence (AI), machine learning (ML) and deep learning (DL).

(GPUs) advance and grow. As shown in **Table 1**, there are some significant differences between DL and ML based on the descriptions of both concepts above [7–10].

A component of AI called ML, teaches programs how to spot patterns in datasets and use those patterns to infer conclusions and predictions. The use of datasets, features, and algorithms in ML software is essential. Datasets are used to train machine learning programs to identify patterns and correlations, so they are necessary. Images, numbers, words, and other types of data are included in such databases. Features, often known as variables, draw attention to important data points that the program should concentrate on. In order to teach the program how to make the best judgments, the appropriate features and algorithms must be chosen. These features and algorithms are the tools for data analysis. The speed and precision of getting the results may vary, even when utilizing multiple methods for the same task may produce equal solutions. It's also crucial to remember that the accuracy of the results is determined by the caliber of the data. Therefore, gathering reliable data will help in reaching the intended results [11, 12].

Forecasting and prediction of data, such as stock market prices, are frequent applications of machine learning. Other straightforward applications that employ it include email spam filters and social media connection suggestions. While DL is used in more complex situations like autonomous vehicles. The network is able to identify traffic lights, find barriers, and more. AI approaches have been recently used in a number of modeling methodologies based on soft computing systems. In the fields of nuclear physics, high-energy physics, materials science, and other sciences, these evolution algorithms have a strong physical existence. The Interaction behavior is complicated by the non-linear relationship between the interaction parameters and the result. AI approaches are essential for the multi-part data processing required to understand the interactions of fundamental particles. These techniques act as alternatives to more traditional methods. In this regard, AI methods like genetic algorithms, genetic programming, and Genetic Expression Programming (GEP) can be used as substitute tools to imitate these interactions. The learning algorithm of AI approaches serves as a driving force behind their use, as it discovers the links between variables in datasets. Then builds models to account for those associations (mathematically dependent). To evaluate experimental data and gain a better knowledge of many physics processes, a new computer science approaches are needed. Experimental data were acquired and described by a mathematical equation.

The success of modeling enables us to anticipate areas where experimental data are lacking. Due to its generalization, noise tolerance, and fault tolerance, AI has become increasingly popular in recent years as a potent tool for creating data correlations and

	Machine learning (ML)	**Deep learning (DL)**
Approach	Requires structured data	Does not require structured data
Human Involvement	Requires human Involvement for mistakes	Does not require human Involvement for mistakes
Hardware	Can function on CPU	Requires GPU/ significant computing power
Time	Takes seconds to hours	Takes weeks
Uses	Forecasting, predicting, and other simple applications	More complex applications like autonomous vehicles

Table 1.
Comparison between DL and ML.

has been successfully applied in materials science. Alaa F. Abd El-Rehim et al., used ANN to Simulate and forecast the Vickers Hardness [13–16]. H A M Ali and D M Habashy used ANN model for calculating the electrical impedance, AC conductivity and dielectric properties [17]. D. M. Habashy et al., simulate and forecast Entropy per Rapidity using ANN model at LHC Energies [18]. D. M. Habashy et al., used ANN model for training Particles multiplicity per rapidity for different charged particles observed in Au + Au heavy-ion collisions with energies varying between 2 and 200 GeV [19]. D.M. Habashy et al., train and forecast micro-hardness of nano-crystalline TiO2 using ANNs [20].

2. Applications in industry

Our GPS navigation systems' traffic predictions are made using machine learning algorithms [1–3]. Based on location- and velocity-related data acquired from our daily data feedback, they indicate busy routes. In order to suggest people we might know, ML software analyze our social media usage data, including the people we have connected with, the profiles we have visited, and our hobbies. Email spam filters use ML to deliver a more dependable and resilient solution. It guarantees that the filter is continually updated to recognize the most recent spammers' tricks. The AlphaGo program from Google DeepMind is among the most well-known applications of deep learning. Go is a board game that the system learned to play by competing against expert players. By making plays without the aid of a human, it was finally able to play at a level above that of the world champion Lee Sedol. One business adopting deep learning for self-driving cars is Tesla. Their deep learning program is utilized for semantic segmentation, road item detection and avoidance, monocular depth estimation, and image analysis. Amazon developed Alexa Conversations to give people a more organic engagement experience. Deep learning is used to ensure more natural interactions rather than ones that are forced or inflexible. On the basis of the presented photos, a deep learning (DL) system is asked to distinguish between dogs and cats. These photos will first be sent via the neural network's various levels. The subsequent layers will individually identify the distinctive features of dogs and cats, ultimately establishing the proper characteristics of each species. Finally, it will generate a result that accurately separates the photos into those of dogs and cats. Unlike the preceding machine learning example, the deep learning system in this case does not need structured data to categories the animals.

3. Artificial neural networks (ANNs) and their constituent parts

Data inputs, weights, biases, activation functions, and outputs make up ANNs [4, 13, 21].

1. These are the data that you wish to process, or data inputs.

2. Weights: These specify how significant each input is to the final result.

3. Biases: These reflect the degree to which assumptions about the outcome are made. More Inferences are made when the bias is larger, while less Inferences are made when the prejudice is smaller.

4. The weighted average of the inputs and the bias used make up the activation functions. They make the decision on whether or not the data will move on to the network's next layer.

Artificial neurons are present in every layer of a NN and transmit data through them. There is bias in these neurons. Weighted channels are used to convey inputted data via the layers. The weighted value of the data is added to the bias and used within the activation mechanism when it reaches the neurons. The neuron may or may not be triggered according to the outcome of the function. If the neuron is active, the information will be transmitted to the following layer. The data will not move on to the next layer, though, if it's not active. Until an output is created, this process will be repeated across the layers. Decisions produced by the deep learning software are the outputs. DL system does not require human involvement to learn from their mistakes because of the various layers in the neural network. It represents a meager step toward creating an artificial general intelligence system that is capable of autonomous decision-making.

4. Adaptive neuro-fuzzy inference system (ANFIS)

As is well known, the fuzzy logic system has some limitations, such as the need to determine the rule base, which is usually solved by referring to expert knowledge. It is obvious that the membership functions required for the formation of fuzzy sets must be determined. ANFIS allows you to use data sets to determine the rule base and membership functions for this purpose [22–25]. The ANFIS system employs two approaches: NNs and FL, if these two systems are merged, they may obtain a successful outcome that involves either fuzzy intelligence or neural network analytical abilities. The ANFIS structure, like that of other fuzzy systems, is divided into two parts: introductory and concluding, which have been connected by a set of rules. ANFIS training includes determining the parameters associated with these parts and using an optimization algorithm. During training, ANFIS makes use of the existing input–output data pairs. And after that, IF-THEN fuzzy rules are created to determine how these parts are related to one another (Jang 1993).

4.1 ANFIS architecture

ANFIS structure is formed of nodes and the bonds that link them. Because some or all of the nodes have influence over the end nodes, it is adaptive. An algorithm is applied to discover the relationship between input and output nodes. We can see 5 different layers in the ANFIS network structure, denoting that it is a multi-layer network. The structure has 2 inputs and one output, as well as four membership functions and two rules. The layer structure of ANFIS is explained below in accordance with the ANFIS structure shown in **Figure 2**.

The input values are obtained by the first layer, which defines the membership functions that apply to them. The "fuzzification layer" is also called. This layer's outputs are the inputs' fuzzy membership grades, which are determined using the following equations:

$$O_{1,i} = \mu_{A_i}(x), i = 1, 2 \tag{1}$$

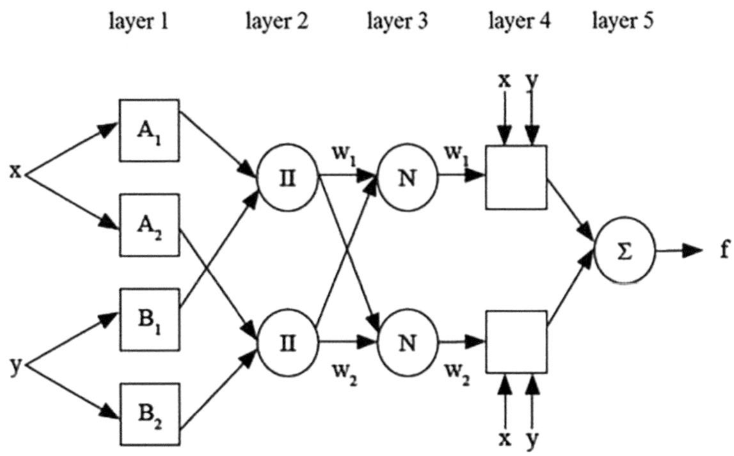

Figure 2.
ANFIS general structure in five layers.

$$O_{1,i} = \mu_{B_i}(y), i = 3, 4 \quad (2)$$

Where x and y are the node's inputs, and A_i and B_i are the verbal marks linked with this node function. $\mu_{A_i}(x)$ and $\mu_{B_i}(y)$ can be assigned any fuzzy membership function. Due to function, the 2nd layer is called the rule layer because it multiplies the input signal values into each node and determines the rule firing strength. This layer employs fuzzy operators to fuzzify the inputs, and also the AND operator. The output of the 2nd Layer is as follows:

$$O_{2,i} = \omega_i = \mu_{A_i}(x) * \mu_{B_i}(y), i = 1, 2 \quad (3)$$

The third layer's role is to normalize the computed firing strengths by dividing each value by the total firing strength, and it is referred to as the normalization layer, the output of Layer 3 $\overline{\omega}_i$ is as follows:

$$O_{3,i} = \overline{\omega}_i = \frac{\omega_i}{\omega_1 + \omega_2}, i = 1, 2 \quad (4)$$

The fourth layer will receive the normalized input values combined with the result parameter set and is called as defuzzification layer.

$$O_{4,i} = \overline{\omega}_i f_i = \overline{\omega}_i (p_i x + q_i y + r_i), i = 1, 2 \quad (5)$$

Where p_i, q_i, and r_i are the consequent parameters.

In the network's final layer, a single fixed node labeled ANFIS calculates the total output as the cumulative of all input variables. The model's overall output is given by

$$O_{5,i} = \sum_i \overline{\omega}_i f_i = \frac{\sum_i \omega_i f_i}{\sum_i \omega_i}, i = 1, 2 \quad (6)$$

During the training process, the best values of MF such as (Triangular.—Trapezoidal.—Piecewise linear.—Gaussian.—Singleton) and subsequent parameters for

ANFIS model experiences are noted as a training data set and training algorithms, see **Figure 3**. The ANFIS model was trained to utilize backpropagation and hybrid algorithms. The backpropagation algorithm calculates output errors for each layer and uses them to update layer parameters. The hybrid training algorithm is so named because it employs two gradient descent and least-squares optimization techniques.

ANFIS was created using MATLAB software, and various membership functions were used to train it. To achieve the optimal membership function parameters, the ANFIS approach's inputs are fuzzified with membership functions and trained using training data measured under normal and abnormal conditions.

4.2 The proposed hybrid ANFIS modeling

The present work proposed a hybrid model combined of ANN and FL (called ANFIS model). This model optimize D mesons ratios production cross-section (D^+/D^0, D^{*+}/D^0, $D_s^+/D^0 and\ D_s^+/D^+$) and differential production cross section of prompt D^0, D^+, D^{*+}, D_s^+ mesons as a function of Transverse momentum distribution (P_T) in pp. collisions at different the total center of mass energy, $\sqrt{s}$ = 5.02 and 7 TeV [26, 27]. Eight ANFIS models are designed to achieve this goal using MATLAB ANFIS editor. ANFIS (1–4) models simulate and predict Ratios of D—meson D^+/D^0, D^{*+}/D^0, D_s^+/D^0 *and* D_s^+/D^+ respectively. The inputs of these models are Transverse momentum distribution (P_T) and $\sqrt{s}$, while the output is Ratios of D—meson. ANFIS (5–8) models simulate differential production cross section of prompt D^0, D^+, D^{*+}, D_s^+ mesons respectively. The inputs of these models are Transverse momentum distribution (P_T) and $\sqrt{s}$, while the output is differential production cross section of prompt D^0, D^+, D^{*+}, D_s^+ mesons. The data collected from experiments are divided into two sets, namely, training set and testing set. The training set is used to train the ANFIS hybrid model. The testing data set is used to confirm the accuracy of the proposed model. It ensures that the relationship between inputs and outputs, based on the training and test sets are real. The data set is divided into two groups, 70% for training and 30% for testing. As a neural network, ANFIS must also be taught over a predetermined number of training cycles (epochs). **Figure 4** displays a flowchart of the ANFIS. To determine the ideal architecture parameters, the ANFIS model was run using experimental data. Different training epoch counts were used in several simulations to test how closely the check error (the difference between the output of the ANFIS and the validation data) and training error (the difference between the output of the ANFIS and the training data) were

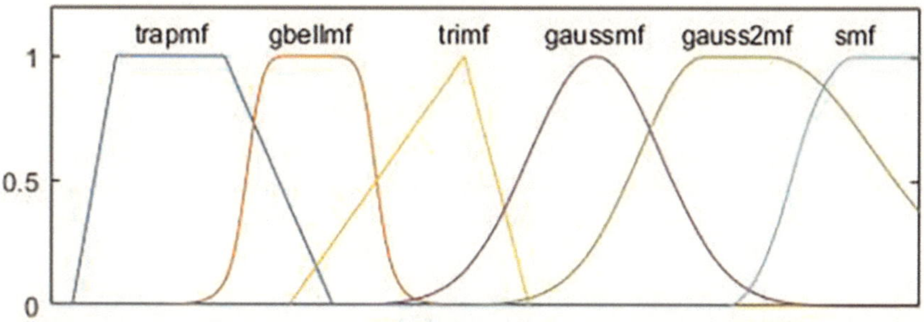

Figure 3.
Membership functions.

related (so that ANFIS would have generalization capability). The root mean squared error (RMSE) and coefficient of correlation (R^2), both of which are provided in Eqs. (7) and (8) respectively, are statistical measures that are used to evaluate error.

$$\text{RMSE} = \sqrt{\frac{\sum_{i=1}^{n}\left(Opre_i - Oexp_i\right)^2}{n}} \tag{7}$$

$$R^2 = 1 - \left(\frac{\sum_i\left(Opre_i - Oexp_i\right)^2}{\sum_i\left(Opre_i\right)^2}\right) \tag{8}$$

The predicted and experimental outputs are represented by $Opre_i$ and $Oexp_i$, respectively, and n is the number of paired input/output pairs.

5. Results and discussion

The goals of this chapter are to use the ANFIS model to model, simulate and predict the behavior of D mesons ratios production cross-section ($D^+/D^0, D^{*+}/D^0, D_s^+/D^0$ *and* D_s^+/D^+) and differential production cross section of prompt (D^0, D^+, D^{*+} and D_s^+ mesons) as a function of P_T in pp. collisions at ($\sqrt{s}$ = 5.02 and 7 TeV).

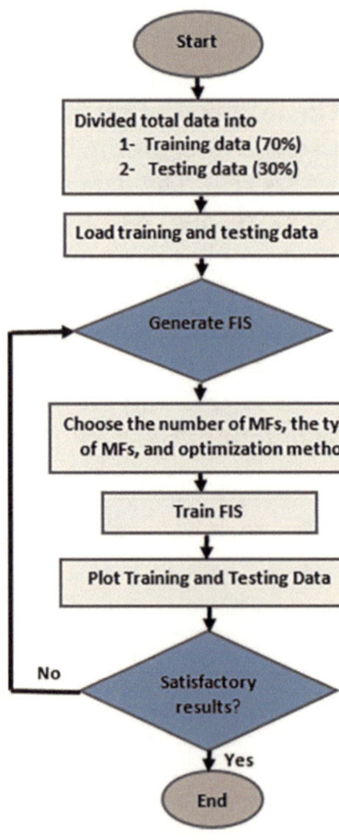

Figure 4.
Flowchart of ANFIS model.

Specific steps are taken in order to achieve the chapter's objectives. The ANFIS model is loaded with experimental data. MATLAB is utilized to process the modeling (R2017 a). Eight ANFIS networks have been generated to finish the modeling process forD mesons ratios production cross-section (D^+/D^0, D^{*+}/D^0, D_s^+/D^0 and D_s^+/D^+) and differential production cross section of prompt (D^0, D^+, D^{*+} and D_s^+ mesons) as a function of P_T in pp. collisions at ($\sqrt{s}$ = 5.02 and 7 TeV). For each ANFIS network, the number and the type of membership functions (MF) for both inputs and outputs are determined. The number of epochs is also adjustable to obtain the lowest training error. After re-training multiple ANFIS networks with various specifications, the optimal ANFIS network with the lowest possible training error is discovered. The input patterns of the designed ANFIS hybrid have been trained to produce target patterns that modeling D mesons ratios (D^+/D^0, D^{*+}/D^0, D_s^+/D^0 and D_s^+/D^+) and differential production cross section of prompt (D^0, D^+, D^{*+}, D_s^+ mesons) as a function of P_T in pp. collisions at ($\sqrt{s}$ = 5.02 and 7 TeV).

The best ANFIS network for this dataset was found within 50 epochs and two hidden layers with 6 and 6 neurons as specified in **Figure 5**. Membership functions (MFs) also play a fundamental role in Fuzzy Inference Systems modeling. It is worth mentioning that these system modeling are chosen carefully. The performance of ANFIS is very sensitive to the amount of MFs in the system. As a result, it is expected that complex systems with a great quantity of MFs will perform poorly because of the amount of the premise parameters that need to be estimated. MFs can have many shapes. Accordingly, the results of the experiments show that ANFIS with Generalized bell-shaped (gbellmf) has attained the best performance in all simulated and predicted experimental data (see **Figure 6**). The generalized bell-shaped MF (gbellmf) has a

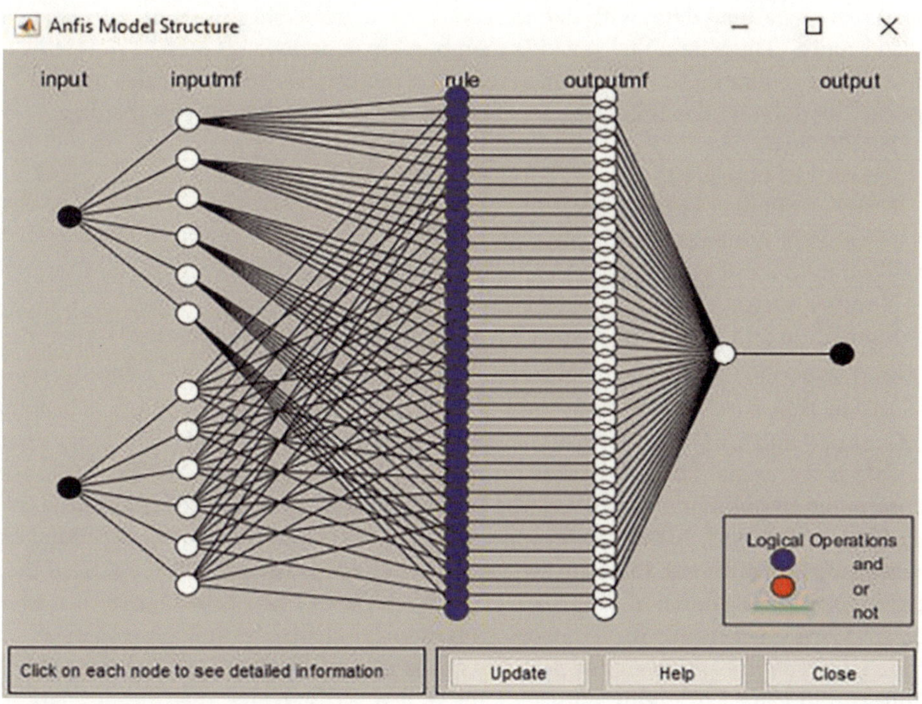

Figure 5.
The ANFIS architecture of the best network.

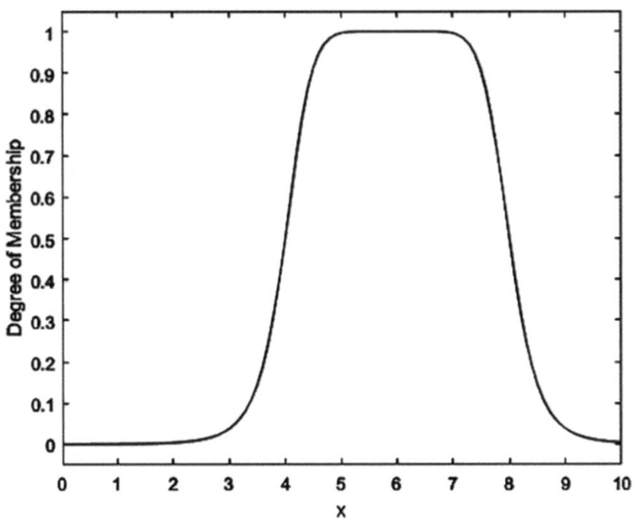

Figure 6.
The generalized bell-shaped membership function (gbellmf).

bell-like symmetrical shape. According to: $f = (x, a, b, c) = \frac{1}{1+\left|\frac{x-c}{a}\right|^{2b}}$. This function has three parameters: a determines the width of the bell-shaped curve, with a larger value resulting in a wider membership function, b is a positive integer that defines the shape of the curve on either side of the central plateau, and c determines the center of the curve in a universe of discourse. The function is used to generate an initial output FIS matrix from training data, with default values for membership function numbers "6 6" and types "These defaults" provide membership functions on each of the two inputs. The generated fuzzy inference system structure has 36 fuzzy rules of type "linear" by default. The following are the final parameters that best fit the data.

ANFIS info.

Number of nodes: 101

Number of linear parameters: 108

Number of nonlinear parameters: 36

Total number of parameters: 144.

Number of fuzzy rules: 36

Simulation and prediction results based on ANFIS used to model the$D$—meson ratios production cross section (D^+/D^0, D^{*+}/D^0, D_s^+/D^0 *and* D_s^+/D^+) and differential production cross section of prompt (D^0, D^+, D^{*+}, D_s^+ mesons) for pp. collision at different P_T and $\sqrt{s}$ (5.02 and 7TeV) are given in **Figures 7–10(a–d)**. The experimental data is represented by solid circle symbols, while the simulated ANFIS results are represented by solid line curves and the predicted ANFIS results are represented by solid square symbols. After training, the ANFIS models have been used to predict D meson ratios at different values of P_T and $\sqrt{s}$ (5.02 and 7 TeV) that are not used in training session as shown in **Figures 7–10(a–d)**. **Figures** 7 and **8(a–d)**, they introduce the ANFIS trained results for Transverse momentum distribution measured at $\sqrt{s}$ (5.02 and 7TeV) respectively for particles ratios and prediction for experimentally unmeasured values. Studying **Figures 7** and **8**, it was discovered that the simulation ANFIS curves and the experimental data symbols nearly matched, indicating excellent simulation results and the prediction data symbols are clear and produce acceptable

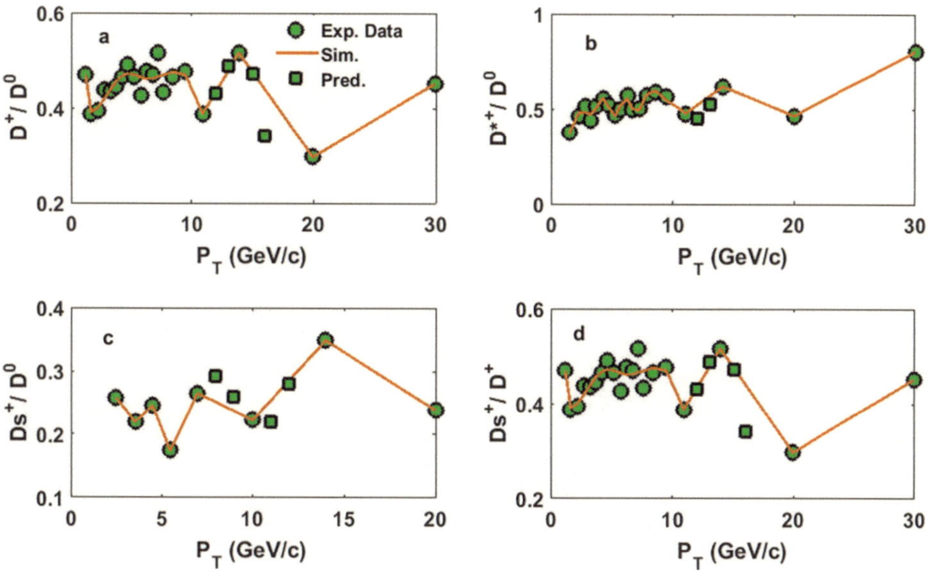

Figure 7.
Transverse momentum distribution measured at $\sqrt{S}$*=5.02 TeV for particles ratios* (D^+/D^o(a), D^{*+}/D^o(**b**), D_s^+/D^o(**c**) and D_s^+/D^+(**d**) *are compared with ANFIS simulation and prediction.*

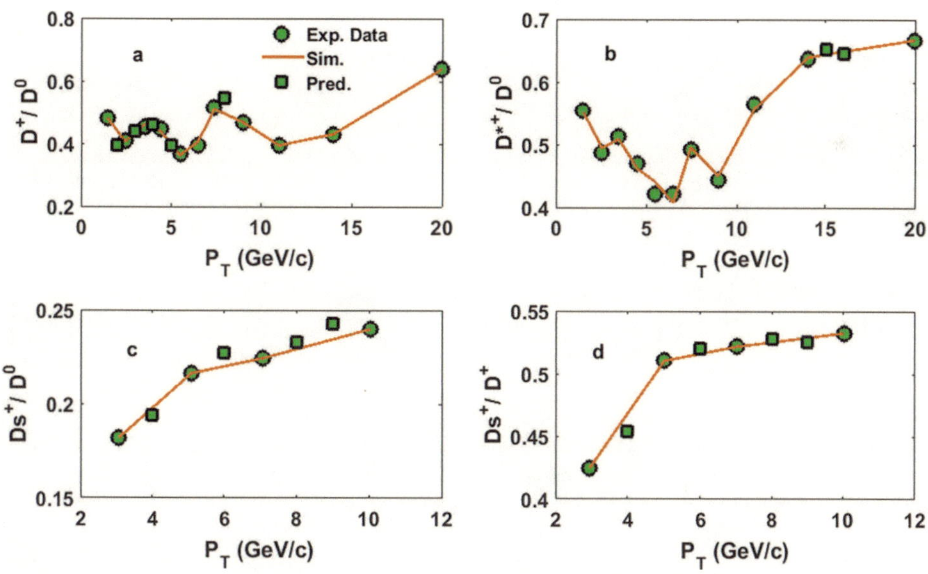

Figure 8.
*Same as in **Figure 7** but at* $\sqrt{S}$ *= 7 TeV.*

results. The ratios of the P_T-differential cross sections of prompt mesons in pp. collisions at $\sqrt{s}$ = 5.02 TeV, the transverse momentum interval $0 < P_T < 30$ GeV/c are reported in **Figure 7(a, b, d)**, and $0 < P_T < 20$ in **Figure 7(c)** but in **Figure 8** at $\sqrt{s}$ = 7 TeV, the transverse momentum interval $1 < P_T < 20$ GeV/c are reported in **Figure 8(a, b)**, and $3 < P_T < 12$ GeV/c in **Figure 8(c, d)**. In **Figure 9**, the production cross sections were

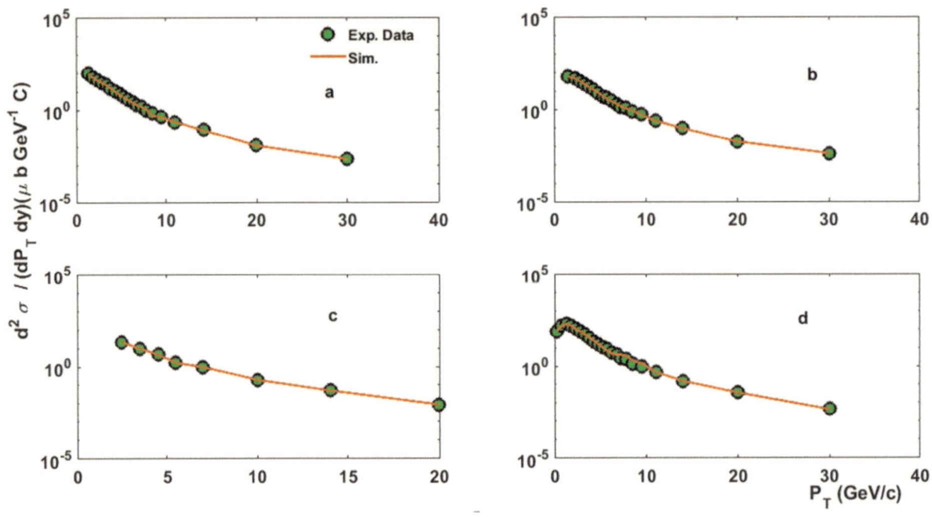

Figure 9.
$P_T$differential production cross section of prompt D^0(a), D^+(b), D_s^+(c), D^{*+}(d) mesons in pp *collisions at $\sqrt{s}$ = 5.02 TeV using ANFIS.*

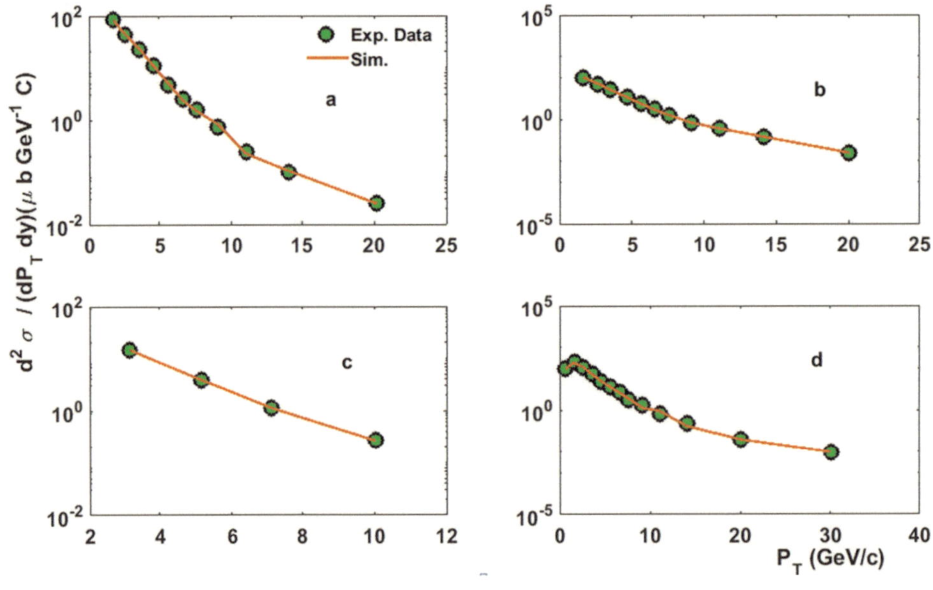

Figure 10.
*Same as in **Figure 9** but at $\sqrt{S}$ = 7 TeV.*

measured at $\sqrt{s}$ = 5.02 TeV in the transverse momentum interval $0 < P_T < 30$ GeV/c for D^0, $1 < P_T < 30$ GeV/c for D^+ and D^{*+}, and in $2 < P_T < 24$ GeV/c for D_s^+ mesons and were measured at$\sqrt{s}$ = 7 TeV, in the transverse momentum interval $0 < P_T < 30$ GeV/c for D^0, $1 < P_T < 24$ GeV/c for D^+ and D^{*+}, and in $2 < P_T < 12$ GeV/c for D_s^+ mesons as shown in **Figure 10**. **Figures 9** and **10(a–d)**, they introduce the ANFIS trained results for P_T differential production cross section of prompt mesons in pp. collisions at $\sqrt{s}$ (5.02 and 7TeV) respectively. The ANFIS simulation results show a

ANFIS parameters type	ANFIS 1 D^+/D^0	ANFIS 2 D^{*+}/D^0	ANFIS 3 D_S^+/D^0	ANFIS 4 D_S^+/D^+
R^2	0.9327	0.9795	1	1
RMSE	0.151	0.012	2.07×10^{-7}	4.95×10^{-7}

Table 2
RMSE and R^2 for D- meson ratios (D^+/D^o, D^{*+}/D^o, D_s^+/D^o and D_s^+/D^+) in pp *collision.*

ANFIS parameters type	ANFIS 5 D^0	ANFIS 6 D^+	ANFIS 7 D^{*+}	ANFIS 8 D_S^+
R^2	0.998	0.999	1	1
RMSE	1.0819	0.2626	0.1426	1.86×10^{-5}

Table 3
RMSE and R^2 for differential production cross section of prompt D^o, D^+, D^{*+}, D_s^+ mesons in pp *collision.*

high level of agreement with the experimental data. We notice that the curves obtained by ANFIS hybrid model show the best fitting to the experimental data for simulation and prediction. This is agreed with Khajeh [28]; and Tortum, Yayla [29]; and G. M. Behery [30]. The Minimum root mean squared error (0.151, 0.012, 2.07×10^{-7}, 4.95×10^{-7}, 1.0819, 0.2626, 0.1426 and 1.86×10^{-5}) and higher coefficient of correlation (0.9327, 0. 9795, 1, 1, 0.998, 0.999, 1 and 1) for ANFIS (1–8) respectively were obtained as shown in **Tables 2** and **3** respectively.

6. Conclusion

In the area of theoretical high-energy physics, the suggested ANFIS system has gained a good reputation. Due to variations in the kind, quantity, and number of membership functions as well as the number of epochs, the system is intended to identify the best ANFIS that can do the best test and prediction. As a result, numerous efforts are performed to identify the best ANFIS that makes use of a few epochs and membership functions. On the pp. interaction, ANFIS was used and put to the test. The pp. -based ANFIS model calculates D mesons ratios ($D^+/D^0, D^{*+}/D^0, D_s^+/D^0$ *and* D_s^+/D^+) and differential production cross section of prompt D^0, D^+, D^{*+}, D_s^+ mesons as a function of P_T at $\sqrt{s}$ = 5.02 and 7 TeV. ANFIS system reached the optimal solution using 50 epochs and 'gbellmf' membership function. The training simulation results demonstrated flawless fitting to the experimental data. With data points not used in training, the ANFIS's prediction ability is tested, and it performs well. The outcomes convincingly show the viability and efficacy of such a method for obtaining collision information. The proposed ANFIS is a powerful mechanism for forecasting the behavior of pp. interaction.

Acknowledgements

The authors are grateful to Artificial Intelligence Group- Ain Shams University, for it fruitful discussions.

Conflict of interest

The authors declare that they have no conflicts of interest regarding the publication of this chapter.

Notes/thanks/other declarations

The authors would like to thank anonymous for their insightful comments and suggestions for improving the chapter.

Author details

Doaa Mahmoud Habashy[1*], Mahmoud Yaseen El-Bakry[1], El-Sayed Ahmed El-Dahshan[2] and Hanem Ibrahim Lebda[1]

1 Faculty of Education, Department of Physics, Artificial Intelligence Group- Ain Shams University, Cairo, Egypt

2 Faculty of Science, Department of Physics, Artificial Intelligence Group- Ain Shams University, Cairo, Egypt

*Address all correspondence to: doaamahmoud@edu.asu.edu.eg

References

[1] Haykin S. Neural Networks and Learning Machines. 3rd ed. Prentice Hall, Upper Saddle River: Pearson; 2008

[2] Ali-A, Fathalla-A, Salah-A, Bekhit-M, Eldesouky-E. Marine Data Prediction: An Evaluation of Machine Learning, Deep Learning, and Statistical Predictive Models, Computational Intelligence and Neuroscience. 2021;**2021**:8551167. DOI: 10.1155/2021/8551167

[3] Kim P. Matlab deep learning With machine learning, neural networks, and artificial intelligence, In MATLAB Deep Learning. Apress; 2017. DOI: 10.1007/978-1-4842-2845-6

[4] Website: https://www.simplilearn.com/tutorials/artificial-intelligence-tutorial/ai-vs-machine-learning-vs-deep-learning [Accessed: December 10, 2022].

[5] EL-Sayed AE, El-Bakry MY. Modeling of Transverse Momentum Spectra for Charged Particles in Proton-Proton Collisions Based on Soft Computing Approaches. Journal of Computational and Theoretical Transport. 2017;**46**:410-426. DOI: 10.1080/23324309.2017.1405272

[6] Breiman L. Random Forests. Machine Learning. 2001:**45**:5-32. DOI: 10.1023/A:1010933404324

[7] Ciresan D, Meier U, Masci J, Gambardella L, Schmidhuber J. Flexible, High Performance Convolutional Neural Networks for Image Classification. In: 22nd International Joint Conference on Artificial Intelligence, Barcelona, Catalona, Spain. July 16-22, 2011; pp. 1237-1242. DOI: 10.5591/978-1-57735-516-8/IJCAI11-210

[8] Krizhevsky A, Sutskever I, Hinton G. Proceedings of the Advances in Neural Information Processing Systems. Vol. 25. NV, USA: Lake Tahoe; 2012. pp. 1090-1098

[9] Ciresan D, Meier U, Schmidhuber J. Multi-Column Deep Neural Networks for Image Classification. In: IEEE Conference on Computer Vision and Pattern Recognition. 2012. pp. 3642-3649.DOI: 10.1109/CVPR.2012.6248110

[10] Nauck D, Nürnberger A, Neuro-fuzzy Systems: A Short Historical Review A. Studies in Computational Intelligence. 2013;**445**:91-109. DOI: 10.1007/978-3-642-32378-2_7

[11] Mohd J, Abid H, Ravi PS, Rajiv S. Artificial Intelligence Applications for Industry 4.0: A Literature-Based Study. Journal of Industrial Integration and Management. 2022;**07**:83-111. DOI: 10.1142/S2424862221300040

[12] Tiruneh GG, Fayek AR, Sumati V. Neuro-fuzzy systems in construction engineering and management research. Automation in Construction. 2020;**119**: 103348. DOI: 10.1016/j.autcon.2020.103348

[13] El-Bakry SY, El-Bakry MY. Neural Network Representation For Electron and Positron Collisions with Sodium and Potassium. Indian Journal of Physics. 2004;**78**:1313-1318

[14] Alaa FA, Habashy DA, Zahran HY, Soliman HN. Mathematical Modelling of Vickers Hardness of Sn-9Zn-Cu Solder Alloys Using an Artificial Neural Network. Metals and Materials International. 2021;**27**:4084-4096. DOI: 10.1007/s12540-020-00940-1

[15] Alaa FA, Heba YZ, Habashy DA, Hana MA. Simulation and Prediction of

the Vickers Hardness of AZ91 Magnesium Alloy Using Artificial Neural Network Model, Crystals. 2020;**10**:290. DOI: 10.3390/cryst10040290

[16] Abd El-Rehim AF, Zahran HY, Al-Masoud HM, Habashy-DM. Microhardness and microstructure characteristics of AZ91 magnesium alloy under different cooling rate conditions, Materials Research Express. 2019;**6**: 086572-086587. DOI: 10.1088/2053-1591/ab1ad6

[17] Ali HAM, Habashy DM. The electrical impedance, AC conductivity and dielectric properties of phenol red compound investigated and modeled by an artificial neural network. Communications in Theoretical Physics. 2020;**72**:105701-105711. DOI: 10.1088/1572-9494/aba24d

[18] Habashy DM, El-Bakry-Mahmoud Y, Scheinast W, Hanafy M. Entropy per Rapidity in Pb-Pb Central Collisions using Thermal and Artificial Neural Network (ANN) Models at LHC Energies. Chinese Physics C. 2022;**46**:073103-073114. DOI: 10.1088/1674-1137/ac5f9d

[19] Habashy DM, El-Bakry Mahmoud Y, Tawfik A, Abdel Rahman RM, Hanafy M. Particles multiplicity based on rapidity in Landau and artificial neural network (ANN) models. International Journal of Modern Physics A. 2022;**37**: 2250002. DOI: 10.1142/ S0217751 X22500026

[20] Habashy DM, Mohamed HS, El-Zaidia EFM. A simulated neural system (ANNs) for micro-hardnessof nano-crystalline titanium dioxide. Physica B: Condensed Matter. 2019;**556**:183-189. DOI: 10.1016/j.physb.2018.12.007

[21] Jang JSR. ANFIS: Adaptive-network-based fuzzy inference system. IEEE Transactions on Systems, Man, and Cybernetics: Systems. 1993;**23**:665-684. DOI: 10.1109/21.256541

[22] Zadeh LA. Fuzzy sets. Information and Control. 1965;**8**:338-353. DOI: 10.1016/S0019-9958(65)90241-X

[23] Jang JSR, Sun CT. Neuro-fuzzy modeling and control. In: Proceedings of the IEEE. 1995;**83**:378-406

[24] Akkoyun S, Torun Y. Neuro-fuzzy modeling of deformation parameters for fusion-barriers. Nuclear Engineering and Technology. 2021;**53**:1612-1618. DOI: 10.1016/j.net.2020.10.017

[25] El-Bakry-M Y. A Study of K-P Interaction at High Energy using Adaptive Fuzzy Interface System. International Journal of Modern Physics C (IJMPC). 2004;**15**:1013-1020. DOI: 10.1142/S0129183104006467

[26] ALICE Collaboration. Measurement of $\boldsymbol{D^0}$, $\boldsymbol{D^+}$, $\boldsymbol{D^{*+}}$ and $\boldsymbol{D_s^+}$, production in pp collisions at $\sqrt{s}$ = 5.02 TeV with ALICE. European Physical Journal C: Particles and Fields. 2019;**79**:388. DOI: 10.1140/epjc/s10052-019-6873-6

[27] ALICE Collaboration. Measurement of D-meson production at mid-rapidity in pp collisions at $\sqrt{s}$ =7 TeV. European Physical Journal C. 2017;**77**:550. DOI: 10.1140/epjc/s10052-017-5090-4

[28] Khajeh A, Modarress H. Prediction of solubility of gases in polystyrene by Adaptive Neuro-Fuzzy Inference System and Radial Basis Function Neural Network. Neural Network Expert Systems with Applications. 2010;**37**: 3070-3074. DOI: 10.1016/j.eswa.2009.09.023

[29] Tortum A, Yayla N, Gökdag M. The modeling of mode choices of intercity freight transportation with the artificial

neural networks and adaptive neuro-fuzzy inference system. Expert Systems with Applications. 2009;**36**(3 Part 2): 6199-6217

[30] Behery GM, El -Harby AA, El -Bakry MY. ANFIS and Neural Networks Systems For Multiplicity Distributions in Proton-Proton Interactions. Applied Artificial Intelligence. 2013;**27**:304-322. DOI: 10.1080/08839514.2013.774212

Chapter 10

Study of Approaches to Predict Personality Using Digital Twin

Vrinda Tandon and Ritika Mehra

Abstract

With a growing proportion of online activities on social networking sites on different mediums like Facebook, Instagram, Twitter, LinkedIn the requirement for personality prediction associated with this online mediated behavior has also increased significantly. The user generated content on social media can be effectively leveraged to record, analyze and predict personality through different psychological approaches like MBTI, Big Five, and DISC. Predicting personality has displayed an intrinsic influence in multifarious domains like career choice, political influence, brand inclination, customized advertising, improvising learning outcomes, recommender system algorithms and so on. The objective of this paper is to stipulate an overview of different strategies used by researchers to predict personality based on the social media usage and user generated content across prominent social media platforms. It was observed that the personality traits can be accurately inferred from user behavior reflected on social media through attributes like status posted, pictures uploaded, number of friends, groups joined, network density, liked content. As of now, Facebook followed by Twitter are the most prominent social media platforms for conducting the study however, the use other social media platforms like Instagram, LinkedIn are expected to increase exponentially for carrying out personality prediction study.

Keywords: digital twin, AI, personality traits, social networking, personality prediction

1. Introduction

A digital twin is an imitation of a physical entity that enables cost effective stimulation. It assists organizations to identity optimization opportunities and improves operational performance. By stating it a replica does not mean that it's simply an object, to be more specific digital twin is a process which is entwined with today's leading technologies like AI (Artificial Intelligence), IoT (Internet of Things), XR (extended reality) and Cloud. It follows a bi-directional approach i.e. the flow of information is from both the digital entity and physical entity [1].

Digital twin is broadly classified in three types i.e. DT Prototype (DTP), Digital Twin Instance (DTI), Digital Twin Aggregation (DTA). DTA is simply the prototype of the physical asset which is to be created. It consists of designing, processing and analyzing physical entity. (DTI) is the digital twin of each instance of the manufactured product and is used to run tests under different scenarios. Meanwhile DTA as the name suggests is the aggregation of DTIs to determine product capabilities, prognostics and learning [1]. Azad M. Madni et al. in his study classified digital twins in four levels i.e. Pre-Digital Twin, Digital Twin, Adaptive Digital Twin and Intelligent Digital Twin. These four levels were classified based on the level of sophistication and scope [2].

1.1 Digital twins and AI

Both digital twin and AI are mutually beneficial for each other. AI can help digital twin to make more informed decisions in order to make improvement across various operations for e.g. design process can be accelerated by quickly evaluating many possible design patterns. On the other hand, digital twin generates stimulated data that can be used to train AI models. As of today, most of the advanced reinforcement learning is happening in the gaming sector. Implementing digital twin assists ML test scenarios by creating a virtual environment for e.g. in reinforcement learning, it is feasible to create multiple scenarios or perform multiple tests which would be impossible in physical environment [3]. In terms of industrial benefit, digital twin along with AI can help organizations to improve speed, quality as well as efficiency of working operations. Digital Twins is a proficient element to unlatch advanced forms of Artificial Intelligence.

1.2 Digital twin nowadays

In a survey conducted by Gartner during covid-19 pandemic, 31% of the respondents mentioned that they are leveraging digital twin technology to ensure safety of customers and employees by enabling remote asset monitoring to minimize in-person monitoring [4, 5]. For the same, as illustrated in **Figure 1**, 47% (i.e. 28% and 19%) of the organizations focus to increase their priority in IoT implementations in spite of the problematic effects of covid-19 [4]. Whereas, 35% of the organizations reduced their IoT investment as an impact of Covid-19. This leads to the fact that majority of the organizations seek to invest more in IoT for cost reduction. With IoT being implemented Garter predicted that the one-third of medium to large size organizations will carry out at least one DT by the end of 2023.

Digital twins are changing the way of work across different industries. It encourages cost effective model driven decisions by considering every possible scenario instead of relying on direct implementation of real world stimulations. It allows

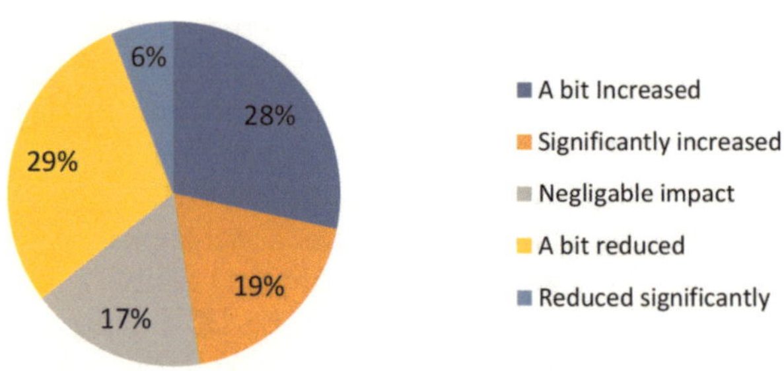

Figure 1.
Covid-19 impact on IoT implementation [4].

predicting scenarios, testing before manufacturing, foreseeing maintenance failures and many other benefits such as-

- Fault detection
- Improves performance hence productivity via optimization
- Lowers maintenance cost and reduce down time with the help of predictive maintenance
- Increases efficiency during the process of making decision.

1.3 Implementing digital twin

Digital twin begins with procuring both static and dynamic knowledge. After gathering relevant information, a conceptual model or say, mathematical model is designed considering exchange of attributes between physical environment and virtual model. This model virtually represents the physical processes on which tests are performed with the help of experimental data. It is very important to analyze which data should come in and accumulated in the digital twin. For this purpose, physical attributes of the real world are collected with the help of sensors to measure critical inputs and execution of the desired adjustment takes place with the help of accurators. After specifying the data requirement, communication protocols (TCP/IP, Pub/Sub, ERP) are determined for smooth transmission of data. These connections are affirmed with the help of technologies like cloud computing, network communication and security. Data coming from different sources is collected, processed and then aggregated in the database along with the real time data. Then, the virtual model is created by modeling all the input and output factors. This virtual model is capable of getting real time data inputs from the physical environment. By effectively visualizing, monitoring and analyzing data, it enables stimulations to generate a feasible output leading to an efficient decision making [6, 7] (**Figure 2**).

2. Application in distinct domains

By offering its efficiency and competitive advantage, this technology would benefit wide range of industries like manufacturing, aerospace, health care, personality [8, 9] prediction, automotive and construction sector [1, 10, 11] (**Figure 3**).

a. Manufacturing: DT technology is used in manufacturing industries to a great extent. Some of the applications are as follows:

    - Product customization: Ensuring effective product customization as per the need of the customer and its impact on production process.

Figure 2.
Flow chart of basic steps of DT implementation [5].

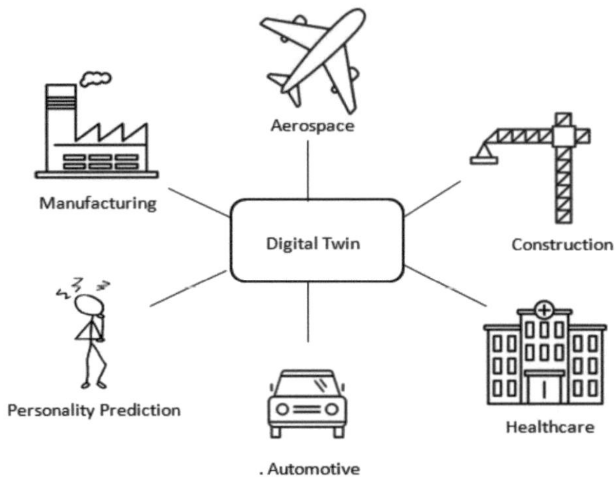

Figure 3.
Digital twin across distinct sectors.

- Product development: Minimize development cost of the product by testing feasibility before manufacturing.
- Avoiding heavy failures by predictive maintenance.

b. Automotive Sector: Automotive industries utilize this technology to produce a feasible prototype of the product. All possible scenarios regarding safety, customer interaction, vehicle maintenance are analyzed to diagnose any issue beforehand by creating a virtual model.

c. Construction Sector:

- Predictive approach for resource allocation and management.
- Continuous updation of data gives an added advantage over static 3D models.
- Optimize designs by providing more visibility and considering operational needs in advance.
- Reduce the chance of delay in construction with the help of stimulation.
- Uphold deadlines and budget by improvising communication and logistics in supply chain.

d. Aerospace: In 2002, the concept of digital twins was addressed by John Vickers from NASA and today the aerospace sector is already relying on this technology for safe stimulation environment, mapping and interrogating transport system.

e. Healthcare Sector: digital twin technology can help to ensure whether a particular treatment or medical equipment would be suitable for patient undergoing treatment for example, CardioInsight [7] mapping system creates 3D map of heart's electrical activity. This diagnostic method also got approval from FDA.. It can contribute a lot to move forward the medical sector by:

- Stimulating dosage effects
- Creating models for optimized medical equipment
- Building efficient care delivery models
- Creating models for evaluating possible risks during surgery
- Designing care coordination models
- Virtual surgery can be performed beforehand

f. Personality traits recognition: Analyzing personality helps to predict an assumed unique way in which an individual responds and makes decision in a particular situation and since services and products are customer-driven, it's important to place customer requirements at the center of business strategy. This idea has evolved industrial mass production to personalized production by understanding human personality traits.

3. Personality assessment fundamentals

The personality model used and prediction algorithm employed to calculate personality trait results are two of the most vital factors for personality assessment.

3.1 Types of personality model

1. Enneagram: The Enneagram consists of nine comprehensive key types that interact in a unique way.

   After analyzing Enneagram number, Enneagram wing comes into picture which determines how Enneagram number uniquely relates to the personality. Enneagram wing generally emphasizes other aspects of individuality where it is common to exhibit different Enneagram type characteristics [12, 13].

   The nine Enneagram types are-The Reformer; The Helper; The achiever; The individualist; The investigator; The loyalist; The enthusiast; The challenger; The peacemaker.

2. DISC: it is an acronym that represents four personality types i.e. D (Dominance), I (Influence), S (Robustness), C(Caution).

3. MBTI: Myers Briggs Type Indicator determines personality of an individual in four dimensions where each dimension is further divided into two contradictory key factors like- Extraversion (E) – Introversion (I), Sensing (S) – Intuition (N), Thinking (T) – Feeling (F), Judging (J) – Perceiving (P).

Sr. no.	Tag	Personality type
1	ISTJ	The Inspector
2	ISTP	The Crafter
3	ISFJ	The Protector
4	ISFP	The Artist
5	INFJ	The Advocate
6	INFP	The Mediator
7	INTJ	The Architect
8	INTP	The Thinker
9	ESTP	The Persuader
10	ESTJ	The Director
11	ESFP	The Performer
12	ESFJ	The Caregiver
13	ENFP	The Champion
14	ENFJ	The Giver
15	ENTP	The Debater
16	ENTJ	The Commander

Table 1.
MBTI personality types.

These key factors are embedded together to form 16 different personality types as shown in the table above (**Table 1**):

4. Big five: also known as OCEAN (openness, conscientiousness, extraversion, agreeableness, and neuroticism), describes human personality [14] by considering five features where these five features are closely interrelated where every individual has a score in these five personality feature. Therefore, this inter-relatability of the features in the Big five personality makes it suitable for multitask learning [12, 15].

4. Approaches to determine personality traits

Different methodologies can be utilized for determining personality traits of an individual. One of the simplest approach is personality questionnaires where various personality inventories are developed in order to create concise psychological scales like- TIPI (Ten Item Personality Inventory), 28-item questionnaire, (BFI) Big Five Inventory which consists of 44 items, NEO PI, NEO PI-R personality inventory (the Revised NEO Personality Inventory) and NEO-FFI-3 (NEO Five Factor Inventory-3) [12, 15]. Apart from these questionnaires, social media usage questionnaire in the form of survey along with other approaches were also utilized by many researchers to determine personality traits.

Linear Discriminate Analysis (LDA), PCA (Principal Component Analysis), SVM (Support *vector machines*), Multinomial Naive Bayes and AdaBoost are some of the most utilized classification algorithms for determining the text emotion. Linear Discriminate Analysis is a robust model used for classification and dimension reduction tool which can also be leveraged in data pre-processing. LDA assumes that data is distributed normally and each class has identical covariance matrices. As the classes are assumed to be linearly separable, multiple discrimination functions addressing several hyper-planes are made to recognize the classes [16]. AdaBoost is one of the most efficient ensemble methods that endeavor to approximate Bayes classifier by combining distinct weak classifiers after adding up the probabilistic predictions. It ultimately gives one single strong classifier. SAMME.R that is a variety of AdaBoost is used for multiclass classification problems [17, 18].

5. Background and previous studies

A survey of previous work review is carried out based on research papers on distinct approaches like questionnaires, profile information, status updates, groups or communities joined on platforms like Facebook, Twitter and LinkedIn. Schrammel et al. [19] reported the outcomes of web-based survey to examine the relationship between information shared by the individuals on online platforms, their usage patterns and personality attributes. 162 completed questionnaires were gathered from different countries around 44% of the people who participated in the study were graduated. Five factor personality scale questionnaire was used. In order to determine behavioral patterns, community specific questions were also included. The results displayed that individuals with high extraversion scores tends to have a greater number of connections. Meanwhile, agreeableness was not related to the number of connections. It was also observed that individuals investing more time online share more personal information on profile [19]. T. Ryan et al. conducted research on self-selected 1158 Facebook users and 166 Facebook non users where participants were aged between 18 and 44. A bundle of 124 questions comprising Big Five Inventory, SELSA-S and NPI-29 along with Facebook usage survey consisting 18 questions was conducted specifically for Facebook users. The data gathered from both Facebook users and non-users was divided into two groups in order to differentiate which characteristics will more likely belong to Facebook users and non-users [20]. B. Zhong et al. [21] in his study explored the relation between personality characteristics and time spent on internet for different purposes like social media, study and other activities. 436 University students participated in this study by answering survey questions on the total internet usage, multitasking experience, need for cognition, Information and Communication Technology (ICT) innovativeness. The one of the two main factors that can be considered in this study is that an individual might not be much prompted to handle and process information on social networking sites. Another factor is that an individual with high NFC could be more efficient and hence gives less time to social networking sites and vice versa [21].

Mohammad Dalvi-Esfahani et al. utilized personality traits as moderators to evaluate the impact of Perspective Taking (PT) and Empathic Concern (EC) on social media addiction (SMA) [22]. 592 high school students between 15 and 18 yrs. participated in this study. The data gathered was analyzed using Partial Least Squares Structural Equation Modeling (PLS-SEM). It was observed that behavior associated SMA was reflected more among high income schools [22]. Pavica Sheldon et al. aimed to determine whether Big five traits, contextual age indicators and fear of missing out are significant indicators of addictions associated with social networking sites. A survey of 337 students consisting of 193 women was conducted. In this study the social activity, interpersonal interaction and life satisfaction were estimated by The Rubin and Rubin life position scale. Big Five Inventory-10 was utilized to measure personality attributes and Bergen Facebook Addiction Scale (BFAS) was used to evaluate social media addiction. As a result, a positive correlation was realized between social activity and Snapchat. It was also observed that FOMO and social media addiction were highly correlated [23]. In a recent study conducted by J. Brailovskaia et al. the consequences of Covid-19 pandemic and its association with the social media usage were examined. A total 550 individuals participated where 76.2% were women. 65.3% of individuals belonged to the student category whereas the percentage of employed individuals was 33.5%. It was discovered that higher the burden of the pandemic, the lower will be the self-control which will ultimately lead to excessive social media usage [24] (**Table 2**).

Publication	Purpose	Input data	Methodology	Conclusion
T. Ryan [19, 20]	• To identify personality traits associated with Facebook users and non-users. • To determine whether these traits are connected the way individuals utilize Facebook.	1635 self-chosen Australian Internet users out of which 1324 completed the online questionnaire (1158 FB user and 166 FB non users)	FB usage questionnaire and other including BFI.	Facebook users are less conscientious and socially desolate than non-users but were quite often found to be more extraverted and narcissistic as opposed to the non-users.
B. Zhong et al. [21]	To explore the relation between personality characteristics and social media	436 University students consisting 118 males and 318 females (average age = 20)	Survey questions to determine the relationship between the Big Five personality traits and internet usage.	• There is a negative correlation between Need For Cognition (NFC) and SNS usage. • People with high NFC also tend to add less people to their SNS accounts.
Mohammad Dalvi-Esfahani [22]	To determine the impact of EC and PT on SMA where personality traits were utilized as moderators.	Questionnaire from 592 high school students where 42.1% were male students.	• PLS-SEM approach was used to analyze data. • Characteristics of personality were estimated with the help of NEO-Five Factor Inventory which consisted in total of 60 items. • A series of ANOVA analysis was utilized	• EC and PT can negatively predict SMA. • Extraversion adversely directed the relationships between EC and SMA and PT and SMA. • ANOVA unveiled that SMA was reflected irrespective of gender but was reflected more on high income schools.
Pavica Sheldon [23]	To examine significant indicators of social media addictions.	Survey of 337 undergraduate students (141 men and 193 women)	• Pearson product–moment correlations and hierarchical linear	Positive correlation was observed between FOMO and social media addiction. Social activity

Publication	Purpose	Input data	Methodology	Conclusion
		was conducted with average age = 23.35 and SD = 8.08	regressions were conducted where life-position indicators and Big Five personality characteristics with the social media addiction attributes were correlated first. • Predictors of social networking sites addictions were analyzed with the help of three hierarchical linear regressions.	and Snapchat were also found to be positively correlated with each other.
Cecilie Schou Andreassen [25]	To identify association between narcissism, self-esteem and habit-forming usage of social networking sites.	Total 23,532 participants aged between 16 and 88 years with mean age = 38.5 years and SD = 13.3.	Open web-based survey consisting- • Bergen Social Media Addiction Scale (BSMAS) • Narcissistic Personality Inventory-16 • Rosenberg Self-Esteem Scale	• It was observed that scores for addictive usage of social platforms were higher among young. • Self-esteem and addictive use of social networking platforms were found to be positively correlated with narcissism. Whereas, negative correlation was there between self-esteem and social media usage.
J. Brailovskaia [24]	To analyze the burden caused by the pandemic on excessive usage of social networking platforms and mental health.	550 individuals from Germany participated in this research out of which, 76.2% were women. Mean age = 27.08 and SD = 6.74.	Participation survey link was sent to 600 randomly selected people between March – May 2020	• Burden caused by the pandemic was positively correlated with excessive social media usage and anxiety whereas sense of control was negatively correlated. • Anxiety symptoms were also observed to be correlated positively with addictive social media usage.

Table 2.
Personality Prediction using Questionnaire approach.

Content generated by the user on Facebook was associated with the responses given by them in the questionnaire. The Facebook features used to determine personality traits in this research are- Total status updated, Photos uploaded, Total number of groups joined, Number of Facebook friends, Number of Facebook likes, Number of times the user is tagged in photos by other profiles.

Questionnaires are time consuming and there are high chances of a person not giving accurate answers to avoid being judged. Whereas, the usage-based approach only focuses on the time given by the user rather than the activity type. Hence in order to mitigate these limitations, some other aspects of social media usage were included. Another popular approach utilized to determine personality traits is to study user behavior on social media through certain level of activities like posting status, commenting, following certain groups/ communities.

In order to bridge the gap between social networking sites and personality traits with the help of machine learning, J. Golbeck et al. [26] utilized BFI (Big Five Personality Inventory) and Facebook profile information i.e. "About me" and status update of the individuals. Data was accumulated from 279 individuals meanwhile; linguistic analysis was performed on data gathered from only 167 individuals where at least 10 words in the text fields were present. Golnoosh Farnadi et al. [27] in his study utilized Facebook user generated data from myPersonality app to predict personality traits of an individual. 9917 status updates from 250 Facebook users, frequency as well as time of posting were collected from myPersonality project. One or more personality traits were assigned to the user based on the answers given by them in questionnaire. The four features leveraged in this study are LIWC feature (from status updates and other text), time related feature, social network feature (network size and density) and other features like number of statues per user, number of urls and words occurring more than once [27].

Amichai-Hamburger et al. and Vinitzky et al. in their research suggested a strong connection between user behavior on Facebook and personality for this purpose they targeted participants who tend to use Facebook often. The data was gathered from 237 Israeli university students comprising 136 female and 101 male with mean age 22 years. In order to conduct study in accordance with the five-factor model, students were asked to NEO-PI-R and a self-report measure. Apart from this, user generated content on Facebook was also measured and encoded. Four dimensions of users aimed were: basic information, personal information, contact information & education and work information. It was observed that highly extroverted people tends to share less personal information as they are more confident with their social skills than introverted people also, extroversion was positively correlated to the number of friends instead of groups joined. It was also found that highly neurotic individuals less likely post their pictures as compared to less neurotic people [28].

Michael Tadsse et al. conducted a research to predict the personality traits of Facebook users using Big Five model. myPersonality dataset was used with 250 Facebook users and 9917 Facebook updates. Features selected for analysis, in accordance with the personality attributes were- network size, density, transitivity, betweenness and brokerage. It was found that the extraversion personality trait along with SNA (Social Network Analysis) resulted in best prediction accuracy [29]. Y Bachrach et al. attempted to predict personality traits by analyzing Facebook activities of a user by their actions (posts, groups and likes) and their friends for e.g. tagging and size & density of the friendship network [30].

Content generated by the user on Facebook was associated with the responses given by them in the questionnaire. The Facebook features used to determine personality traits in this research are:

- Total status updated.

- Photos uploaded.
- Total number of groups joined.
- Number of Facebook friends.
- Number of Facebook likes.
- Number of times the user is tagged in photos by other profiles.

Correlations were determined between these features utilized by producing plots known as 'Clustered Scatter Plots'. By including more Facebook features the accuracy of the model improvised. And a clear picture was observed to determine the correlation between each personality trait and Facebook feature for e.g., agreeableness was found to be positively correlated with groups joined, posts liked and total friends. This study primarily focused on the data of individuals who are active on social media and agreed to use the application for personality prediction. This sort of data might lead to selection bias. Also, the total number of posts liked, groups joined were taken into consideration but not the type of content liked or group joined (**Table 3**).

Publication	Purpose	Input data	Methodology	Conclusion
J. Golbeck et al. [26]	Predicting user personality through information available on their FB profile.	Performed linguistic analysis on 167 out of 279 individuals. (avg. Age = 31.2 years)	Utilized BFI-45 and FB information "About me" and status updates using Facebook API.	Each of the five personality traits were predicted by training m5sup'Rules and Gaussian Processes ML algorithms.
GolnooshFarnadi et al.[27]	To predict personality traits from Facebook status updates, their time stamps and social network attributes.	9917 status updates from 250 Facebook users, frequency as well as time of posting were collected from myPersonality project.	User generated data on Facebook, answers given by the users in questionnaire, 81 features extracted using LIWC SVM, KNN and NB are the three ML techniques that were explored in this study.	ML based approaches generalize across domains. • Accurate models can be build where no training data is present by utilizing training examples from other sources.
Michael M. Tadesse [29]	To predict personality traits via user generated posts on facebook.	myPersonality dataset consisting of 250 Facebook users and 9917 posts.	• OpenNLP was employed for data pre-processing. • LIWC and SPLICE were used to analyze content of posts. • Pearson correlation analysis was	• Neurotic users less likely use words relating to positive emotions. • Negative correlation was found between conscientious individuals and words with

Publication	Purpose	Input data	Methodology	Conclusion
			performed to inspect significant features for personality traits prediction.	negative emotions. • The best prediction performance was observed for extraversion personality trait by SNA.
Chris Sumner et al. [31]	To determine whether there is any relationship between Big Five Personality traits and Facebook activity.	Facebook profile information from 537 participants from 15 countries. (mean age = 30 years with 349 female and 174 male participants)	• BFI-44 questionnaire along with 79 Facebook data points consisting age, total number of friends, biography and gender. • LIWC tool was used to analyze user generated content i.e. post and photo caption/ description. • Zero-order Spearman's correlation was used for linguistic analysis.	• People with high extraversion and Conscientious tend to use words with positive emotion. A negative correlation was also observed between the use of swear word and • Conscientiousness while Neuroticism is positively correlated to it. • A positive correlation was also observed between number of words used in one sentence and Neuroticism as well as Agreeableness.
Y. Bachrach et al. [30]	Determined the aspects of profile by bifurcating Facebook activities in two categories- • Solely based on user's activity (posting status, posts liked or groups joined) • With respect to actions of user and their connections (size and density of the network, total number of posts tagged)	Records of Facebook profile features of 180,000 users.	• Five Factor Model questionnaire. • Prediction algorithm used- Multivariate Linear Regression	• Claimed that the prediction accuracy of the model is increased by adding more Facebook features for analysis. • All the Facebook features used were positively correlated with Extraversion but negatively correlated with Conscientiousness • A positive correlation was also found between Openness and Neuroticism with total status updates, likes and groups joined by an individual.

Table 3.
Personality Prediction using content generated on Facebook in association with questionnaire.

5.1 Synthesis of previous work carried on twitter data

In a research conducted by Daniel Ricardo Jaimes Moreno et al., twitter data was utilized to predict personality traits. On twitter users generally post content in form of text i.e., tweets. For training purpose, the author utilized PAN CLEF dataset of 152 users containing 14,166 tweets. The prediction of personality was seen as a classification problem where after creating TFIDF matrix, different dimensionality reduction techniques were applied like PCA (Principal Components Analysis), LDA (Linear Discriminant Analysis) and NMF (Non-negative Matrix Factorization) for extracting latent features. The results displayed that for extraverted traits, TFIDF and PCA performed better whereas, LDA technique gave better results for agreeableness, conscientiousness and stability. Therefore, the best performance was performed by LDA technique [32].

Aditi V.Kunte et al. [33] made hypothesis based on twitter dataset using Twitter API. Data pre-processing was done by conversion to lower case, removing stop-words and special characters. Classification algorithms were then applied to the pre-processed data in order to classify user personality in the class labels of Big Five personality prediction model. It was observed that accuracy, precision recall and F1-score were highest for Multinomial NB as compared to other classification algorithms [34].

Pavan Kumar K. N. and Marina L. Gavrilova [34] aimed to determine MBTI personality traits by taking user generated data in form of latest 50 tweets. For this purpose, the author used a combination of TFIDF, GloVe word embedding technique and SVM classifier. The TFIDF document term matrix and GloVe embeddings of the tweets posted by an individual is utilized to construct decision tree ensembles composed of CART (Classification and Regression tress).

It was observed that the prediction accuracy changed across the MBTI dimensions where S-N (Sensation-Intuition) and E-I (Extraversion-Introversion) dimensions were considerably more reliable. Alexia Katrimpouza et al. [35] used questionnaire and educational activities of students on Twitter in order to determine how learning outcomes are correlated to twitter usage. In total, three studies were conducted where twitter activities of students were analyzed during each study. It was seen that the personality characteristics Openness and Conscientiousness were related with the twitter usage in one of the studies. The tools to implement this study are- SML scale (Social Media Learning), Big Five personality test, TAS (Technology Affinity Survey) and ICTL (Communications Technology Learning). Fabio R. Galloa et al. [36] has made hypothesis that NKB (Network Knowledge Base) model can be associated with personality prediction to develop a hybrid model to predict actions and reactions made by an individual in their social networking feeds. A specimen of NKB and stream of news items for each individual was used to train classifier in order to predict if the user will make a move i.e. a certain action in a certain amount of time. For tuning hyper- parameters the different algorithms utilized were- Logistic Regression, One Class SVM, Random Forests, Decision Trees, Multinomial Naive Bayes, and Complement Naive Bayes. The bigger scope of this study is to reduce the pathogenic feeds by analyzing the information flow on social media [36] (**Table 4**).

Utku Pamuksuz et al. and Joseph T. Yun et al. [38] targeted on three brands i.e. McDonald's, Harley-Davidson and Tom's Shoes, to determine personality of the brand's twitter accounts and the followers associated with it. The main objective of the author was to obtain the connection between human personality and brand's personality on social networking platforms. Crimson Hexagon was utilized to accumulate

Publication	Purpose	Input data	Methodology	Conclusion
Daniel Ricardo Jaimes Moreno [32]	This study focused on the extraction of features by exploring distinct dimensionality reduction techniques.	Utilized PAN CLEF 2015 data set of 152 different users consisting of 14,166 tweets (approx. 100 tweets per user).	• Matrix was transformed using TFIDF • For extracting latent feature, different dimensionality reduction techniques were utilized- PCA, LDA and NMF.	• Extroverted tweets carried more words on an average in comparison ofTweets from other characteristics. • LDA displayed better results for agreeableness, conscientiousness and stability. • Non-negative Matrix Factorization[37] technique gave inefficient results for each trait as compared to other methods.
Aditi V. Kunte [33]	Whether LDA, AdaBoost or Multinomial NB has higher relevance for personality prediction over twitter dataset.	Twitter data was retrieved using Twitter API consisting 2 columns- index and status.	• Big Five personality prediction model was utilized. • Focused on real time data using Twitter API.	Multinomial NB displayed highest accuracy, precision, recall and F1-score than AdaBoost and LDA.
Pavan Kumar K. N. [34]	To associate MBTI personality traits in accordance with the latest 50 tweets posted by the user.	• Latest 50 tweets made by the user. • Twitter MBTI dataset was obtained from the PersonalityCafe forum.	• To extract relevant text features TFIDF and GloVE word embeddings technique was utilized. • SVM was used for text classification.	The prediction accuracy changed across the MBTI dimensions where S-N (Sensation-Intuition) and E-I (Extraversion-Introversion) dimensions were considerably more reliable.
Alexia Katrimpouza [35]	The educational activities of students were utilized throughout the study in order to obtain an interaction model between social media and participants.	• Inputs were taken from different number of students during each phase of the study. • First study: 19 students; 19 female (aged between 18 and 28) • Second study: 80 students; (aged between 17 and 47) • Third study: 46 students aged between 18 and 33; 1 man and 45 women	• Twitter activity of the students was analyzed with reference to the total number of tweets posted in each study. • Big Five personality questionnaire along with TAS, ICTL and SML scale was used.	• It was observed that the students who took part in the weekly tasks were more efficacious in class as compared with the ones who did not take part. • Openness and Conscientiousness were associated with the twitter usage in one of the studies. • It was found that the proper consumption of twitter activity amplified the learning experience of students by sharing and

Publication	Purpose	Input data	Methodology	Conclusion
				searching relevant information.
Fabio R. Galloa [36]	Introduced NKB framework for predicting user actions and reactions on the post.	Tweets and follow network of users.	• Big Five model (from the Personality Insights service offered by IBM Cloud) • Time of the day when prediction is carried out Sentiments present in the post, percentage of positive items in the feed, percentage of negative items	• ML algorithms are efficient to make predictions about how a person responds to the content in the feed. • The best execution was produced by a variant based on Complement Naive Bayes. • Personality type proved to be single most impactful feature for displaying significant drops in recall.

Table 4.
Predicting Personality using content posted on Twitter.

brand related user generated dated between 13 July 2009 until 1 October 2015. To attain relation between users and brands, cosine similarity measure was leveraged. The results displayed that the personality type of twitter followers of Harley and Tom's was closer to the brand personality but it was not the same for McDonald's and its followers. However, this difference was mitigated on removing neuroticism from the analysis [38].

5.2 Synthesis of previous work carried on multifarious social networking platforms

Niels et al. [39] focused on determining whether job related social platforms can be leveraged to predict personality of a person. Both organization and individual seeking for opportunities can benefit with such sort of assessment as personality affects the job performance. In first study, the respondents were bifurcated into two samples i.e. student (mean age = 23.1, SD = 2.78) and employed individuals (mean age = 36.8 and SD = 10.87) whereas in second study, a homogonous sample of consisting total of 97 participants from one organization took part. A different set of Big Five questionnaire was utilized. To some extent, self-rated extraversion and self-presentation were found to be predicted via LinkedIn profile. However, Big five traits were not evaluated significantly beyond extraversion [39]. Shyron Qianyun Ma et al. [40] conducted a study to examine personality traits of LinkedIn users. A sample of 301 LinkedIn users with average age of 33 years was utilized. The dataset consisted of 43.2% of individuals with 11 or more years of experience and 62.8% post graduated individuals. Snowball technique was used where each LinkedIn individual was asked to approach other five connections to join via mail. The results displayed a close relation between extraversion and agreeableness with LinkedIn use. Meanwhile no significant relation was observed between openness and LinkedIn use. It was also found that individuals with high agreeableness were capable of gaining more social bridging capital [40].

Sumer S. Vaid et al. [41] attempted to determine the linkage between personality traits and frequency of using distinct social networking platforms. Big Five and Dark Triad personality was assessed in association with social network specific usage patterns. For this purpose, relation between different types of social media was analyzed followed by personality traits recognition in accordance of individual's inclination towards social media usage. The data gathered from exploratory and confirmatory studies were then put together to run a progression of ad-hoc analysis on pooled dataset [41] (**Table 5**).

Publication	Purpose	Input data	Methodology	Conclusion
Shyron Qianyun Ma [40]	To understand the means in which personality assessment impacts the LinkedIn use and features use.	301 LinkedIn users with average age of 33 years participated in the study.	• Big Five 20 item inventory was utilized. • Internet Social Capital Scale (ISCS) was used to measure perceived social capital. • Zero order correlation analysis and five parallel multiple regression analysis were performed to test the hypothesis.	• Extraversion and agreeableness were found to be closely related to LinkedIn use. • Individuals with agreeableness gained more social bridging capital. • No significant correlation was found between openness and LinkedIn use or feature use.
Niels [39]	To analyze whether user data generated on job related social platforms is efficient to predict personality traits of an individual.	• In first study, respondents were divided into 2 samples i.e. student sample (age range: 19–35) and working sample (age range: 22–65). • 97 people from one homogonous organization participated in second study.	Different set of questionnaires based on Big five factor model was utilized in both the studies.	Big five traits could not be analyzed significantly beyond extraversion.
Sumer S. Vaid [41]	To examine multiplatform social media and the personality characteristics linked with the use of different sort of social networking sites.	• The data for exploratory (N: 1803) and confirmatory study (N: 2317 was collected from young university students in the US. • 10 different types of social media platforms were analyzed. .	Regression analysis was done on pooled dataset. • PCA was used to characterize use of various types of social networking platforms. • Cross sectional approach was followed.	• Extraversion as well as Openness were closely related with the Facebook and messaging platforms usage. • Young adults higher in Neuroticism used image-content sharing and micro-blogging platforms more frequently. • Conscientiousness was connected to

Publication	Purpose	Input data	Methodology	Conclusion
				more noteworthy usage of the YouTube platform. • Media ecologies of young adults was categorized in 4 use dimensions- • Blogs and Virtual Worlds • Social Networking Sites and Messaging Platforms • Forums and Media-Sharing Platforms • Facebook.

Table 5.
Predicting personality using LinkedIn and multifarious social media platforms.

6. Conclusion

Personality can be predicted on the basis of behavioral characteristics or linkage and content-based data available. In this study we have attempted to explore distinct approaches to determine personality traits with respect to the text posted by the user or their activities on social networking sites. In behavior or say activity-based analysis, an individual is susceptible to display different behaviors at different time intervals i.e. a person can act differently in different situations which will ultimately lead to activation of a specific personality characteristic and will assert expression for it at that specific time. Features like groups joined, total number of friends were also utilized by researchers to accurately determine personality traits. Facebook due to myPersonality application was one of the most prominent platforms used. But in 2018 ban was announced for improper control of data and harvesting personal information of users. After this suspension, Twitter became the most utilized platform to gather data and determine personality traits commonly using Myers Briggs and Big Five approaches.

Author details

Vrinda Tandon and Ritika Mehra*
Dev Bhoomi Uttarakhand University, Uttarakhand, India

*Address all correspondence to: riti.arora@gmail.com

References

[1] Fuller A, Fan Z, Day C, Barlow C. Digital twin: Enabling technologies, challenges and open research. IEEE Access. 2020;**8**:108952-108971

[2] Madni AM, Madni CC, Lucero SD. Leveraging digital twin technology in model-based systems engineering. System. 2019;**7**(1):7

[3] Lv Z, Xie S. Artificial intelligence in the digital twins: State of the art, challenges, and future research topics. Digital Twin. 2021;**1**(12):12

[4] https://www.gartner.com/en/newsroom/press-releases/2020-10-29-gartner-survey-reveals-47-percent-of-organizations-will-increase-investments-in-iot-despite-the-impact-of-covid-19-

[5] https://www.marketsandmarkets.com/Market-Reports/digital-twin-market-225269522.html

[6] Latif HH. A Methodology for Developing Digital Twins in the Manufacturing Plant. North Carolina State University; NC state University library, USA: ProQuest Dissertations Publishing; 2020. Available from: https://repository.lib.ncsu.edu/bitstream/handle/1840.20/38132/etd.pdf?sequence=1&isAllowed=y

[7] Corral-Acero J, Margara F, Marciniak M, Rodero C, Loncaric F, Feng Y, et al. The 'digital Twin'to enable the vision of precision cardiology. European Heart Journal. 2020;**41**(48): 4556-4564

[8] John OP. The "Big Five" Factor Taxonomy: Dimensions of Personality in the Natural Language and in Questionnaires. In: Pervin L, Editor. Handbook of Personality: Theory and Research. NewYork, US: Guiltford Press; 1990. pp. 66-100. Available from: https://psycnet.apa.org/record/1990-98135-003

[9] Schwartz HA, Eichstaedt JC, Kern ML, Dziurzynski L, Ramones SM, Agrawal M, et al. Personality, gender, and age in the language of social media: The open-vocabulary approach. PLoS One. 2013;**8**(9):e73791

[10] Medium. What is Digital Twin Technology — Use Cases, Solutions, and Examples: https://gramener.medium.com/digital-twin-technology-f2312d64ca36

[11] Rasheed A, San O, Kvamsdal T. Digital twin: Values, challenges and enablers from a modeling perspective. Ieee Access. 2020;**8**:21980-22012

[12] Retorio What's the Difference Between the Big 5, the Enneagram, and MBTI?. https://www.retorio.com/blog/pre-employment-assessment-big-5

[13] Brown A, Bartram D. Relationships between OPQ and Enneagram Types. London: SHL Group plc; 2005

[14] Sun J, Tian Z, Fu Y, Geng J, Liu C. Digital twins in human understanding: A deep learning-based method to recognize personality traits. International Journal of Computer Integrated Manufacturing. 2021;**34**(7–8):860-873

[15] Gosling SD, Rentfrow PJ, Swann WB Jr. A very brief measure of the big-five personality domains. Journal of Research in Personality. 2003;**37**(6):504-528

[16] https://www.sciencedirect.com/topics/computer-science/linear-discriminant-analysis

[17] Freund Y, Schapire RE. A decision-theoretic generalization of on-line

learning and an application to boosting. Journal of Computer and System Sciences. 1997;**55**(1):119-139

[18] Vergara D, Hernández S, Jorquera F. Multinomial naive Bayes for real-time gender recognition. In: 2016 XXI Symposium on Signal Processing, Images and Artificial Vision (STSIVA). Bucaramanga, Colombia: IEEE; 2016, Aug. pp. 1-6. DOI: 10.1109/STSIVA.2016.7743331

[19] Schrammel J, Köffel C, Tscheligi M. Personality traits, usage patterns and information disclosure in online communities. People and Computers XXIII Celebrating People and Technology. 2009;**2009**:169-174

[20] Ryan T, Xenos S. Who uses Facebook? An investigation into the relationship between the Big Five, shyness, narcissism, loneliness, and Facebook usage. Computers in Human Behavior. 2011;**27**(5):1658-1664

[21] Zhong B, Hardin M, Sun T. Less effortful thinking leads to more social networking? The associations between the use of social network sites and personality traits. Computers in Human Behavior. 2011;**27**(3):1265-1271

[22] Dalvi-Esfahani M, Niknafs A, Alaedini Z, Ahmadabadi HB, Kuss DJ, Ramayah T. Social media addiction and empathy: Moderating impact of personality traits among high school students. Telematics and Informatics. 2021;**57**:101516

[23] Sheldon P, Antony MG, Sykes B. Predictors of problematic social media use: Personality and life-position indicators. Psychological Reports. 2021;**124**(3):1110-1133

[24] Brailovskaia J, Margraf J. The relationship between burden caused by coronavirus (Covid-19), addictive social media use, sense of control and anxiety. Computers in Human Behavior. 2021;**119**:106720

[25] Andreassen CS, Pallesen S, Griffiths MD. The relationship between addictive use of social media, narcissism, and self-esteem: Findings from a large national survey. Addictive Behaviors. 2017;**64**:287-293

[26] Golbeck J, Robles C, Turner K. Predicting personality with social media. In: CHI'11 Extended Abstracts on Human Factors in Computing Systems. New York, NY, United States: Association for Computing Machinery; 2011. pp. 253-262. DOI: 10.1145/1979742.1979614

[27] Farnadi G, Zoghbi S, Moens MF, De Cock M. Recognising personality traits using facebook status updates. In: Proceedings of the International AAAI Conference on Weblogs and Social Media. Cambridge, MA, USA. 2013;7(2):14-18. DOI: 10.1609/icwsm.v7i2.14470

[28] Amichai-Hamburger Y, Vinitzky G. Social network use and personality. Computers in Human Behavior. 2010;**26**(6):1289-1295

[29] Tadesse MM, Lin H, Xu B, Yang L. Personality predictions based on user behavior on the facebook social media platform. IEEE Access. 2018;**6**:61959-61969

[30] Bachrach Y, Kosinski M, Graepel T, Kohli P, Stillwell D. Personality and patterns of Facebook usage. In: Proceedings of the 4th Annual ACM Web Science Conference. New York, NY, United States. 2012, Jun. pp. 24-32. DOI: 10.1145/2380718.2380722

[31] Sumner C, Byers A, Shearing M. Determining personality traits & privacy

concerns from facebook activity. Black Hat Briefings. 2011;**11**(7):197-221

[32] Moreno DRJ, Gomez JC, Almanza-Ojeda DL, Ibarra-Manzano MA. Prediction of personality traits in twitter users with latent features. In: 2019 International Conference on Electronics, Communications and Computers (CONIELECOMP). Cholula, Mexico: IEEE; 2019. pp. 176-181. DOI: 10.1109/CONIELECOMP.2019.8673242

[33] Kunte AV, Panicker S. Using textual data for personality prediction: A machine learning approach. In: 2019 4th International Conference on Information Systems and Computer Networks (ISCON). Mathura, India: IEEE; 2019. pp. 529-533. DOI: 10.1109/ISCON47742.2019.9036220

[34] Kumar KP, Gavrilova ML. Personality traits classification on twitter. In: 2019 16th Ieee International Conference on Advanced Video and Signal Based Surveillance (Avss). Taipei, Taiwan: IEEE; 2019. pp. 1-8. DOI: 10.1109/AVSS.2019.8909839

[35] Katrimpouza A, Tselios N, Kasimati MC. Twitter adoption, students' perceptions, Big Five personality traits and learning outcome: Lessons learned from 3 case studies. Innovations in Education and Teaching International. 2019;**56**(1):25-35

[36] Gallo FR, Simari GI, Martinez MV, Falappa MA. Predicting user reactions to twitter feed content based on personality type and social cues. Future Generation Computer Systems. 2020; **110**:918-930

[37] Strömqvist Z. Matrix Factorization in Recommender Systems: How Sensitive Are Matrix Factorization Models to Sparsity?. Sweden: Department of Statistics Uppsala University; 2018. Available from: http://uu.diva-portal.org/smash/get/diva2:1214390/FULLTEXT01.pdf

[38] Pamuksuz U, Yun JT, Humphreys A. A brand-new look at you: Predicting brand personality in social media networks with machine learning. Journal of Interactive Marketing. 2021;**56**(1):1-15

[39] Van de Ven N, Bogaert A, Serlie A, Brandt MJ, Denissen JJ. Personality perception based on LinkedIn profiles. Journal of Managerial Psychology. 2017

[40] Ma SQ, Leung L. The impacts of personality traits, use intensity and features use of LinkedIn on bridging social capital. Applied Research in Quality of Life. 2019;**14**(4):1059-1078

[41] Vaid SS, Harari GM. Who uses what and how often?: Personality predictors of multiplatform social media use among young adults. Journal of Research in Personality. 2021;**91**:104005

Chapter 11

Scaling Subspace-Driven Approaches Using Information Fusion

Sally Ghanem and Hamid Krim

Abstract

In this work, we seek to exploit the deep structure of multi-modal data to robustly exploit the group subspace distribution of the information using the Convolutional Neural Networks (CNNs) formalism. Upon unfolding the set of subspaces constituting each data modality, and learning their corresponding encoders, an optimized integration of the generated inherent information is carried out to yield a characterization of various classes. Referred to as deep Multimodal Robust Group Subspace Clustering (DRoGSuRe), this approach is compared against the independently developed state-of-the-art approach named Deep Multimodal Subspace Clustering (DMSC). Experiments on different multimodal datasets show that our approach is competitive and more robust in the presence of noise.

Keywords: sparse learning, computer vision, unsupervised classification, subspace clustering, multi-modal sensor data

1. Introduction

Unsupervised learning is a very challenging topic in Machine Learning (ML) and involves the discovery of hidden patterns in data for inference with no prior given labels. Reliable clustering techniques will save time and effort required for classifying/labeling large datasets that might have thousands of observations. Multi-modal data, increasingly in need for complex application problems, have become more accessible with recent advances in sensor technology, and of pervasive use in practice. The plurality of sensing modalities in our applications of interest, provides diverse and complementary information, necessary to capture the salient characteristics of data and secure their unique signature. A principled combination of the information contained in the different sensors and at different scales is henceforth pursued to enhance understanding of the distinct structure of the various classes of data. The objective of this work is to develop a principled multi-modal framework for object clustering in an unsupervised learning scenario. We extract key class-distinct features-signatures from each data modality using a CNNs encoder, and we subsequently non-linearly combine those features to generate a discriminative characteristic feature. In so doing, we work on the hypothesis that each data modality is approximated by a Union of low dimensional Subspaces which highlights underlying hidden features. The UoS structure is unveiled by pursuing sparse self-representation of the given data modality. The subsequent aggregation of the multi-modal subspace structures yields a jointly unified characteristic subspace for each class.

1.1 Related work

Subspace clustering has been introduced as an efficient way for unfolding union of low-dimensional subspaces underlying high dimensional data. Subspace clustering has been extensively studied in computer vision due to the vast availability of visual data as in [1–4]. This paradigm has broadly been adopted in many applications such as image segmentation [5], image compression [6], and object clustering [7].

Uncovering the principles and laying out the fundamentals for multi-modal data has become an important topic in research in light of many applications in diverse fields including image fusion [8], target recognition [9–12], speaker recognition [13], and handwriting analysis [14]. Convolutional neural networks have been widely used on multi-modal data as in [15–19]. A multi-modal subspace clustering-inspired approach was also proposed in [20]. The emphasis of our formulation results in a different optimization problem, as the multi-modal sensing seeks to not only account for the private information which provides the complementarity of the sensors, but also the common and hidden information. This yields, as an end result, a different network structure than that of [20] with a different application space inspiration. In addition, the robustness of fusing multi-modal sensor data each with its distinct intrinsic structure, is addressed along with a potential scaling for viability. A thorough comparison of our results to the multimodal fusion network in [21] is carried out, with a demonstration of resilient fusion under a variety of limiting scenarios including limited sensing modalities (sensor failures). In [22], the authors proposed a deep multi-view subspace clustering approach that combined global and local structures to help achieve a small distance between samples of the same cluster and make samples in different clusters of different views farther. To that end, they used a discriminative constraint between different views. The discriminative constraint is based on the Hadamard product between the features extracted by the convolutional auto-encoder for the different views. In contrast, our approach is based on the minimizing the group norm, which we proved with a derivation in earlier work [23] and entails a smaller angle between the different subspaces across all modalities, thus promoting the goal of obtaining a common latent space. Moreover, minimizing the group norm also provides as well as group sparse solution along data modalities. Sun et al. [24] proposed a deep trainable multi-view subspace clustering method, named self-supervised deep multi-view subspace clustering (S2DMVSC) that learns the common latent subspace using two losses: spectral clustering loss and classification loss in order to denoise the imperfect correlations among data points.

In this paper, we prove that our formulation, which is based on the group norm of the self-representation matrices and the commutation loss between them, provides a natural way to fuse multi-modal data by employing the self-representation matrix as an embedding for each data modality, making our approach robust under different types of potential limitations. It is good to note that our proposed approach secures the individual sensor data-points relations resulting in more flexibility for each data modality.

1.2 Contributions

Building on the work of Deep Subspace Clustering (DSC) [25], we propose a new and principled multi-modal fusion approach which accounts for a sensor' capacity to house private and unique information about some observed data as well as that information which is likely also captured and hence common to other sensors. This is accounted for in our robust fusion formulation for multi-modal sensor data. Unveiling the complex UoS of multi-modal data also requires us to account for scaling in our proposed formulation and solution, which in turn invokes the learning of multiple/deep scale Convolutional Neural Networks. Our proposed Multi-modal fusion approach, by virtue of each sensor information structure (i.e., private plus shared) seeks to enhance and robustify the subspace approximation of shared information for each of the sensors, thus yielding a parallel bank of UoS for each of the sensors. The robust Deep structure effectively achieves scaling while securing structured representation for unsupervised inference. We compare our approach to a well-known deep multimodal network [21] which was also based on [25].

In our proposed approach, we thus define the latent space in a way that safeguards the individual sensor private information which hence dedicates more degrees of freedom to each of the sensors. In contrast to the approach in [21]. In our evaluation, we use two recently released data sets each of which we partition into learning and validation subsets. The learned UoS structure for each of the data sets is then utilized to classify new observed data points, which illustrates the generalization power of the proposed approach. Different scenarios with corresponding additive noise to either the training set or the testing set, or both, were used to thoroughly investigate the robustness, and resilience of the clustering approach performance. Experimental results confirm a significant improvement for our Deep Robust Group Subspace Recovery network (DRoGSuRe) under numerous limiting scenarios and demonstrate robustness under these conditions.

The balance of the paper is organized as follows, in Section 2, we provide the problem formulation, background along with the derivation for our proposed approach, Deep Robust Group Subspace Recovery (DRoGSuRe). In Section 3, we describe the attributes of the proposed approach and contrast it to Deep Multimodal Subspace Clustering algorithm (DMSC). In Sections 4 and 5, we present a substantiative validation along with experimental results of our approach, while Section 6 provides concluding remarks.

2. Deep robust group subspace clustering

2.1 Problem formulation

We assume having a set of data observations, each represented as a $\boldsymbol{m}$ dimensional vector $\boldsymbol{x_k}(\boldsymbol{t}) \in \mathbb{R}^m$, where $\boldsymbol{k} = \mathbf{1}, \mathbf{2}, \dots, \boldsymbol{n}$. Moreover, we consider having $\boldsymbol{T}$ data modalities, indexed by $\boldsymbol{t} = \mathbf{1}, \mathbf{2}, \mathbf{3}, \dots, \boldsymbol{T}$. Each data modality can then be described as $\boldsymbol{x_k}(\boldsymbol{t}) \in \mathbb{R}^m$, where $\boldsymbol{X}(\boldsymbol{t}) = [\boldsymbol{x_1}(\boldsymbol{t})\ \boldsymbol{x_2}(\boldsymbol{t}) \dots \boldsymbol{x_n}(\boldsymbol{t})]$. Our objective is to assign each set of data observations into clusters that can be efficiently represented by a low-dimensional subspace. This is equivalent to finding a partitioning $\{\boldsymbol{X}^1(\boldsymbol{t}), \boldsymbol{X}^2(\boldsymbol{t}), \dots, \boldsymbol{X}^P(\boldsymbol{t})\}$ of $[\boldsymbol{n}]$ observations, where $\boldsymbol{P}$ is the total number of clusters underlying each data modality indexed by $\boldsymbol{p}$. Furthermore, each linear subspace can be described as $S^p(t) \subset \mathbb{R}^m$ with $\dim S^p(t) \ll m$.

We will exploit the self-expressive property presented in [1, 26], which entails that each data observation $\boldsymbol{x}_i(\boldsymbol{t})$ can be represented as a linear combination of all features from the same subspace $\boldsymbol{S}(\boldsymbol{x}_i(\boldsymbol{t}))$ as follows,

$$\boldsymbol{x}_i(\boldsymbol{t}) = \sum_{i \neq j,\, x_j(t) \in S(x_i(t))} w_{ij}(t) x_j(t). \tag{1}$$

If we stack all the data points $\boldsymbol{x}_i(\boldsymbol{t})$ into columns of the data matrix $\boldsymbol{X}(\boldsymbol{t})$. The self-expressive property can be written in a matrix form as follows,

$$\boldsymbol{X}(\boldsymbol{t}) = \boldsymbol{X}(\boldsymbol{t})\boldsymbol{W}(\boldsymbol{t})\ s.t.\boldsymbol{W}_{ii} = \mathbf{0}. \tag{2}$$

The important information about the relations among data samples is then recorded in the self-representation coefficient matrix $\boldsymbol{W}(\boldsymbol{t})$. Under a suitable arrangement/permutation of the data realizations, the sparse coefficient matrix $\boldsymbol{W}(\boldsymbol{t})$ is an $\boldsymbol{n} \times \boldsymbol{n}$ block-diagonal matrix with zero diagonals provided that each sample is represented by other samples only from the same subspace. More precisely, $\boldsymbol{W}_{ij}(\boldsymbol{t}) = \mathbf{0}$ whenever the indexes $\boldsymbol{i}, \boldsymbol{j}$ correspond to samples from different subspaces. As a result, the majority of the elements in $\boldsymbol{W}$ are equal to zero. A diagram showing our algorithm is depicted in **Figure 1**.

Our algorithm consists of three main stages; the first stage is the encoder which encodes the input modalities into a latent space. The encoder consists of $\boldsymbol{T}$ parallel CNNs, where $\boldsymbol{T}$ is the number of data modalities. Each modality data is fed into one network, and the output of each network represents the modality data projection into its corresponding hidden/latent space. The second component of the auto encoder is $\boldsymbol{T}$ self-expressive layers, the goal of which is to enforce the self-expressive property among the data observations of each data modality. Each self-expressive layer is a fully connected layer which independently operates on the output of each encoder. The last stage is the decoder which reconstructs input data from the self-expressive layers' output. The objective function sought through this approximation network is reflected in Eq. (5). The group sparsity introduced in [23] requires the minimization of the group norm of matrices $\boldsymbol{W}(\boldsymbol{t})$, which in turn, entails a smaller angle between the different spaces across all modalities, thus promoting the goal of obtaining a common latent space. Note that minimizing group norm provides a group sparse

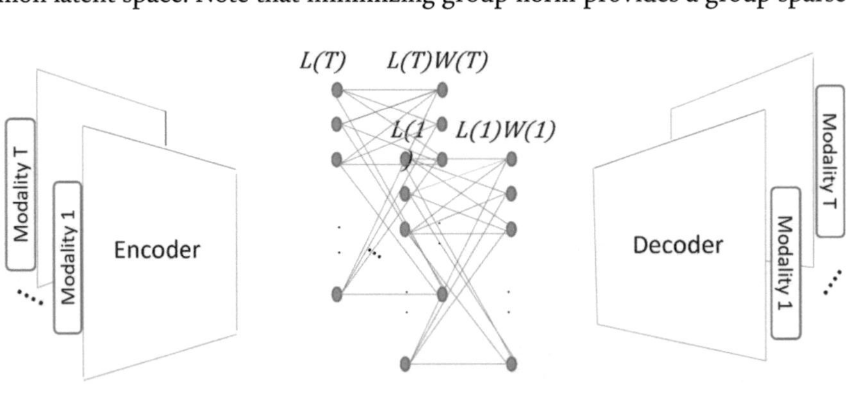

Figure 1.
Deep robust group subspace clustering diagram.

solution along data modalities. If we in addition, constrain the coefficient matrices corresponding to each data modality to commute, therefore, we ensure their sharing the same eigen vectors. The idea of commutation has been used in [27–29]. We define $\boldsymbol{\Omega} = \{\boldsymbol{W}(\boldsymbol{t})\}_{t=1}^{T}$, where $\boldsymbol{W}(\boldsymbol{t}) = [w_{kj}(\boldsymbol{t})]_{k,j}$ and the group l-norm $\|\boldsymbol{\Omega}\|_{1,2}$ as:

$$\|\Omega\|_{1,2} = \sum_{k,j}\sqrt{\sum_{t=1}^{T} w_{k,j}^2(t)}. \tag{3}$$

We also define $[\boldsymbol{W}(\boldsymbol{t_1}), \boldsymbol{W}(\boldsymbol{t_2})]$ as,

$$[\boldsymbol{W}(t_1), \boldsymbol{W}(t_2)] = \boldsymbol{W}(t_1)\boldsymbol{W}(t_2) - \boldsymbol{W}(t_2), \boldsymbol{W}(t_1) = 0 \tag{4}$$

The loss function is then rewritten as,

$$\begin{aligned} min_{\,\boldsymbol{W}(t)/w_{kk}(t)=0} &\sum_{t_1,t_2}^{T} \|[\boldsymbol{W}(t_1)\boldsymbol{W}(t_2)]\|^2 + \|\boldsymbol{\Omega}\|_{1,2} \\ &+\frac{\gamma}{2}\sum_{t=1}^{T} \|\boldsymbol{X}(t) - \boldsymbol{X}_r(t)\|_F^2 \\ &+\rho\sum_{t=1}^{T} \|\boldsymbol{W}(t)\|_1 + \frac{\mu}{2}\sum_{t=1}^{T} \|\boldsymbol{L}(t) - \boldsymbol{L}(t)\boldsymbol{W}(t)\|_F^2 \end{aligned} \tag{5}$$

where $\boldsymbol{X_r}(\boldsymbol{t})$ represent the reconstructed data corresponding to modality $\boldsymbol{t}$, and $\boldsymbol{L}(\boldsymbol{t})$ is the output of the $\boldsymbol{t}^{th}$ encoder with input $\boldsymbol{X}(\boldsymbol{t})$. $\boldsymbol{W}(\boldsymbol{t})$ is the sparse weight function that ties the data observation for modality $\boldsymbol{t}$. Solving DRoGSuRe in Tensorflow and using the adaptive momentum based gradient descent method (ADAM) [30] results in minimizing the loss function. For each data modality, the weights of the encoder, the self-expressive layer and the decoder are individually calculated, however, fine-tuning the weights is based on the loss function, which is a function of the group norm and the pairwise product difference between sparse coefficient matrices. $\|\|_1$ denotes the $\boldsymbol{l_1}$ norm, i.e., the sum of absolute values of the argument. The Lagrangian objective functional may be rewritten as,

$$\begin{aligned} L(\boldsymbol{W}(t)) = &\sum_{t_1,t_2}^{T} \|[\boldsymbol{W}(t_1)\boldsymbol{W}(t_2)]\|^2 + \|\boldsymbol{\Omega}\|_{1,2} \\ &+\rho\sum_{t=1}^{T} \|\boldsymbol{W}(t)\|_1 + \frac{\gamma}{2}\sum_{t=1}^{T} \|\boldsymbol{X}(t) - \boldsymbol{X}_r(t)\|_F^2 \\ &+\sum_{t=1}^{T}\frac{\mu}{2} \|\boldsymbol{L}(t)\boldsymbol{W}(t) - \boldsymbol{L}(t)\|_F^2 \\ &+\sum_{t=1}^{T} <\boldsymbol{L}(t)\boldsymbol{W}(t) - \boldsymbol{L}(t), \boldsymbol{Y}(t)> \end{aligned} \tag{6}$$

Assume $\hat{\boldsymbol{W}}(\boldsymbol{t}) = \boldsymbol{I} - \boldsymbol{W}(\boldsymbol{t})$, we update $\boldsymbol{W}(\boldsymbol{t})$ as follows,

$$\begin{aligned} \boldsymbol{W}_{\boldsymbol{k}+1}(t) = arg\,min_{\,\boldsymbol{W}(t)} \sum_{t_1,t_2}^{T} \|[\boldsymbol{W}(t_1)\boldsymbol{W}(t_2)]\|^2 + \|\boldsymbol{\Omega}\|_{1,2} + \rho\|\boldsymbol{W}(t)\|_1 + \\ < \boldsymbol{L}_{\boldsymbol{k}+1}(t)\boldsymbol{W}(t) - \boldsymbol{L}_{\boldsymbol{k}+1}(t), \boldsymbol{Y}_{\boldsymbol{k}}(t) \\ > + \frac{\mu_k}{2}\|\boldsymbol{L}_{\boldsymbol{k}+1}(t)\boldsymbol{W}(t) - \boldsymbol{L}_{\boldsymbol{k}+1}(t)\|_F^2 \end{aligned} \tag{7}$$

Similar to [4], we utilize linearized ADMM [31] to approximate the minimum of Eq. (7) since the algorithmic solution is complicated and yields a non-convex optimization functional. It has been shown that linearized ADMM is very effective for $\boldsymbol{l_1}$ minimization problems and the augmented Lagrange multiplier (ALM) method can take care of the non-convexity of the problem [32, 33]. Therefore, utilizing an appropriate augmented Lagrange multiplier $\boldsymbol{\mu_k}$, we can compute the global optimizer by solving the dual problem. The solution to Eq. (7) can be approximated, using linearized soft thresholding, as follows,

$$\begin{aligned} \boldsymbol{W}_{\boldsymbol{k}}^{+}(t) = prox_{\frac{\rho}{\mu\eta_1}}(\boldsymbol{W}_{\boldsymbol{k}}(t) + \frac{\boldsymbol{L}_{\boldsymbol{k}+1}^{T}\left(\boldsymbol{L}_{\boldsymbol{k}+1}\hat{\boldsymbol{W}}_{\boldsymbol{k}}(t) - \frac{\boldsymbol{Y}_{\boldsymbol{k}}(t)}{\mu_k}\right)}{\eta_1} \\ + \sum_{t_1,t_2=1,t_1\neq t_2}^{T} \{(\boldsymbol{W}_{\boldsymbol{k}}(t_1)\boldsymbol{W}_{\boldsymbol{k}}(t_2) \\ -\boldsymbol{W}_{\boldsymbol{k}}(t_2)\boldsymbol{W}_{\boldsymbol{k}}(t_1))W_k^T(t_2) + W_k(m)(\boldsymbol{W}_{\boldsymbol{k}}(t_1)\boldsymbol{W}_{\boldsymbol{k}}(t_2) \\ -\boldsymbol{W}_{\boldsymbol{k}}(t_2)\boldsymbol{W}_{\boldsymbol{k}}(t_1))\} \end{aligned} \tag{8}$$

$$\boldsymbol{W}_{\boldsymbol{k}+1}(t) = \gamma_{\frac{\rho}{\mu\eta_2}}(\boldsymbol{W}_{\boldsymbol{k}}^{+}(t)) \tag{9}$$

where $\boldsymbol{\eta_1} \geq \|\boldsymbol{L}\|_2^2$. We alternatively update $\boldsymbol{L}(\boldsymbol{t})$ as,

$$\boldsymbol{L}_{\boldsymbol{k}+1}(t) = \boldsymbol{L}_{\boldsymbol{k}}(t) + \mu_k\left(\boldsymbol{L}_{\boldsymbol{k}}(t)\hat{\boldsymbol{W}}_{k+1}(t) - \frac{Y_k(t)}{\mu_k}\right)\hat{W}_{k+1}^{T}(t). \tag{10}$$

where $\boldsymbol{prox}_{\beta}(\boldsymbol{A}_{i,j}(\boldsymbol{t})) = \boldsymbol{A}_{i,j}(\boldsymbol{t}) * \frac{\max\left\{\left(\sqrt{\sum_{t=1}^{T}\boldsymbol{A}_{i,j}(\boldsymbol{t})^2 - \beta}\right), \boldsymbol{0}\right\}}{\sqrt{\sum_{t=1}^{T}\boldsymbol{A}_{i,j}(\boldsymbol{t})^2}}$ and $\gamma_{\tau}(\boldsymbol{B}_{i,j}) = \boldsymbol{sign}(\boldsymbol{B}_{i,j}) * \max\{(|\boldsymbol{B}_{i,j}| - \boldsymbol{\tau}), \boldsymbol{0}\}$. The Lagrange multipliers are updated as follows,

$$\boldsymbol{Y}_{\boldsymbol{k}+1}(t) = \boldsymbol{Y}_{\boldsymbol{k}}(t) + \mu_k(\boldsymbol{L}_{\boldsymbol{k}+1}(t)\boldsymbol{W}_{\boldsymbol{k}+1}(t) - \boldsymbol{L}_{\boldsymbol{k}+1}(t)) \tag{11}$$

$$\mu_{k+1} = \epsilon\mu_k \tag{12}$$

After computing the gradient of the loss function, the weights of each multi-layer network, that corresponds to one modality, are updated while other modalities' networks are fixed. In other words, after constructing the data during the forward pass, the loss function determines the updates that back-propagates through each layer. The encoder of the first modality is updated, afterwards, the self-expressive layer of that modality gets updated and finally the decoder. Since the weights corresponding to each modality are dependent on other modalities, we update each part of the network corresponding to each modality with the assumption that all other networks' components corresponding to other modalities are fixed. The resulting sparse coefficient matrices $\boldsymbol{W}(\boldsymbol{t})$'s, for $\boldsymbol{t} = \mathbf{1}, \mathbf{2}, \dots, \boldsymbol{T}$ are then integrated as follows,

$$\boldsymbol{W}_{Total} = \sum_{t=1}^{T} \boldsymbol{W}(t) \tag{13}$$

Integrating the sparse coefficient matrices helps reinforcing the relation between data points that exist in all data modalities, thus establishing a cross-sensor consistency. Furthermore, adding the sparse coefficient matrices reduces the noise variance introduced by the outliers. A similar approach was introduced in [34] for Social Networks community detection, where an aggregation of multi-layer adjacency matrices was proved to provide a better Signal to Noise ratio, and ultimately better performance. To proceed with distinguishing the various classes in unsupervised manner, we construct the affinity matrix as follows,

$$\boldsymbol{A} = \boldsymbol{W}_{Total} + \boldsymbol{W}_{Total}^{T} \tag{14}$$

where $\boldsymbol{A} \in \mathbb{R}^{n \times n}$. We subsequently use the spectral clustering method [35] to retrieve the clusters in the data using the above affinity matrix as input.

2.2 Theoretical discussion

In order to justify the multiple banks of self-expressive layers, we assume that each modality $\boldsymbol{X}(\boldsymbol{t})$ may be expressed as a private information contribution $\boldsymbol{X}_p(\boldsymbol{t})$ and a shared information $\boldsymbol{X}_s(\boldsymbol{t})$ such that,

$$\boldsymbol{X}(t) = \boldsymbol{X}_s(t) + \boldsymbol{X}_p(t) \tag{15}$$

The shared information can be represented as follows,

$$\boldsymbol{X}_s(t) = \sum_{t=1}^{T} F(\boldsymbol{W}(t)(\boldsymbol{\Pi}_s \boldsymbol{X}(t))) \tag{16}$$

where $\boldsymbol{\Pi}_s = \cap_{t=1,\dots,T} \boldsymbol{\Pi}_s^t$. $\boldsymbol{X}_s(\boldsymbol{t})$ and $\boldsymbol{X}_p(\boldsymbol{t})$ are distinct and will hence lie in different subspaces, which will hence be mapped to different components in $\boldsymbol{W}(\boldsymbol{t})$. Similarly for the subspaces spanned by $\boldsymbol{X}_p(\boldsymbol{t}_i)$ and $\boldsymbol{X}_p(\boldsymbol{t}_j)$, $\boldsymbol{i} \neq \boldsymbol{j}$, the corresponding components of $\boldsymbol{W}(\boldsymbol{t}_i)$ and $\boldsymbol{W}(\boldsymbol{t}_j)$ will almost surely not coincide. On the other hand, the components of $\boldsymbol{W}(\boldsymbol{t}_i)$ and $\boldsymbol{W}(\boldsymbol{t}_j)$ corresponding to $\boldsymbol{X}_s(\boldsymbol{t}_i)$ and $\boldsymbol{X}_s(\boldsymbol{t}_j)$ will almost surely coincide, thus justifying the construction of a layered $\boldsymbol{W}_{Total}$, and thereby improving the SNR. In addition, the decoder will help protect and maintain the private information corresponding to each modality $\boldsymbol{X}_p(\boldsymbol{t})$ by ensuring that data can be reconstructed again from the latent space with minimal loss. In the following, we will elaborate more on how aggregating affinity matrices should impact the overall clustering performance. The idea of aggregating affinity matrices is not new, in fact, it has been used extensively in clustering and community detection field.

For example, in [36], the authors proposed a method that combines the self-similarity matrices of the eigenvectors after applying a Singular Value Decomposition on clusters. In [37], they proposed merging the information provided by the multiple modalities by combining the characteristics of individual graph layers using tools from subspace analysis on a Grassmann manifold. In [38], they propose a multilayer spectral graph clustering (SGC) framework that performs convex layer aggregation.

Proposition: The persistent differential scaling of $\boldsymbol{m}$-modal Group Robust Subspace Clustering Fusion yields an order $\boldsymbol{m}$-improvement resilience over the singly differential scaling fusion.

The proof of the proposition can be found in Appendix A. We basically show that by perturbing one or more data modalities, our proposed approach introduces less error to the overall affinity matrix as compared to DMSC. Hence, preserving the performance and yielding a graceful degradation of the clustering accuracy as an increasing number of modalities get corrupted by noise.

3. Affinity fusion deep multimodal subspace clustering

For completeness, we provide a brief overview of the Deep Multimodal Subspace Clustering algorithm which was proposed in [4]. As noted earlier for DRoGSuRe and similarly for Affinity Fusion Deep Multimodal Subspace clustering (AFDMSC), the network is composed of three main parts: a multimodal encoder, a self-expressive layer, and a multimodal decoder. The output of the encoder contributes to a common latent space for all modalities. The self-expressiveness property applied through a fully connected layer between the encoder and the decoder results in one common set of weights for all the data sensing modalities. This marks a divergence in defining the latent space with DRoGSuRe. Our proposed approach, as a result, safeguards the private information $\boldsymbol{X_p(t)}; \boldsymbol{t} = \mathbf{1}, \ldots, \boldsymbol{T}$ individually for each of the sensors, i.e., dedicating more degrees of freedom for each of the sensors. This contrasts with AFDMSC. The reconstruction of the input data by the decoder, can yield the following loss function to secure the proper training of the self-expressive network,

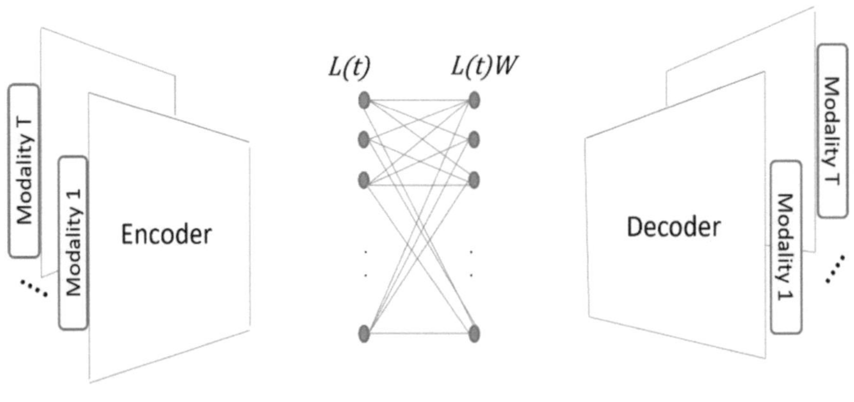

Figure 2.
Deep multimodal subspace clustering diagram.

$$min_{\,W|w_{kk}=0}\|\boldsymbol{W}\|_2+\frac{\gamma}{2}\sum_{t=1}^{T}\|\boldsymbol{X}(t)-\boldsymbol{X}_r(t)\|_F^2+\frac{\mu}{2}\sum_{t=1}^{T}\|\boldsymbol{L}(t)-\boldsymbol{L}(t)\boldsymbol{W}\|_F^2, \tag{17}$$

where $\boldsymbol{W}$ represents the parameters of the self expressive layer, $\boldsymbol{X}(\boldsymbol{t})$ is the input to the encoder, $\boldsymbol{X}_r(\boldsymbol{t})$ denote the output of the decoder and $\boldsymbol{L}(\boldsymbol{t})$ denotes the output of the encoder. $\boldsymbol{\mu}$ and $\boldsymbol{\gamma}$ are regularization parameters. An overview for the DMSC approach is illustrated in **Figure 2**.

4. Experimental results

4.1 Dataset description

We will evaluate our approach on two different datasets. The first dataset we will use is the Extended Yale-B dataset [39]. The same dataset has been used extensively in subspace clustering as in [1, 40]. The dataset is composed of 64 frontal images of 38 individuals under different illumination conditions. In this work, we will use the augmented data used in [4], where facial components such as left eye, right eye, nose and mouth have been cropped to represent four additional modalities. Images corresponding to each modality have been cropped to a size of 32×32. A sample image for each modality is shown in **Figure 3**. The second validation dataset we use is the ARL polarimetric face dataset [41]. This consists of facial images for 60 individuals in the visible domain and in four different polarimetric states. All the images are spatially aligned for each subject. We have also resized the images to 32×32 pixels. Sample images from this dataset are shown in **Figure 4**.

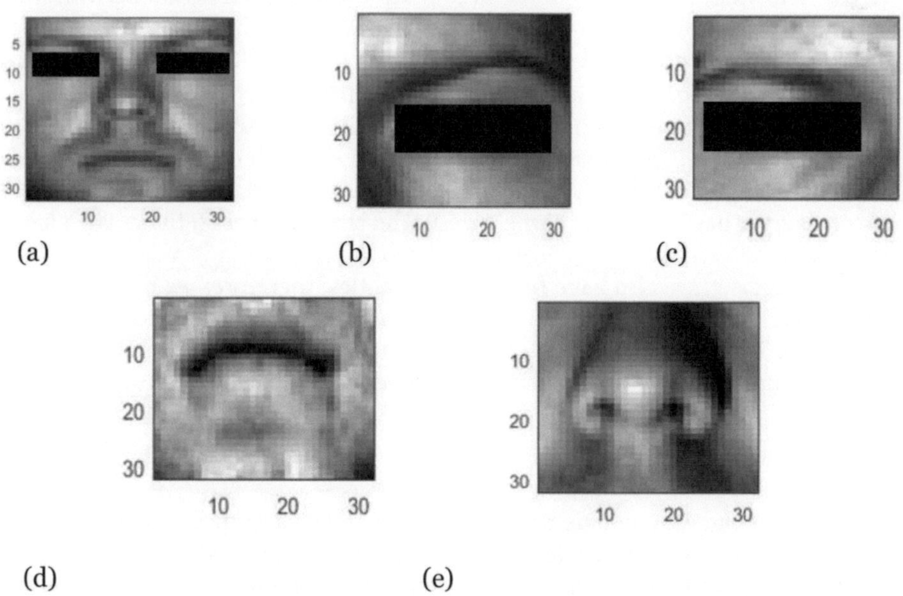

Figure 3.
Sample images from the augmented extended Yale-B Dataset. (a) Face. (b) Left eye. (c) Right eye. (d) Mouth. (e) Nose.

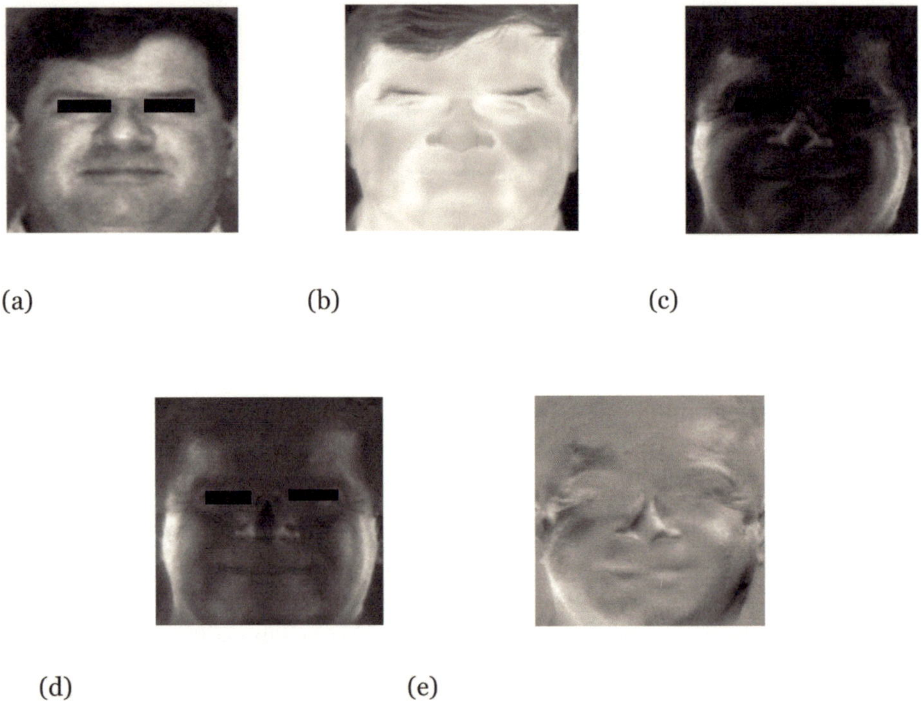

Figure 4.
Sample images from the ARL polarimetric dataset. (a) Visible. (b) DoLP. (c) S0. (d) S1. (e) S2.

4.2 Network structure

In the following, we will elaborate on how we construct the neural network for each dataset. Similar to [4], we implemented DRoGSuRe with Tensorflow and used the adaptive momentum based gradient descent method (ADAM) [30] to minimize the loss function in Eq. (5) with a learning rate of $\mathbf{10^{-3}}$.

In case of ARL dataset, we have five data modalities and will therefore have 5 different encoders, self-expressive layers and decoders. Each encoder is composed of three neural layers. The first layer consists of 5 convolutional filters of kernel size 3. The second layer has 7 filters of kernel size 1. The last layer has 15 filters with kernel size equals 1.

For EYB dataset, we also have five data modalities, therefore, we have 5 different encoders, self-expressive layers and decoders. Each encoder is composed of three neural layers. The first layer consists of 10 convolutional filters of kernel size 5. The second layer has 20 filters of kernel size 3. The last layer has 30 filters of kernel size 3.

4.3 Noiseless results

In the following, we compare the performance of our approach versus the DMSC approach when learning the union of subspaces structure of noise-free data. First, we divide each dataset into training and validation sets to be able to classify a newly observed dataset, using the structure learned through the current unlabeled data. The ARL expression dataset used for training consists of 2160 images per modality. The validation baseline images include 720 images total per modality. For the EYB, we

	Learning	Validation
DMSC	97.59%	98.33%
DRoGSuRe	100%	100%

Table 1.
Performance comparison for ARL dataset.

	Learning	Validation
DMSC	98.82%	98.89%
DRoGSuRe	98.42%	98.76%

Table 2.
Performance comparison for EYB dataset.

randomly selected 1520 images per modality for training and 904 images for validation. The sparse solution $\boldsymbol{W}(\boldsymbol{t})$ corresponding to each data modality, provides important information about the relations among data points, which may be used to split data into individual clusters residing in a common subspace. Observations from each object can be seen as data points spanning one subspace. Interpreting the subspace-based affinities based on $\boldsymbol{W}(\boldsymbol{t})$ as a layered set of networks, we proceed to carry out what amounts to modality fusion. The $\boldsymbol{T}$ sparse matrices are added to produce one sparse matrix for both modalities, $\boldsymbol{W}_{\boldsymbol{Total}}$, thereby improving performance. Observations associated with one object/individual are clustered as one subspace where the contribution of each sensor is embedded in the entries of the $\boldsymbol{W}_{\boldsymbol{Total}}$ matrix. For clustering by $\boldsymbol{W}_{\boldsymbol{Total}}$, we apply spectral clustering.

After learning the structure of the data clusters, we validate our results on the validation set. We extract the principal components (eigen vectors of the covariance matrix) of each cluster in the original (training) dataset, to act as a representative subspace of its corresponding class. We subsequently project each new test point onto the subspace corresponding to each cluster, spanned by its principal components. The $\boldsymbol{l_2}$ norm of the projection is then computed, and the class with the largest norm is selected to be the class of this test point. For DRoGSuRe, we use the coefficient matrix $\boldsymbol{W}_{\boldsymbol{Total}}$ in Eq. (13) to cluster the test data points coming from all data modalities. We compare the clustering output labels with the ground truth for each dataset. The results for ARL and EYB datasets are depicted in **Tables 1** and **2** respectively. From the results, it is clear that DRoGSuRE technique for the fused data remarkably outperforms DMSC in case of ARL dataset. The reason behind the significant improvement is the layered structure of our proposed approach that constructs the latent space in a way that safeguards the individual sensor private information which hence dedicates more degrees of freedom to each of the sensors. In addition, the ARL dataset structure offers modalities that are different in nature and individually provides new information in contrast to the EYB dataset. However, in case of EYB dataset and in the noiseless case, DMSC performed better than DRoGSuRe.

4.4 Noise training with single and multiple modalities

In the following, we test the robustness of our approach in the case of noisy learning. We distort one modality at a time by shuffling the pixels of all images in that

	DMSC L	DMSC V	DRoGSuRe L	DRoGSuRE V
Mod 0	87.17%	86.67%	95.37%	95%
Mod 1	91.67%	90%	98.29%	98.33%
Mod 2	92.77%	92.78%	99.17%	99.44%
Mod 3	90.55%	90.57%	99.31%	99.44%
Mod 4	92.78%	91.11%	96.44%	96.67%

Table 3.
ARL dataset: Distorting one modality.

	DMSC L	DMSC V	DRoGSuRe L	DRoGSuRE V
Mod 0 & 1	82.22%	82.78%	92.27%	94.58%
Mod 1 & 2	91.11%	91.11%	97.22%	97.36%
Mod 0 & 3	85.51%	82.56%	93.01%	95.42%
Mod 1 & 4	91.67%	89.44%	97.22%	97.36%
Mod 2 & 3	90%	89.72%	97.69%	97.78%

Table 4.
ARL dataset: Distorting two modalities.

	DMSC L	DMSC V	DRoGSuRe L	DRoGSuRE V
Mod 0	87.96%	88.5%	93.29%	94.69%
Mod 1	91.84%	91.15%	95.79%	97.46%
Mod 2	89.01%	88.72%	98.03%	97.57%
Mod 3	92.69%	91.81%	95.59%	96.68%
Mod 4	91.45%	91.59%	97.17%	97.35%

Table 5.
EYB dataset: Distorting one modality.

modality during the training phase. By doing so, we are perturbing the structure of the sparse coefficient matrix associated with that modality, thus impacting the overall W matrix for both DRoGSuRe and DMSC. Testing with clean data, i.e., no distortion, demonstrates the impact of perturbing the training and hence performing an

	DMSC L	DMSC V	DRoGSuRe L	DRoGSuRE V
Mod 0 & 2	86.64%	85.18%	96.84%	96.13%
Mod 0 & 4	87.83%	89.16%	94.54%	95.8%
Mod 1 & 4	86.38%	86.06%	94.21%	95.8%
Mod 2 & 3	88.22%	84.96%	91.58%	93.92%
Mod 3 & 4	88.03%	86.28%	94.08%	95.35%

Table 6.
EYB dataset: Distorting two modalities.

inadequate training, e.g., insufficient data or non-convergence. This can also be considered as augmenting the training data with new information or a new view for one modality which might not necessarily contained in the testing or the validation data. Moreover, we repeat the same experiment with the distortion of two modalities before learning the sparse coefficient matrices for both DMSC and DRoGSuRe. The results for the ARL dataset are depicted in **Tables 3** and **4**, while results for the EYB dataset are shown in **Tables 5** and **6**. For ARL dataset, we refer to Visible, S0, S1, S2 and DoLP as Mod 0, 1, 2, 3 and 4 respectively. For EYB Dataset, we refer to Face, left eye, nose, mouth and right eye as mod 0, 1, 2, 3, and 4. We refer to each modality as Mod, where L denotes learning and V denotes validation results. From the results, it is clear that DRoGSuRe is showing a significant improvement in the clustering accuracy as compared to DMSC for both learning and validation set. The reason for that, is again, due to the fact that perturbing one or two modalities would have less impact on the overall performance for DRoGSuRe in comparison to DMSC.

4.5 Testing with limited noisy testing data

In the following, we study the effect of using noiseless data for training while validating with noisy and missing data. We add Gaussian noise to one data modality in the validation set and vary the SNR by varying the noise variance. We subsequently assume that we only have one modality available at testing. Then, we keep increasing the number of available noiseless data modalities beside the noisy modality. We

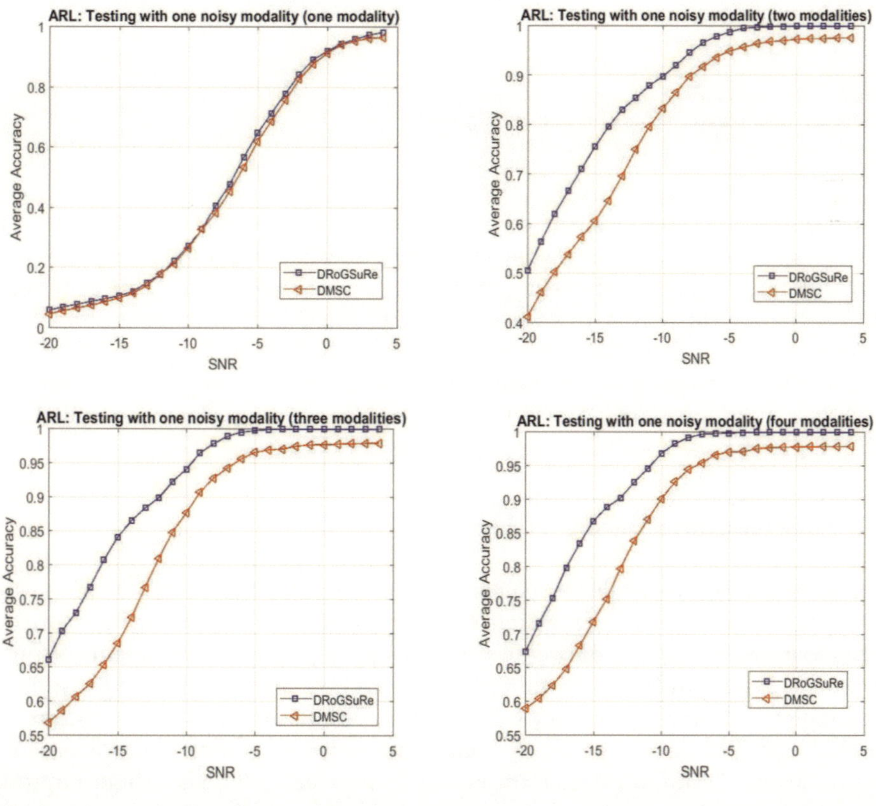

Figure 5.
ARL noiseless training and validating on limited noisy data.

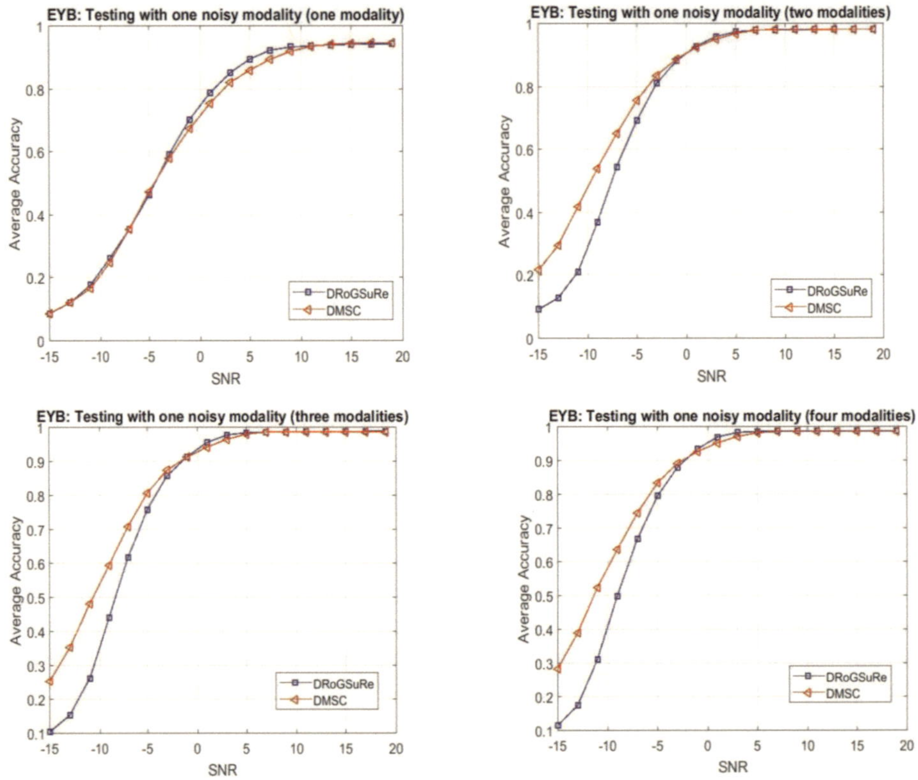

Figure 6.
EYB noiseless training and validating on limited noisy data.

average the results considering all different combinations of data modalities for ARL and EYB datasets. The results are depicted in **Figures 5** and **6** respectively. For the ARL dataset, we note the increasing gap between DMSC and DRoGSuRe as we augment the sensing capacity with noise-free modalities. On the other hand, for the EYB dataset and at lower SNR, the performance of DRoGSuRe is slightly worse than DMSC which might be explained by the results in **Table 2**; as the training accuracy for DMSC is slightly better than DRoGSuRe in the case of clean training. However, at higher SNR, the performance of the two approaches is very close.

4.6 Missing modalities during testing

In the following, we evaluate the performance of DRoGSuRe and DMSC in case of missing data modalities during testing. It is not uncommon to have one or more sensors that might be silent during testing, thus justifying this experiment for further assessment. We try different combinations of available modalities during testing, and we average the clustering accuracy for each trial. Results are depicted in **Figures 7** and **8** for ARL and EYB data respectively. Again, we notice a significant improvement for DRoGSuRe over DMSC for ARL Dataset. For EYB dataset, there is a slight improvement for DRoGSuRe over DMSC. The reason behind the slight improvement is because our approach introduces less error to the overall affinity matrix as compared

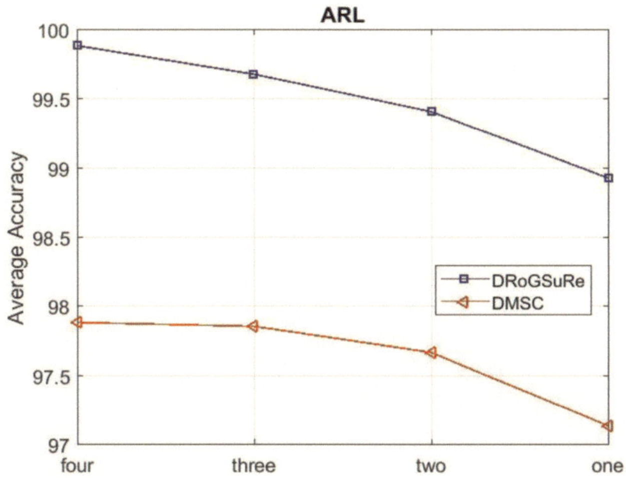

Figure 7.
Missing modalities during testing for ARL dataset.

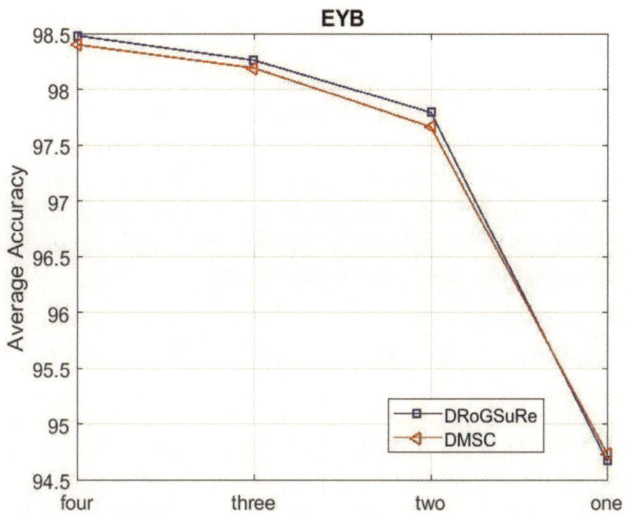

Figure 8.
Missing modalities during testing for EYB dataset.

to DMSC. Hence, preserving the performance and yielding a graceful degradation of the accuracy although DMSC was the state of the art for the EYB dataset.

5. Feature concatenation

Here we propose a rationale along with an alternative solution for enhancing the performance for EYB multi-modal data. Due to the specific structure of the EYB multi-modal data, the concatenation of the features corresponding to each modality is a reasonable alternative. By doing so, we are adjoining together the features representing each part of the face. Since the four modalities correspond to non-overlapping partitions of the face, the feature set corresponding to each partition will

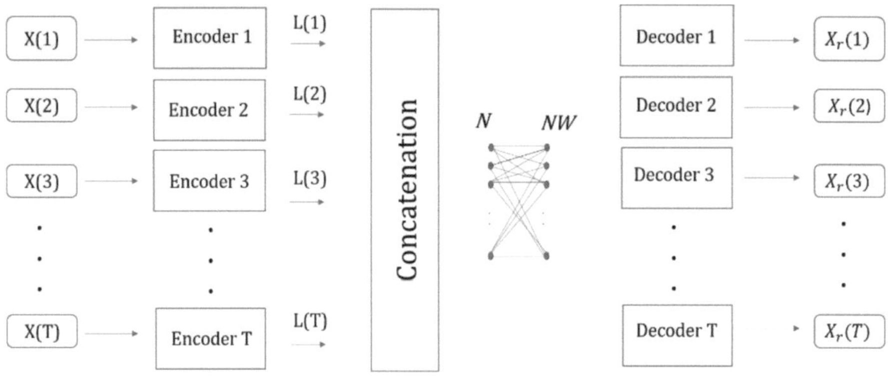

Figure 9.
CNN Concatenation Network.

solely provide complementing information. A similar idea is proposed in [4] and is referred to as Late concatenation, where the multi-modal data is integrated in the last stage of the encoder. Their resulting decoder structure remains the same for either affinity fusion or late concatenation. This entails de-concatenating the multi-modal data prior to decoding it. Our proposed approach on the other hand, results in a self-expressive layer being driven by the concatenated features from the $\boldsymbol{M}$ encoder branches. Afterwards, we feed the self-expressive layer output to each branch of the decoder. The concatenated information results in a more efficient code for the data, thereby resulting in an overall parsimonious with a sparse structure of the decoder, results in a decoder composed of three neural layers. The first layer consists of 150 filters of kernel size 3. The second layer consists of 20 layers of kernel size 3. The third layer consists of 10 layers of kernel size 5. Our approach is illustrated in **Figure 9**. We optimize the weights of the auto-encoder as follows,

$$min_{\,W|w_{kk}=0}\rho\|\boldsymbol{W}\|_1+\frac{\gamma}{2}\sum_{t=1}^{T}\|\boldsymbol{X}(t)-\boldsymbol{X}_r(t)\|_F^2+\frac{\mu}{2}\sum_{t=1}^{T}\|\boldsymbol{N}-\boldsymbol{N}\boldsymbol{W}\|_F^2, \tag{18}$$

where $\boldsymbol{N}=[\boldsymbol{L}(\mathbf{1})\|\boldsymbol{L}(\mathbf{2})\|\boldsymbol{L}(\mathbf{3})\|\boldsymbol{L}(\mathbf{4})\|\boldsymbol{L}(\mathbf{5})]$.

We compared the performance of our proposed approach against the late concatenation approach in [4] and the results are depicted in **Table 7** for the EYB dataset.

From the previous table, we can conclude that concatenating the features from the encoder and feeding the concatenated information to each decoder branch achieves a better performance for this type of multi-modal data structure. The reason behind this enhancement is the combination of efficient extraction of the basic features from the whole face and finer features from each part of the face. Promoting more efficiency as noted, this concatenation may also be intuitively viewed as adequate mosaicking, in which different patterns complement each other. In the following, we will show how

	Learning	**Validation**
DMSC Late Concatenation	95.66%	94.7%
CNN Concatenation Network	99.28%	99.3%

Table 7.
Concatenation performance for EYB dataset.

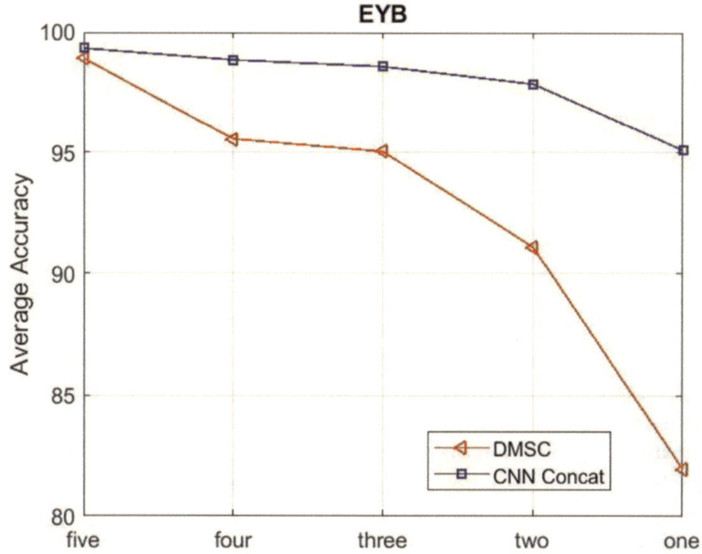

Figure 10.
Missing modalities during testing for EYB dataset.

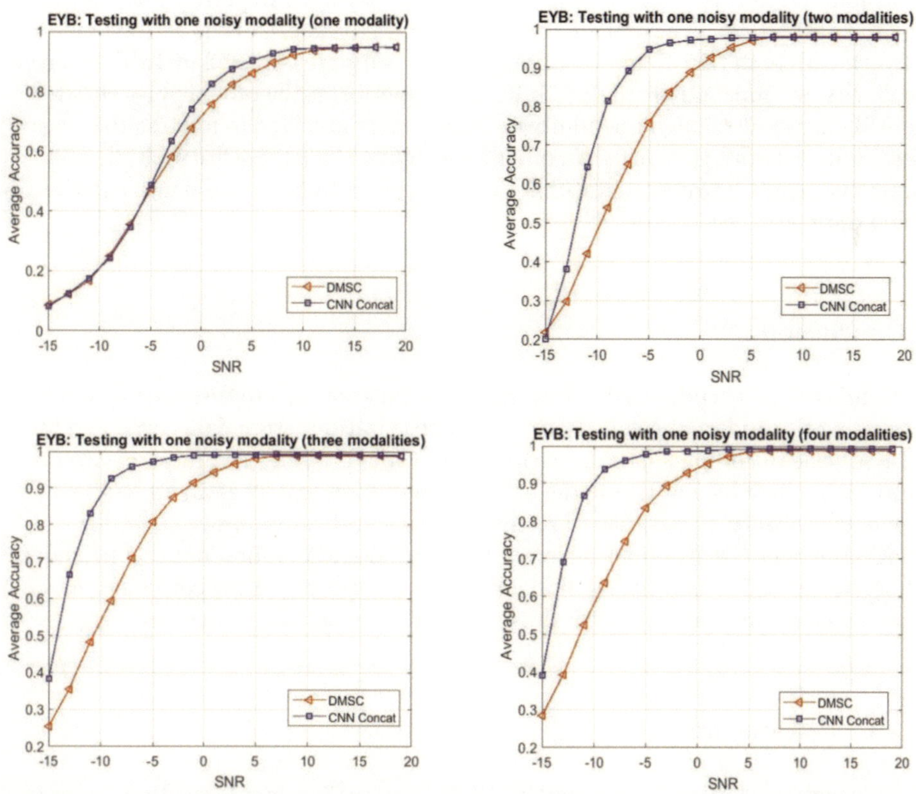

Figure 11.
EYB noiseless training and validating on limited noisy data.

	Learning	Validation
DMSC	97.59%	98.33%
DRoGSuRe	100%	100%
CNN Concat	99.44%	99.17

Table 8.
Concatenation performance for ARL dataset.

our proposed approach performs in two cases: missing and noisy test data. The results of the new proposed approach, which we refer to as CNNs concatenation network, is compared to the state-of-the-art DMSC network [4]. We start by training the auto-encoder network using 75% of the data and then we test on the rest of the data. In **Figure 10**, we show how the performance degrades by decreasing the number of available modalities at testing from five to one. From the results, it is clear how the CNNs concatenation network outperforms the DMSC network. Additionally, we repeated the same experiment we performed in subsection 4.5. We train the network with noiseless data and then add Gaussian noise to one data modality at the testing. Additionally, we vary the number of available modalities at testing from one to four. The results are depicted in **Figure 11**. From the results, it is clear how the concatenated CNNs is more robust to noise than DMSC.

In addition, we have utilized the Concatenation network to perform object clustering on the ARL data. We compare the clustering performance of the concatenation network with both DMSC and DRoGSuRe. The results are depicted in **Table 8**. From the results, we conclude that DRoGSuRe still outperforms the other approaches for the ARL dataset. Although the number of parameters involved in training the DRoGSuRe network is higher than other approaches, since there are multiple self-expressive layers, however, DRoGSuRe is more robust to noise and limited data availability during testing.

6. Conclusion

In this paper, we proposed a deep multi-modal approach to fuse data through recovering the underlying subspaces of data observations from data corrupted by noise to scale to complex data scenarios. DRoGSuRe provides a natural way to fuse multi-modal data by employing the self-representation matrix as an embedding for each data modality. Experimental results show a significant improvement for DRoGSuRe over DMSC under different types of potential limitations and provides robustness with limited sensing modalities. We also proposed the concatenated CNNs model, which can work better for different multi-modal data structures.

Acknowledgements

This work was in part supported by DOE-National Nuclear Security Administration through CNEC-NCSU under Award DE-NA0002576. The first author was also in part supported by DTRA.

Conflict of interest

The authors declare no conflict of interest.

A. Appendix. Parameter perturbation analysis

To theoretically compare our proposed variational scaling fusion approach DRoGSuRe to DMSC, we proceed by way of a first order perturbation analysis on the parameter set $\boldsymbol{W}^i$ of respectively either technique $\boldsymbol{i} = \mathbf{1}, \mathbf{2}$. This will, in turn impact the associated affinity matrix $\boldsymbol{A}^i$, which as we will later elaborate directly impacts the subspace clustering procedure which is central to the inference following the fusion procedure.

Adopting the original formulation for the first persistently differential scaling approach, namely that $\boldsymbol{T}$ modalities are jointly exploited, results in, $\boldsymbol{X}^1(\boldsymbol{t}) = \left[\boldsymbol{x}_1^1(\boldsymbol{t})\boldsymbol{x}_2^1(\boldsymbol{t}) \ldots \boldsymbol{x}_n^1(\boldsymbol{t})\right]$, where $\boldsymbol{x}_k^1(\boldsymbol{t}) \in \mathbb{R}^m, \boldsymbol{t} = \mathbf{1}, \mathbf{2}, \ldots, \boldsymbol{T}$ represents the $\boldsymbol{k}^{th}$ observation. The second approach only effectively uses only one subspace structure of the fused modalities $\boldsymbol{X}^2(\boldsymbol{t}) = \left[\boldsymbol{x}_1^1\boldsymbol{x}_2^1 \ldots \boldsymbol{x}_n^1\right]$.

A first order perturbation on the data may be due to noise or to a degradation of a given sensor, and results in a perturbation of the UoS parameters,

$$\tilde{\boldsymbol{W}}_1^i = \boldsymbol{W}_1^i + \delta^i \tag{A1}$$

For the first method, each modality will have an associated subspace cluster parameter set $\{\boldsymbol{W}_t^1\}_{t=1,\ldots,T}$, with $\boldsymbol{W}_t^1 \in \mathbb{R}^{n\times n}$. The overall parameter set for DRoGSuRe can then be written as,

$$\tilde{\boldsymbol{W}}^1 = \tilde{\boldsymbol{W}}_1^1 + \boldsymbol{W}_2^1 + \ldots + \boldsymbol{W}_m^1 \tag{A2}$$

Where the unperturbed overall sparse coefficient matrix is written as follows, $\boldsymbol{W}_{tot}^1 = \boldsymbol{W}_1^1 + \boldsymbol{W}_2^1 + \ldots + \boldsymbol{W}_m^1$. A similar development follows for method 2, with the difference that the contributing modalities are fused a priori.

Proof. We first write the affinity matrix associated with DRoGSuRE as,

$$\tilde{\boldsymbol{A}}^1 = \tilde{\boldsymbol{W}}_{tot}^1 + \left(\tilde{\boldsymbol{W}}_{tot}^1\right)^T \tag{A3}$$

$$\tilde{\boldsymbol{A}}^1 = \tilde{\boldsymbol{W}}_1^1 + \boldsymbol{W}_2^1 + \ldots + \boldsymbol{W}_m^1 + \left(\tilde{\boldsymbol{W}}_1^1 + \boldsymbol{W}_2^1 + \ldots + \boldsymbol{W}_m^1\right)^T \tag{A4}$$

where the superscript $\boldsymbol{T}$ denotes transpose. This is equivalent to,

$$\tilde{\boldsymbol{A}}^1 = \tilde{\boldsymbol{A}}_1^1 + \sum_{i=2}^{T} \boldsymbol{A}_i^1 \tag{A5}$$

Where $\mathbf{0} \leq \tilde{\boldsymbol{A}}_1^1(\boldsymbol{i}, \boldsymbol{j}) \leq \mathbf{1} + \boldsymbol{\delta}^1$. The unperturbed collective affinity matrix $\boldsymbol{A}^1$ can be similarly written $\boldsymbol{A}^1 = \sum_{i=1}^{T} \boldsymbol{A}_i^1$ with the unity constraint on each entry of all matrices. We may also write the magnitude of the difference as,

$$\left|A^1 - \tilde{A}^1\right| = \delta^1 + \left(\delta^1\right)^T \tag{A6}$$

Letting $\Delta = \boldsymbol{\delta}^1 + \left(\boldsymbol{\delta}^1\right)^T \in \mathbb{R}^{n\times n}$, and assuming $\boldsymbol{\epsilon} = \boldsymbol{max}_{i,j}[\Delta]_{i,j}$, we can write,

$$\left\|A - \tilde{A}\right\|_F \leq n\epsilon \tag{A7}$$

Given the Δ matrix individual entry bounds, we conclude that,

$$0 \leq \epsilon \leq \frac{1}{t} \tag{A8}$$

Since DMSC assumes having one sparse coefficient matrix $\boldsymbol{W}$ for all data modalities, which is equivalent to only one subspace structure of the fused modalities $\boldsymbol{X}^2(\boldsymbol{t}) = \left[\boldsymbol{x}_1^2 \ldots \boldsymbol{x}_n^2\right]$. Therefore, the UoS parameters will be perturbed by $\boldsymbol{\delta}^2$ as follows,

$$\tilde{W}^2 = W^2 + \delta^2 \tag{A9}$$

The affinity matrix associated with DMSC can be written as follows, $\tilde{A}^2 = \tilde{W}^2 + \left(\tilde{W}^2\right)^T$, which is equivalent to,

$$\tilde{A}^2 = W^2 + \delta^2 + \left(W^2\right)^T + \left(\delta^2\right)^T \tag{A10}$$

Similarly, the unperturbed affinity matrix will be as follows,

$$A^2 = W^2 + \left(W^2\right)^T \tag{A11}$$

From Eqs. (A10) and (A11), the magnitude of the difference can be written as follows,

$$|A^2 - \tilde{A}^2| = \delta^2 + \left(\delta^2\right)^T \tag{A12}$$

Letting $\gamma = \delta^2 + \left(\delta^2\right)^T \in \mathbb{R}^{n\times n}$, i.e., $\left|A^2 - \tilde{A}^2\right| = \gamma$, and assuming $\boldsymbol{\Psi} = \boldsymbol{max}_{i,j}[\gamma]_{i,j}$, we can write $\left\|A^2 - \tilde{A}^2\right\|_F \leq \boldsymbol{n\Psi}$. Given the γ matrix individual entry bounds, we conclude $\boldsymbol{0 \leq \Psi \leq 1}$. If we only perturb one modality, knowing that $\boldsymbol{0 \leq A(i,j) \leq 1}$, therefore the error could lie between $\boldsymbol{0 \leq \Psi \leq 1}$, which entails either creating a fake relation between two data points or erasing an existing relation. $\boldsymbol{\epsilon}$ and $\boldsymbol{\Psi}$ are random variables that do not have to follow a specific distribution, however, in any case $\boldsymbol{E}(\epsilon^2) \ll \boldsymbol{E}(\Psi^2)$ and therefore $\boldsymbol{SNR_{DRoGSuRe}} \gg \boldsymbol{SNR_{DMSC}}$.

In light of the above two bounds, and the results of [42], where it is shown that the spectral clustering dependent on the respective projection operators $\boldsymbol{P}_{W^1}$ and $\tilde{\boldsymbol{P}}_{\tilde{W}^1}$ onto the vector subspaces spanned by the principal eigenvectors of $\boldsymbol{W}_{tot}^1$ and $\tilde{\boldsymbol{W}}_{tot}^1$ of may be written as,

$$\left\|P_{W^1} - \tilde{P}_{\tilde{W}^1}\right\|_F \leq \frac{\sqrt{2}}{\alpha^1}\left\|A^1 - \tilde{A}^1\right\|_F \tag{A13}$$

where $\boldsymbol{\alpha}^1$ is the spectral gap between the $\boldsymbol{k}^{th}$ and $(\boldsymbol{k}+\mathbf{1})^{st}$ eigen value of $\boldsymbol{A}^1$, $|\lambda_k^1 - \lambda_{k+1}^1|$. Similarly, for DMSC, the bound on the projection operators is,

$$\left\| P_{W^2} - \tilde{P}_{\tilde{W}^2} \right\|_F \leq \frac{\sqrt{2}}{\alpha^2} \left\| \boldsymbol{A}^2 - \tilde{\boldsymbol{A}}^2 \right\|_F \tag{A14}$$

where $\boldsymbol{\alpha}^2 = |\lambda_k^1 - \lambda_{k+1}^1|$. Since $\boldsymbol{W}_1^1, \boldsymbol{W}_2^1, \ldots, \boldsymbol{W}_T^1$ happen to commute and if they happen to be diagonalizable, therefore, they share the same eigenvectors. As a result, the eigenvectors of $\boldsymbol{W}_1^1 + \boldsymbol{W}_2^1 + \ldots + \boldsymbol{W}_T^1$ are also the same and the corresponding eigenvalue that is the sum of the corresponding eigenvalues of $\boldsymbol{W}_1^1, \boldsymbol{W}_2^1, \ldots$ and $\boldsymbol{W}_T^1$.Therefore, $\lambda_k^1 \gg \lambda_k^2$ From all the above, we can conclude that smaller error yielding to better clustering, hence preserving the performance, yields the improvement by the T-factor noted in the proposition and shown in the two perturbation developments.

Author details

Sally Ghanem and Hamid Krim*
North Carolina State University, Raleigh, USA

*Address all correspondence to: ahk@ncsu.edu

References

[1] Elhamifar E, Vidal R. Sparse subspace clustering: Algorithm, theory, and applications. IEEE Transactions on Pattern Analysis and Machine Intelligence. 2013;**35**:2765-2781

[2] Favaro P, Vidal R, Ravichandran A. A closed form solution to robust subspace estimation and clustering. In: CVPR 2011. Colorado springs, Colorado, USA: IEEE; 2011. pp. 1801-1807

[3] Li CG, Vidal R. Structured sparse subspace clustering: A unified optimization framework. In: Proceedings of the IEEE Conference on Computer Vision and Pattern Recognition. Boston, MA, USA: IEEE; 2015. pp. 277-286

[4] Bian X, Panahi A, Krim H. Bi-sparsity pursuit: A paradigm for robust subspace recovery. Signal Processing. 2018;**152**: 148-159

[5] Yang AY, Rao SR, Ma Y. Robust statistical estimation and segmentation of multiple subspaces. In: 2006 Conference on Computer Vision and Pattern Recognition Workshop (CVPRW'06). New York, NY, USA: IEEE; 2006. p. 99

[6] Hong W, Wright J, Huang K, Ma Y. Multiscale hybrid linear models for lossy image representation. IEEE Transactions on Image Processing. 2006;**15**:3655-3671

[7] Ho J, Yang MH, Lim J, Lee KC, Kriegman D. Clustering appearances of objects under varying illumination conditions. In: 2003 IEEE Computer Society Conference on Computer Vision and Pattern Recognition, 2003. Proceedings. Madison, Wisconsin, USA: IEEE; 2003. p. I

[8] Hellwich O, Wiedemann C. Object extraction from high-resolution multisensor image data. In: Third International Conference Fusion of Earth Data. France: Sophia Antipolis; 2000

[9] Korona Z, Kokar MM. Model theory based fusion framework with application to multisensor target recognition. In: 1996 IEEE/SICE/RSJ International Conference on Multisensor Fusion and Integration for Intelligent Systems (Cat. No. 96TH8242). Tokyo, Japan: IEEE; 1996. pp. 9-16

[10] Ghanem S, Panahi A, Krim H, Kerekes RA, Mattingly J. Information subspace-based fusion for vehicle classification. In: 2018 26th European Signal Processing Conference (EUSIPCO). Rome, Italy: IEEE; 2018. pp. 1612-1616

[11] Ghanem S, Roheda S, Krim H. Latent code-based fusion: A volterra neural network approach. 2021. arXiv preprint arXiv:2104.04829

[12] Wang H, Skau E, Krim H, Cervone G. Fusing heterogeneous data: A case for remote sensing and social media. IEEE Transactions on Geoscience and Remote Sensing. 2018;**56**:6956-6968

[13] Soong FK, Rosenberg AE. On the use of instantaneous and transitional spectral information in speaker recognition. IEEE Transactions on Acoustics, Speech, and Signal Processing. 1988;**36**:871-879

[14] Xu L, Krzyzak A, Suen CY. Methods of combining multiple classifiers and their applications to handwriting recognition. IEEE Transactions on Systems, Man, and Cybernetics. 1992;**22**: 418-435

[15] Ngiam J, Khosla A, Kim M, Nam J, Lee H, Ng A. Multimodal deep learning.

In: International Conference on Machine Learning (ICML). Bellevue, Washington, USA: International Machine Learning Society (IMLS). The conference. 2011. pp. 689-696

[16] Ramachandram D, Taylor GW. Deep multimodal learning: A survey on recent advances and trends. IEEE Signal Processing Magazine. 2017;**34**: 96-108

[17] Valada A, Oliveira GL, Brox T, Burgard W. Deep multispectral semantic scene understanding of forested environments using multimodal fusion. In: International Symposium on Experimental Robotics. Tokyo, Japan: Springer; 2016. pp. 465-477

[18] Roheda S, Riggan BS, Krim H, Dai L. Cross-modality distillation: A case for conditional generative adversarial networks. In: 2018 IEEE International Conference on Acoustics, Speech and Signal Processing (ICASSP). Calgary, Alberta, Canada: IEEE; 2018b. pp. 2926-2930

[19] Roheda S, Krim H, Luo ZQ, Wu T. Decision level fusion: An event driven approach. In: 2018 26th European Signal Processing Conference (EUSIPCO). Rome, Italy: IEEE; 2018a. pp. 2598-2602

[20] Zhu P, Hui B, Zhang C, Du D, Wen L, Hu Q. Multi-view deep subspace clustering networks. 2019. arXiv preprint arXiv:1908.01978.

[21] Abavisani M, Patel VM. Deep multimodal subspace clustering networks. IEEE Journal of Selected Topics in Signal Processing. 2018;**12**: 1601-1614

[22] Wang Q, Cheng J, Gao Q, Zhao G, Jiao L. Deep multi-view subspace clustering with unified and discriminative learning. IEEE Transactions on Multimedia. 2020;**23**: 3483-3493

[23] Ghanem S, Panahi A, Krim H, Kerekes RA. Robust group subspace recovery: A new approach for multi-modality data fusion. IEEE Sensors Journal. 2020;**20**:12307-12316

[24] Sun X, Cheng M, Min C, Jing L. Self-supervised deep multi-view subspace clustering. In: Asian Conference on Machine Learning, Nagoya, Japan: PMLR. 2019. pp. 1001-1016

[25] Ji P, Zhang T, Li H, Salzmann M, Reid I. Deep subspace clustering networks. Advances in Neural Information Processing Systems. 2017;**30**:24-33

[26] Bian X, Krim H. Bi-sparsity pursuit for robust subspace recovery. In: 2015 IEEE International Conference on Image Processing (ICIP). Québec city, Québec, Canada: IEEE; 2015. pp. 3535-3539

[27] Roheda S, Krim H, Riggan BS. Robust multi-modal sensor fusion: An adversarial approach. IEEE Sensors Journal. 2020b;**21**:1885-1896

[28] Roheda S, Krim H, Riggan BS. Commuting conditional GANS for multi-modal fusion. In: ICASSP 2020–2020 IEEE International Conference on Acoustics, Speech and Signal Processing (ICASSP). Barcelona, Spain: IEEE; 2020a. pp. 3197-3201

[29] Roheda S, Krim H, Luo ZQ, Wu T. Event driven fusion. 2019. arXiv preprint arXiv:1904.11520

[30] Kingma DP, Ba J. Adam: A method for stochastic optimization. 2014. arXiv preprint arXiv:1412.6980

[31] Lin Z, Liu R, Su Z. Linearized alternating direction method with

adaptive penalty for low-rank representation. In: Advances in Neural Information Processing Systems. Granada, Spain: NIPS; 2011. pp. 612-620

[32] Rockafellar RT. Augmented Lagrange multiplier functions and duality in nonconvex programming. SIAM Journal on Control. 1974;**12**: 268-285

[33] Luenberger DG, Ye Y, et al. Linear and Nonlinear Programming. Vol. 2. Springer; 1984

[34] Taylor D, Shai S, Stanley N, Mucha PJ. Enhanced detectability of community structure in multilayer networks through layer aggregation. Physical Review Letters. 2016;**116**: 228301

[35] Ng AY, Jordan MI, Weiss Y. On spectral clustering: Analysis and an algorithm. In: Advances in Neural Information Processing Systems. Vancouver, British Columbia, Canada: NIPS; 2002. pp. 849-856

[36] Gabriel HH, Spiliopoulou M, Nanopoulos A. Eigenvector-based clustering using aggregated similarity matrices. In: Proceedings of the 2010 ACM Symposium on Applied Computing. Switzerland: Sierre; 2010. pp. 1083-1087

[37] Dong X, Frossard P, Vandergheynst P, Nefedov N. Clustering on multi-layer graphs via subspace analysis on Grassmann manifolds. IEEE Transactions on Signal Processing. 2013; **62**:905-918

[38] Chen PY, Hero AO. Multilayer spectral graph clustering via convex layer aggregation: Theory and algorithms. IEEE Transactions on Signal and Information Processing over Networks. 2017;**3**:553-567

[39] Lee KC, Ho J, Kriegman DJ. Acquiring linear subspaces for face recognition under variable lighting. IEEE Transactions on Pattern Analysis and Machine Intelligence. 2005;**27**:684-698

[40] Liu G, Lin Z, Yan S, Sun J, Yu Y, Ma Y. Robust recovery of subspace structures by low-rank representation. IEEE Transactions on Pattern Analysis and Machine Intelligence. 2012;**35**: 171-184

[41] Hu S, Short NJ, Riggan BS, Gordon C, Gurton KP, Thielke M, et al. A polarimetric thermal database for face recognition research. In: Proceedings of the IEEE Conference on Computer Vision and Pattern Recognition Workshops. Las Vegas, NV, USA: IEEE Conference on Computer Vision and Pattern Recognition, CVPR; 2016. pp. 187-194

[42] Hunter B, Strohmer T. Performance analysis of spectral clustering on compressed, incomplete and inaccurate measurements. 2010. arXiv preprint arXiv:1011.0997